AF412885

COSMIC MAGNETIC FIELDS

Magnetic fields are important in the Universe and their effects contain the key to many astrophysical phenomena that are otherwise impossible to understand. This book presents an up-to-date overview of this fast-growing topic and its interconnections to plasma processes, astroparticle physics, high energy astrophysics, and cosmic evolution. The phenomenology and impact of magnetic fields are described in diverse astrophysical contexts within the Universe, from galaxies to galaxy clusters, the filaments and voids of the intergalactic medium, and out to the largest redshifts. The presentation of mathematical formulae is accessible and is designed to add insight into the broad range of topics discussed. Written for graduate students and researchers in physics, astrophysics, and related disciplines, this volume will inspire readers to devise new ways of thinking about magnetic fields in space on galaxy scales and beyond.

PHILIPP P. KRONBERG is Research Professor Emeritus at the University of Toronto, Canada and Visiting Scholar at Los Alamos National Laboratory. He has served on or chaired advisory and management boards of many organisations and facilities and has received numerous awards and distinctions including a Humboldt Award, a Guggenheim Fellowship, and a Killiam Fellowship. For over thirty years Kronberg has pioneered both measurements and physical and mathematical analyses to deduce astrophysical magnetic fields on many scales, from our Milky Way to the most distant quasars.

Cambridge Astrophysics Series

Series editors

Andrew King, Douglas Lin, Stephen Maran, Jim Pringle, Martin Ward, and Robert Kennicutt

Titles available in the series

COSMIC MAGNETIC FIELDS

PHILIPP P. KRONBERG

University of Toronto

CAMBRIDGE
UNIVERSITY PRESS

CAMBRIDGE
UNIVERSITY PRESS

University Printing House, Cambridge CB2 8BS, United Kingdom

Cambridge University Press is part of the University of Cambridge.

It furthers the University's mission by disseminating knowledge in the pursuit of
education, learning and research at the highest international levels of excellence.

www.cambridge.org
Information on this title: www.cambridge.org/9780521631631

© Philipp P. Kronberg 2016

This publication is in copyright. Subject to statutory exception
and to the provisions of relevant collective licensing agreements,
no reproduction of any part may take place without the written
permission of Cambridge University Press.

First published 2016

Printed in the United Kingdom by TJ International Ltd. Padstow Cornwall

A catalogue record for this publication is available from the British Library

Library of Congress Cataloging in Publication Data
Names: Kronberg, Philipp P., 1939-
Title: Cosmic magnetic fields / Philipp P. Kronberg, University of Toronto.
Description: Cambridge : Cambridge University Press, 2016. | Series: Cambridge astrophysics series ; 53 | Includes
bibliographical references and index.
Identifiers: LCCN 2016000896 | ISBN 9780521631631 (hardback : alk. paper)
Subjects: LCSH: Cosmic magnetic fields. | Magnetic fields. | Astrophysics.
Classification: LCC QB462.8 .K76 2016 | DDC 523.01/88–dc23 LC record available at http://lccn.loc.gov/2016000896

ISBN 978-0-521-63163-1 Hardback

While every effort has been made, it has not always been possible to identify the sources of
all the material used, or to trace all copyright holders. If any omissions are brought to our
notice, we will be happy to include the appropriate acknowledgements on reprinting and
in the next update to the digital edition, as applicable.

Cambridge University Press has no responsibility for the persistence or accuracy
of URLs for external or third-party internet websites referred to in this publication,
and does not guarantee that any content on such websites is, or will remain,
accurate or appropriate.

This book is dedicated to my parents Philipp Kronberg and Jean Davidson Kronberg for
their unwavering support of my career in astrophysics and radio astronomy
and
to my sons Paul, Martin, and Michael Kronberg for their constant encouragement in the
writing of this book.

Contents

x   ***Contents***

Preface

Recent observational advances reveal the widespread existence of magnetic fields in the Universe, and produce much firmer estimates of magnetic field *strengths* in both interstellar space and some regions of intergalactic space. Microgauss-level fields are a common component of spiral galaxy disks and halos. Current measurements can detect magnetic fields, directly or indirectly, in the late universe of galaxies as far as "observational reach" can take us. This is at least to $z \sim 3$, and means that they have not globally changed into the past, that covers at least 90% of the lookback time to the origin of the Universe.

They appear to pervade the intracluster medium of clusters of galaxies, and indeed well beyond the cluster core regions. Strengths of ordered magnetic fields in the intracluster medium of cool core galaxy clusters exceed what is typical for the interstellar medium of the Milky Way. This challenges us to explain how they were first generated, and then regenerated to the unexpectedly high measured levels, and to answer the related question of how such widespread magnetic fields influenced the formation of stars and galaxies.

In this book I describe how extragalactic magnetic fields on different scales and in different astrophysical systems can be probed and measured. I also discuss some new methods which could be used to indirectly infer field strengths that are otherwise beyond the reach of direct measurement.

Some basic physical processes responsible for the regeneration of fields are outlined, including mechanisms of magnetic diffusivity and dissipation that influence field amplification. Fast dynamo processes associated with galaxy halo outflow and radio jets can demonstrably magnetise large volumes of intergalactic space; nonetheless, the mechanisms are far from completely understood. Fast-acting dynamo mechanisms appear to operate in galaxies through galaxy outflow, in extragalactic jets and lobes, and possibly in cooling flow clusters. Magnetic reconnection and related magnetohydrodynamic (MHD) processes may also be important.

Whether the intergalactic fields we now observe were originally produced in stars and galaxies, or in the pre-Recombination early Universe is a question with many ramifications, and increasingly discussed. It seems clear that stars and galaxies can seed and regenerate magnetic fields in various ways, post-Recombination. But magnetic fields may have also emerged earlier during Inflation, and perhaps integral to creation of the first subatomic particles as they emerged from the primordial "soup" in the course of enormous energy transfers.

Additional or colour-enhanced illustrations can also be found at the URL www.cambridge .org/9780521631631.

Acknowledgements

I am especially grateful to Sharon Hanna for her invaluable advice and assistance in the preparation and checking of the manuscript. For figure drawing and improvement I am also indebted to Raul Cunha for his competent and cheerful assistance with the large and diverse range of illustrations.

I am indebted to the following colleagues, students, and friends for valued suggestions, checking, and assistance, and additional material: Felix Aharonian, Harjit S. Ahluwalia, Rainer Beck, Julia Becker-Tjus, Gregory Benford, Martin Bernet, Elly Berkhuijsen, Peter Biermann, Guido Birk, Robert Braun, Alan Bridle, Jo-Anne Brown, David Clarke, Stirling Colgate, Raul Cunha, William Daughton, Rod Davies, Eric Donovan, James Drake, Quentin Dufton, Andrew Fletcher, Ernst Fürst, Bryan Gaensler, Götz Golla, Salman Habib, Sharon Hanna, George Heald, Michael Hudson, Norbert Junkes, Karl-Heinz Kampert, Ulrich Klein, Arieh Königl, Roland Kothes, Fritz Krause, Marita Krause, Russell Kulsrud, Harald Lesch, Hui Li, Kathleen Lindeman, Richard Lovelace, Francesco Miniati, Mark Morris, Masanori Nakamura, Andrii Neronov, Phyllis Orbaugh, Frazer Owen, Eugene Parker, Phil Perillat, Rick Perley, Judith Perry, Bharat Ratra, Martin Rees, Patricia Reich, Wolfgang Reich, Brian Reville, Bob Rosner, Vera Rubin, Marcin Sawicki, Michael Scarrott, Gerd Schrade, Herta Schrade, Anwar Shukurov, Martine Simard-Normandin, Dmitry Sokoloff, Todor Stanev, Andrew S. Taylor, John R. Taylor, Chris Thompson, Serap Tilav, Ronald Tucker, Gerrit Verschuur, Heinz Völk, Ievgen Vovk, Richard Wielebinski, Hao Xu, and Ellen Zweibel. The list would be incomplete without acknowledgement of, and gratitude to many graduate students and postdocs who have contributed significantly, directly or indirectly, to my own work on magnetic fields in space.

The writing of this book was made possible with grant support from the Natural Sciences and Engineering Research Council of Canada (NSERC). In addition to NSERC, I am also grateful to the Alexander von Humboldt Foundation, the Canada Council Killam Program, the Guggenheim Foundation, and Los Alamos National Laboratory for enabling many of my own research results over several years.

1

A brief history and background

1.1 Overview of some early results and concepts

The interplay between magnetic fields and ionised gas in space was demonstrated in post-World War II studies of the Earth's aurora, and in attempts to understand the complex and energetic phenomena of the Sun's atmosphere and corona. Much of the basics of magnetoplasma processes in astrophysics have come from solar studies. Here direct measurements of the magnetic field's direction and strength have come through measurement of Zeeman splitting of transitions, and dynamical imaging in the radio, optical, and later Solar X-rays. These studies enabled a theoretical understanding of the role of magnetic fields in particle acceleration in solar flares and prominences. The process of magnetic reconnection, still incompletely understood, was gradually recognised between the late 1940s and early 1960s as an explanation for particle acceleration in solar flares – including reconnection diagrams.

Awareness of magnetic fields in *diffuse* astrophysical plasmas beyond the solar corona began with the discovery of synchrotron radiation (Schwinger 1949). It followed the earlier discovery of cosmic rays (CR) by Hess (1912) and of the polarisation of light from dust-reddened stars. Then came the realisation that supernova remnants emit synchrotron radiation (Shklovskii 1953 and the earlier references therein, Alfvén & Herlofson 1950, Kiepenheuer 1950). Synchrotron radiation requires *both* a magnetic field and relativistic electrons, of density n_e^r, through the following emissivity equation.

$$\varepsilon(v) \approx 10^{-23} n_{r0}^e l\zeta(s)\left(6.3 \times 10^8\right)^{\frac{(s-1)}{2}} \times (B\sin\varphi)^{\frac{(s+1)}{2}} v^{\frac{(1-s)}{2}} \text{ ergs}^{-1}\text{ cm}^{-2}\text{ Hz}^{-1}\text{ sterad}^{-1}. \tag{1.1}$$

The index s is the power law slope of the energy distribution of the relativistic electrons $(n_e^r(E) = n_{e0}^r E^{-s})$, where $\zeta(s)$ is a slowly varying function of s, being unity for $s = 2.5$. l (cm) is the line-of-sight dimension of the emitting region, φ is the pitch angle of the average electron velocity with respect to the local field direction, B is the total field strength in Gauss, and v is in Hz (*cf.* Pacholczyk 1970). The power, L, radiated over all frequencies, is

$$L = 4\pi D_L^2 \int_{v_1}^{v_2} f(v)dv \text{ ergs}^{-1}, \tag{1.2}$$

where D_L is the cosmological luminosity distance, and f is the received spectral density (erg s^{-1} cm^{-2} Hz^{-1}) of the radiation flux. L. Biermann and Schlüter (1951) estimated that interstellar turbulence would infuse the interstellar medium with magenetic fields of order 10^{-6}–10^{-5} G.

A mechanism for the creation of a first seed magnetic field was proposed and applied to stellar magnetic fields by L. Biermann (1950). This process, commonly known as the Biermann battery, has fundamental application to any plasma. In some form it is probably the mechanism that has seeded, i.e. "created", the magnetic fields in many different astrophysical contexts.

During the late 1940s it became clear that the Milky Way contains a distributed magnetic field, and the first estimates of its value, and the notion of interstellar particle acceleration, were established. All were based on sound physical arguments, which I shall briefly review. By today's standards, the data on which sound physical arguments were based might seem impressively limited.

In 1949, Fermi proposed an alternative to Alfvén, Richtmyer, & Teller's (1949) conclusion that CR's were of solar origin and are magnetically confined to the Sun's vicinity. He argued that they could be of interstellar origin. In an insightful paper Fermi proposed that particles above a threshold energy could gain energy by reflection off turbulent motions of a magnetised interstellar medium. Evidence in support of this came from Doppler shift measurements on interstellar absorption lines by Adams (1943), which indicated about 15 kms^{-1} rms scatter in radial velocity.

Fermi's proposal that CRs can be accelerated by such "collisions" with an interstellar magnetic field (now known as Fermi acceleration) also provided a natural explanation for the power law spectrum of cosmic ray particle energies, as is observed. Using Alfvén's relation for the velocity of magnetosonic waves,

$$v_{\mathrm{A}} = \frac{B}{\sqrt{4\pi\rho}}, \tag{1.3}$$

and combining with estimates of the interstellar density of nucleons, Fermi concluded that the interstellar magnetic field strength, $B_{\mathrm{interstellar}}$, is $\approx 5 \cdot 10^{-6}$ G, in effect the consequence of an energy conversion from the kinetic energy of interstellar streaming and turbulence into magnetic energy.

Around the same time the newly discovered properties of CRs, which were at first called *Höhenstrahlung*, also led to early estimates of the interstellar magnetic field strength – using quite different arguments: By the late 1940s it was established that CR nuclei have energies up to $\approx 10^{16}$ eV. The electron component of the galactic CRs was later recognised as associated with interstellar synchrotron radiation from the Milky Way. The observational facts at *ca.* 1949 were: (i) CR energies up to 10^{16} eV, and (ii) isotropy of arrival directions. These were combined with (iii), the proton/electron charge-to-mass ratio, to deduce the strength of the interstellar magnetic field in our region of the Galaxy. Schlüter and L. Biermann (1950) argued that, *if* primary cosmic ray particles were of extragalactic origin (consistent with their isotropy) and if they pervade the Universe, then their energy density, $\varepsilon_{\mathrm{CR}}$, would be $\approx 10^{-12}$ erg cm^{-3}. This would exceed that of extragalactic light by two orders of magnitude. Fermi (1949) and Biermann & Schlüter (1951) argued that this isotropy can only be explained if CRs are *confined* by an interstellar magnetic field. For this to happen, the magnetic field energy density, ε_B, must be comparable with $\varepsilon_{\mathrm{CR}}$, otherwise the cosmic ray particles will leak out of the Galactic disk and destroy the observed isotropy. $\varepsilon_B = 10^{-12}$ erg cm^{-3} corresponds to a magnetic field strength of 5 μG. These arguments were elegant and convincing, and have stood the test of time over the intervening decades. To illustrate with a simple calculation we compare the gyroradius, r_B, of the (then) highest known energy primary CR nuclei ($\approx 10^{16}$ eV), with the thickness of the galactic disk.

$$r_B = \frac{m_\mathrm{p} c^2 \gamma}{eZB_\perp} \approx 10 \left(\frac{E_{16}}{Z \cdot B_{\perp -6}} \right) \ \mathrm{pc} \qquad (1.4)$$

where E_{16} is the energy in units of 10^{16} eV, Z is the atomic number of the CR nucleus, m_p the proton mass and $B_{\perp -6}$ the magnetic field in microgauss. At $B_\perp = 5\ \mu$G and $E_\mathrm{CR} = 10^{16}$ eV, $r_B = 2$ pc, comfortably smaller than the scale height of the galactic disk ($z_\mathrm{G} \approx 1$ kpc).

These deductions of the Milky Way's magnetic field strength from early cosmic ray data and interstellar turbulence predated the first astronomical attempts at measuring B by about 14 years. The latter were to directly measure the Galactic magnetic field strength – by Zeeman-splitting of the 1420.4 MHz interstellar hydrogen line (Chapter 2). Such measurements were technically difficult, and only in the last few years have various Zeeman measurements of $|B|_{\mathrm{interstellar}}$ converged on a result that is close to the original estimates of Fermi, Schlüter, and L. Biermann. Evidence for a relatively large-scale ordered component of the interstellar magnetic field came from observations of the optical polarisation of starlight over large regions of the galactic sky (Hiltner 1951), and later from radio polarimetry of the galactic interstellar radio synchrotron radiation made during the 1950s (Wielebinski & Shakeshaft 1964).

As radio astronomy developed in the 1950s, the discrete radio sources in the newly revealed "radio sky" were thought to be aggregates of radio-emitting stars. The radio star explanation seemed, *prima facie*, a natural extension of the discovery during World War II, that the Sun dominates the radio sky as a single source at metre wavelengths. It was reinforced by the identification of one of the brightest discrete radio sources near the Galactic plane as a supernova remnant, Taurus A, the Crab Nebula. These findings appeared consistent with Grote Reber's earlier discovery that the large-scale radio sky roughly traces the Milky Way, where most stars lie.

The subsequent discovery of a few *extragalactic* radio sources, not associated with any Galactic star, inspired searches for optical identifications for the increasing numbers of discrete radio sources found throughout the 1950s. These included Cygnus A, even brighter than Taurus A, and also lying almost directly in the Galactic plane. The absence of any obvious star or supernova at the newly precise location of Cygnus A, pinpointed with a radio interferometer, now justified a deep optical search at Cygnus A's position using the 5 m Hale telescope at Mount Palomar. Baade & Minkowski (1954) found a very faint, distant galaxy system at a redshift of 0.056. This revolutionary radio-optical correspondence suggested that most of the strongest discrete sources in the radio sky are indeed *extragalactic*, and at great distances.

Their cm wave radio emission is synchrotron radiation (Equation 1.1), and although associated with a galaxy, the emission often originates in large volumes *outside* the parent galaxy. These large, supra-galactic zones – "radio lobes" – have sizes now known to range from kpc-scale to several megaparsecs in the extreme, and are discussed later in the book. The largest radio sources have dimensions that are comparable with entire galaxy clusters and with galaxy–galaxy separation distances. Pertinent to this book is the fact that they are permeated by magnetic fields. Extended radio sources are now found at redshifts $\gtrsim 6$. These systems, having significant magnetic field strengths and magnetic energy, can be found throughout the extragalactic Universe.

It is straightforward to deduce that the largest of these largest systems have been "filled" with magnetic fields over a timespan of $\approx 10^8$ years, a period that is still negligible compared with the age of the Universe. Furthermore, the energy content of these large radio-bright extragalactic clouds, 10^{60}–10^{62} ergs, is a non-negligible fraction of the *rest mass* of the parent galaxy's central collapsed object, which appears to be their energy source. This makes them a remarkable phenomenon of the extragalactic universe. Much of the accumulated energy within these radio source lobes is in the form of magnetic flux – and generated within a relatively short period of cosmic time.

For these galaxy systems, the ultimate energy source must be gravitational, since even a nuclear energy source is insufficient. They have been a subject of considerable analysis since attention was first called to the enormous magnitude of their energies by G. R. Burbidge (1956). Later in the book we elaborate on the consequences of this central energy source, believed to be a supermassive black hole, for galactic and intergalactic magnetic fields. Its formation appears closely associated with the evolution of its host galaxy. The released energy, illustrated by Fig. 1.1, first flows outward in a highly collimated energy pipe, or "jet". Later in the book we show how its conversion from a gravitational form is very efficient. That is, a substantial fraction is magnetic energy, and this is why such systems are important to the theme of this book.

Another possible gravitational energy reservoir for widespread magnetic fields is infall energy that is dissipated in the course of cosmological large scale structure (LSS) formation.

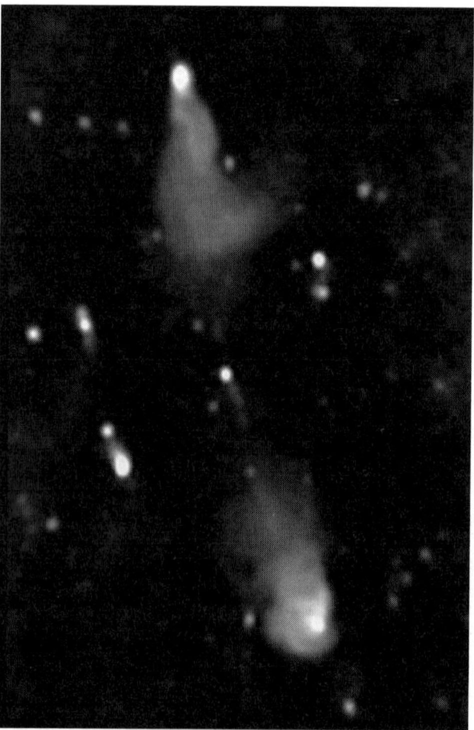

Figure 1.1 Radio image of 2147+816, as imaged by the VLA (NRAO-NVSS survey). Its projected largest dimension is ≈ 2.6 Mpc, comparable to an inter-galaxy separation distance.

As intergalactic matter falls into galaxy groups, clusters, and filaments and sheets of LSS, the accompanying shearing and turbulence create mechanisms for converting gravitational infall energy into thermal and magnetic energy.

Seed fields for galactic and intergalactic magnetic fields have been proposed to originate very early in the timeline of the Universe – before the recombination era at z $\approx$ 1500, during the earlier plasma epoch, or much earlier. Some of these primordial fields might have originated close to the time of the quantum chromo dynamic (QCD) phase transition when the scale of the Universe was as small as ~1 cm and less. This latter possibility also implies interesting connections with particle physics processes in the primordial universe, which are briefly discussed later in the book. Alternatively, or in addition, later field seeding by Biermann battery processes in the first stars was possibly an important contributor. Finally, the black hole-jet processes on various scales might alone have provided much, or most of the magnetic flux of interstellar and intergalactic space. It is possible that the latter field seeding processes overwhelmed the early primordial ones due to stars.

1.2 Observational techniques and results: past, present, and future prospects

Faraday rotation measures (*RM*s) from linear polarisation measurements of extragalactic radio sources began in the 1960s. These soon covered the entire sky with sufficient angular density of sources to indicate that a large-scale, organised magnetic field permeates the disk of our Galaxy (Davies 1968), consistent with Hiltner's (1951) starlight polarisation results. Faraday rotation $\Delta\chi$ of a plane polarised wave is defined as

$$\Delta\chi = 811.9\Delta\lambda^2 \int_0^L n_e B_{\parallel} dl \text{ radians} \tag{1.5}$$

where n_e (cm^{-3}) is the free electron density, $B_{\parallel}$ (μgauss) in an element dl along the line of sight to a linearly polarised radiator at distance L from the observer, and λ is in metres. The Faraday *RM*, defined as $\Delta\chi/\Delta\lambda^2$, is thus the path integral of the electron density-weighted line-of-sight component of the magnetic field. Given some independent estimate of the associated electron density, and the dimensions of the "cloud" containing both the electrons and magnetic field, B can be estimated – as we show later with examples.

Subsequent measurements of the Faraday rotation measures for larger numbers of (polarised) extragalactic radio sources has led to a more refined modelling of the large-scale Galactic magnetic field structure (e.g. Simard-Normandin & Kronberg 1980, Sofue & Fujimoto 1983, Clegg *et al.* 1992, Kronberg & Newton-McGee 2011). The Galactic magnetic field shows an underlying organisation on a grand scale, and appears to have some large-scale field reversal(s). Recent radio studies indicate that virtually all disk galaxies are permeated by large-scale magnetic fields.

The magnetic field in "clouds" containing both synchrotron-emitting relativistic electrons, (n_r, |B|) and thermal electrons, n_e, (Equations 1.1, 1.3) can be diagnosed in a more detailed way by appropriately designed Faraday rotation measurements described in Chapter 2. Generally, each volume element of the cloud will generate polarised synchrotron photons, which can be Faraday rotated along the ray path to the observer. This can happen on their way to us, in any intervening magnetised zone containing n_e and $B_{\parallel}$ – either within the cloud or in front of the cloud. A formalism was developed by Burn (1966) showing a Fourier transform relationship involving the observed Faraday rotation (variation of $\chi(\lambda^2)$), or *RM*. As discussed in Chapter 2, the line of sight distribution of *RM* can be applied to perform a kind of magnetic field "tomography" on a system of "clouds". Only relatively recently, some 40 years later, did it become feasible to extend magnetic field probes of distant extragalactic radio sources by applying variations of Burn's technique to newer observations of extended extragalactic sources (e.g. Brentjens & deBruyn 2005).

In the early 1970s, Faraday rotation measure data were used to probe for widespread intergalactic magnetic fields to large redshifts (Rees & Reinhardt 1972, Nelson 1973, Vallée 1975, Kronberg & Simard-Normandin 1976). The more recent discoveries that (i) the baryonic fraction of the visible universe is only a few percent, and that (ii) intergalactic space consists of cosmic voids and filaments, lead to a refined interpretation of these early $RM(z)$ results. These had produced a global, average present-epoch upper limit $B_{\mathrm{IGM}} \lesssim 10^{-9}$ G and did not distinguish what we now know as filaments, sheets, and giant voids of cosmological LSS.

Positive detections have emerged for magnetic field searches for high electron column density absorption line systems ($N_{\mathrm{e}} \gtrsim 10^{20.5}$ cm^{-2}) in front of radio-loud quasars. The interest in making such measurements is that the absorption line systems can exist at substantial redshifts – hence, significantly into the cosmic past. Thus if an intervenor galaxy at large redshift is detected in absorption, we have the opportunity to compare magnetic field strengths "then", with those "now" in related local-universe systems. Observations of this kind can test for global field strength evolution around galaxy systems over different cosmological epochs (Chapter 12).

Large radio telescopes with polarimetric capability can now image the same interstellar synchrotron emission in nearby galaxies that was detected in our own Milky Way in the early days of radio astronomy. Diffuse synchrotron radiation has recently been found within galaxy clusters, and beyond galaxy clusters in regions of significant galaxy over-density in the nearby universe. In general, searches for large-scale low-frequency (metre-λ) synchrotron radiation constitute a very direct means of detecting the presence of an intergalactic magnetic field. Such searches serve to trace "old" CR electrons and their associated magnetic field, and energy, over significant volumes of intergalactic space.

Satellite-based X-ray telescopes, through imaging and atomic spectroscopy of the intra-cluster intergalactic medium (ICM) within galaxy clusters, have made it possible to combine the consequently derived spatial distribution of the free electrons with Faraday rotation measures. This has led to magnetic field estimates in galaxy clusters, e.g. Kim *et al.* (1990). Energetic electrons in sufficiently large numbers also generate X-rays by inverse-Compton (IC) scattering, either of the 2.7 K CMB (Cosmic microwave background) photons, or, in more extreme cases, of their self-generated synchrotron radiation photons. Where IC-generated X-rays can be distinguished from those of hot gas bremsstrahlung, they provide additional, often important, constraints on the magnetic field.

Observations over the past thirty years have produced magnetic field detections not only in galaxy disks, but also in galaxy halos, clusters of galaxies, and in some very high redshift, i.e. cosmologically early, galaxy systems. These produce both absorption lines and Faraday rotation of the radiation from background quasars. Generally, the more we look for extragalactic magnetic fields, the more ubiquitous we find them to be.

Optical telescopes of 8 m- and higher-class can add very importantly to the power of Faraday *RM* probes, especially in distant Faraday rotating, intervening galaxies in front of background quasars and galaxies. This can be appreciated from Equation (1.5) given the dependence on n_e and B in a Faraday rotating system. High spatial and spectrographic wavelength resolution at the highest possible sensitivity can isolate column densities and temperatures of species in magnetised intervening objects that are associated with an image of Faraday rotation. The evolution of galaxy magnetic fields and gaseous conditions can thus be studied with a combination of optical and radio *RM* measurements over a large range of cosmological epoch. The addition of optical and X-ray observations augment the *RM* synthesis methods outlined in Chapter 2.

Another prospect for measuring magnetic fields is through perturbations of the cosmic microwave background (CMB) by galaxy clusters known as the Sunyaev–Zel'dovich (SZ) effect. The physical process was discovered by Sunyaev & Zel'dovich (1969) and it is largely insensitive to the cluster distance. As instrumental capabilities enable ever more precise detections of the SZ effect, it is possible to test for higher order perturbations of the SZ effect that can be sensitive to magnetic fields. The SZ effect, being largely redshift insensitive, cluster magnetic fields can be probed in this way along with other cosmological parameters, to significant redshifts. Finally, the polarisation properties of the CMB radiation itself can be analysed as an indirect diagnostic of very distant and cosmologically early magnetic fields.

The discovery of "ultra" high energy ($\gtrsim 10^{18.5}$ eV) cosmic rays (UHECR), high energy neutrinos, and extragalactic γ-ray sources provides several additional tools for probing extragalactic magnetic fields. Various interactions among hadrons, leptons, photons, and neutrinos propagating through intergalactic space can potentially be linked to intergalactic magnetic fields at different levels, often at far lower levels than what can be detected by Faraday rotation.

The succeeding chapters of this book elaborate on several topics mentioned in the preceding few pages. They will review some basic relevant theory, current and future measurement methods for diffuse astrophysical magnetic fields, results, and discuss magnetic field origins. Some astrophysical consequences and connections to other astrophysical topics are also explained and discussed.

Many of the earlier theories relating to the detection of magnetic fields (and even some results) date to the 1960s and earlier. From an experimental point of view some were well ahead of their time. More specifically, instrumental development and detections have often had to catch up with theory. In the next chapter and elsewhere in the book, we describe and elucidate methods for probing astrophysical magnetic fields in many contexts, along with discussion of some underlying physical processes.

Articles on astrophysical magnetic fields have been published over the past three decades by Priest (1985), Asseo & Sol (1987), Rees (1987), Wielebinski & Krause (1993), F. Krause (1993), Kronberg (1994), Grasso & Rubenstein (2001), Carilli & Taylor (2002) in the context of galaxy clusters, Han & Wielebinski (2002) for the Milky Way, Widrow (2002), and in a book edited by Wielebinski & Beck (2005).

References

Adams, W. S. 1943, The Structure of Interstellar H and K Lines in Fifty Stars, *Astrophys. J.*, 97, 105
Alfvén, H. & Herlofson, N. 1950, Cosmic Radiation and Radio Stars, *Phys. Rev.*, 78, 616
Alfven, H., Richtmyer, R. D., & Teller, E. 1949, On the Origin of Cosmic Rays, *Phys. Rev.*, 75, 892
Asseo, E. & Sol, H. 1987, Extragalactic Magnetic Fields, *Phys. Rep.*, 148, 307
Baade, W. & Minkowski, R. 1954, Identification of the Radio Sources in Cassiopeia, Cygnus A, and Puppis A, *Astrophys. J.*, 119, 206

Biermann, L. 1950, Über den Ursprung der Magnetfelder auf Sternen und im interstellaren Raum (mit einem Anhang von A. Schlüter), *Zeitschrift für Naturforschung*, 5a, 65

Biermann, L. & Schlüter, A. 1951, Cosmic Radiation and Cosmic Magnetic Fields. II. Origin of Cosmic Magnetic Fields, *Phys. Rev.*, 82, 863

Brentjens, M. A. & de Bruyn, A. G. 2005, Faraday Rotation Measure Synthesis, *Astron. Astrophys.*, 441, 1217

Burbidge, G. R. 1956, On Synchrotron Radiation from Messier 87, *Astrophys. J.*, 124, 416

Burn, B. J. 1966, On the Depolarization of Discrete Radio Sources by Faraday Dispersion, *MNRAS*, 133, 67

Carilli, C. L. & Taylor, G. B. 2002, Cluster Magnetic Fields, *Ann. Rev. Astron. Astrophys.* 40, 319

Carmichael, H. 1964, A Process for Solar Flares in *The Physics of Solar Flares, Proceedings of the AAS-NASA Symposium held 28–30 October, 1963 at the Goddard Space Flight Center, Greenbelt, MD*, ed. W. N. Hess (Washington, DC: National Aeronautics and Space Administration), 451

Clegg, A. W., Cordes, J. M., Simonetti, J. H., & Kulkarni, S. R. 1992, Rotation Measures of Low-latitude Extragalactic Sources and the Magnetoionic Structure of the Galaxy, *Astrophys. J.*, 386, 143

Davies, R. D. 1968, Structure of the Galactic Magnetic Field, *Nature*, 218, 435

Fabian, A. C. & Barçons, X. 1991, Intergalactic matter, *Rep. Prog. Phys.*, 54, 1069

Fermi, E. 1949, On the Origin of the Cosmic Radiation, *Phys. Rev.*, 75, 1169

Grasso, D. & Rubenstein, H. 2001, Magnetic Fields in the Early Universe, *Phys. Rep.*, 348, 163

Han, J. L. & Wielebinski, R.W. 2002, Milestones in the Observations of Cosmic Magnetic Fields, *Chinese J. Astron. Astrophys.*, 2, 293

Hess, V. 1912 – cited in Y. Sekido & H. Elliot, eds. 1985, *Early History of Cosmic Ray Studies* (Dordrecht: D. Reidel)

Hiltner, W. A. 1951, Polarization of Stellar Radiation. III. The Polarization of 841 Stars, *Astrophys. J.*, 114, 241

Kiepenheuer, K. O. 1950, Cosmic Rays as the Source of General Galactic Radio Emission, *Phys. Rev.*, 79, 738

Kim, K. T., Kronberg, P. P., Dewdney, P.E., & Landecker, T. L. 1990, The Halo and Magnetic Field of the Coma Cluster of Galaxies, *Astrophys. J.*, 355, 29

Krause, F. 1993, The Cosmic Dynamo: A Review on History, in *IAU Symp. 157, The Cosmic Dynamo*, eds. F. Krause, K.-H. Rädler, and G. Rüdiger (Dordrecht: Springer), 487

Kronberg, P. P. 1994, Extragalactic Magnetic Fields, *Rep. Prog. Phys.*, 57, 325

Kronberg, P. P. & Simard-Normandin, M. 1976, New Evidence on the Origin of Rotation Measures in Extragalactic Radio Sources, *Nature*, 263, 653

Kronberg, P. P. & Newton-McGee, K. 2011, Remarkable Symmetries in the Milky Way Disk's Magnetic Field, *PASA*, 28, 171

Nelson, A. H. 1973, On the Interpretation of Faraday Rotation Data, *Pub. Astr. Soc. Japan*, 25, 489

Pacholczyk, A. G. 1970, *Radio Astrophysics* (San Francisco: Freeman)

Priest, E. R. 1985, The Magnetohydrodynamics of Current Sheets, *Rep. Prog. Phys.*, 48, 955

Rees, M. J. 1987, The Origin and Cosmogonic Implications of Seed Magnetic Fields, *Q. J. R. Astron. Soc.*, 28, 197

Rees, M. J. & Reinhardt, M. 1972, Some Remarks on Intergalactic Magnetic Fields, *Astron. Astrophys.*, 19, 189

Schlüter, A., & Biermann, L. 1950, Interstellarische Magetfelder Zeitschrift für Naturforschung, 5a, 237

Schwinger, J. 1949, On the Classical Radiation of Accelerated Electrons, *Phys. Rev.*, 75, 1912

Shklovskii, I. S. 1953, The Problem of Cosmic Radio Waves. *Ast. Zhur. SSSR*, 30, 577

Shklovskii, I. S. 1953, On the Nature of the Radiation of the Crab Nebula. *Dokl. Akad. Nauk. SSSR*, 90, 983.

Simard-Normandin, M. & Kronberg, P. P. 1980, Rotation Measures and the Galactic Magnetic Field, *Astrophys. J.*, 242, 74

Sofue, Y. & Fujimoto, M. 1983, A Bisymmetric Spiral Magnetic Field and the Spiral Arms in Our Galaxy, *Astrophys. J.*, 265, 722

Sunyaev, R. A. & Zel'dovich, Y. 1969, Distortions of the Background Radiation Spectrum, *Nature*, 223, 721

Sunyaev, R. A. & Zel'dovich, Y. 1969, The Interaction of Matter and Radiation in a Hot-Model Universe, *Astr. Space Sci.*, 4, 301

Vallée, J. P. 1975, Magnetic Field in the Intergalactic Region, *Nature*, 254, 23

Widrow, L. M. 2002, Origin of Galactic and Extragalactic Magnetic Fields, *Rev. Mod. Phys.*, 74, 775

Wielebinski, R. 1990, *The Interstellar Medium in Galaxies*, ed. J. M. Van der Hulst (Dordrecht: Kluwer), 349

Wielebinski, R. & Beck, R., eds. 2005, *Cosmic Magnetic Fields*, Lecture Notes in Physics (Berlin: Springer), 664

Wielebinski, R. & Krause, F. 1993, Magnetic Fields in Galaxies, *The Astronomy and Astrophysics Review*, 4, 449

Wielebinski, R. & Shakeshaft, J. R. 1964, A Survey of the Linearly Polarized Component of Galactic Radio Emission at 408 Mc/s, *MNRAS*, 128, 19

Zel'dovich, Y. & Sunyaev, R. A. 1980, The Angular Distribution of the Microwave Background, and Its Intensity in the Direction of Galaxy Clusters, *Soviet Astr. Lett.*, 6, 285

2

Methods for probing magnetic fields in diffuse astrophysical plasmas

2.1 Introduction

Synchrotron radiation in radio bands and its polarisation properties can be used in many ways to probe and measure astrophysical magnetic fields (B). Equation (1.1) illustrates that the synchrotron emissivity scales by the local number density of relativistic electrons, n_{er} (and/or positrons). Thus, to quantify B and to fully characterise the state of an interstellar or intergalactic plasma, other companion measurements, sometimes over a range of frequencies, are needed to determine the ionisation fraction, field direction, etc. Magnetic fields play an important role in stars, in early stages of galaxy formation, and in large scale structure evolution and cosmology. This highlights the importance of an appropriate mix of complementary measurements. One example is the independent measurement of $\int n_{th}(l)dl$ along the propagation path to a pulsar to combine with a pulsar's RM to deduce an average $|B|_{interstellar}$. Another is the use of thermal bremsstrahlung X-ray measurements to complement Faraday rotation measurements in a galaxy cluster – with the goal of investigating the intracluster magnetic field, in this case by estimating n_{th} separately.

Different B-probing methods that do not involve synchrotron radiation are described in Section 2.3 below, and in parts of Chapters 9–12 that deal with high energy cosmic rays (CR), X-rays, γ-rays, neutrinos, and the cosmic microwave background (CMB).

2.2 Some basics of polarised EM waves

Most of the techniques discussed in this chapter involve measurement of the polarisation of electromagnetic waves. It is therefore useful to begin by briefly reviewing the parameters that describe the full polarisation state of a stream of radiation.

The four Stokes parameters I, Q, U, and V (Stokes 1852, Chandrasekhar 1950), are a convenient set of scalar quantities which completely specify the state of a stream of incoherent electromagnetic radiation. I specifies the total radiation and V the circularly polarised component. By the IEEE convention, the intensity of the left-handed circularly polarised component is given by $-V$. The linearly polarised component, described by Q and U, is vector-like, having both intensity ($\propto (Q^2 + U^2)^{1/2}$) and orientation ($\chi$). It is important to distinguish the orientation ($0°$–$180°$) of the plane of polarisation of a linearly (or elliptically) polarised wave, from the direction ($0°$–$360°$) for a proper 2-D vector. This explains why Q and U are orthogonal components of a 2-D vector whose argument is 2χ. Thus, in terms of the Stokes parameters, the fractional degree of linear polarisation of an EM wave is given by

$$m_{\mathrm{L}} = \frac{\sqrt{Q^2 + U^2}}{I} \qquad (2.1)$$

and the orientation of the polarisation plane is

$$\chi = \frac{1}{2} \tan^{-1} \frac{U}{Q}. \qquad (2.2)$$

The fractional circular polarisation is

$$m_{\mathrm{C}} = \frac{V}{I}. \qquad (2.3)$$

It is useful to note that a linearly polarised wave can be transmitted (or received) as the instantaneous sum of right- and left-handed circularly polarised waves. Their relative phase determines χ. If $|V_{\mathrm{L}}| \neq |V_{\mathrm{R}}|$, then the wave is elliptically polarised. An elliptically polarised wave can alternatively be decomposed into a purely linearly polarised wave superimposed on a right- or left-handed circularly polarised wave. Most incoherent and thermal sources of cosmic EM waves are unpolarised ($Q = U = V = 0$). This includes black body radiation from stars (if they are not shining through dust), or the optically thin X-ray bremsstrahlung radiation from hot intergalactic gas in clusters of galaxies (Chapter 9). In these cases, Stokes I (always positive) represents purely randomly polarised radiation ($Q = U = V = 0$), and in general

$$I \geq |Q| + |U| + |V|. \qquad (2.4)$$

Further discussion of the Stokes parameters can be found in Chandresekhar (1950) and, especially in the context of radio astronomy, by Cohen (1958). In the following we give a brief overview of observations, mostly radio, that can probe or measure cosmic magnetic fields.

2.3   Zeeman splitting of spectral lines

A direct way of measuring the strength of a uniform magnetic field, independently of other parameters, is to measure Zeeman splitting of a transition in the interstellar gas, stellar atmosphere, etc.

$$v = v_{\mathrm{mn}} \pm \frac{eB}{4\pi mc} \ \mathrm{Hz}, \qquad (2.5)$$

where v_{mn} is the frequency of a single transition and B is in Gauss. The fact that the splitting term is independent of frequency fundamentally favours lower frequency transitions. But this "advantage" is counteracted in situations where B is high, as in dense molecular cloud environments, or in the Sun, where Zeeman splitting can be observed even at optical frequencies.

The first apparently successful Zeeman splitting measurement was made in the late 1960s, for the interstellar 1420.4 MHz neutral hydrogen line in absorption (Davies 1968, Verschuur 1968). The long interval between the suggestion of Bolton & Wild (1957) that an interstellar Zeeman Effect should be observable, and the first detections of the effect is a measure of the technical difficulty of conducting such observations on a radio telescope.

Since those early measurements, interstellar Zeeman measurements have progressed with difficulty. Among principal instruments used are the Jodrell Bank and Hat Creek telescopes, and more recently the Arecibo telescope. Interstellar field strengths of $\sim$1.5–2 μG are a minimum level that can be detected. C. Heiles, T. Troland, and R. M. Crutcher have led much of this work, for which a summary can be found in Heiles & Crutcher (2005). Their results are related in Chapter 5.

Whereas the Zeeman splitting in absorption lines discussed above requires a sufficiently intense background source of small angular size, Zeeman broadening of HI emission lines can in principle be widely observed. But these measurements require a very precise knowledge of the telescope aperture and feed characteristics, that is, precise knowledge of telescope side lobes and spillover effects in each spectral frequency channel. For further reading on this topic, see Heiles & Crutcher (2005) and references therein.

Except for the Sun and in stellar atmospheres, where B can be at gauss-levels and above (e.g. Mestel & Landstreet 2005), Zeeman splitting is so far only detectable in special situations where the transitions are intrinsically narrow – such as maser lines, or narrow absorption lines. In the cold neutral medium (CNM) of our Galaxy, Zeeman splitting in 21 cm absorption lines reveals a median magnetic field strength of $\sim$6 μG (Heiles & Troland 2005). This result has significance for inferring magnetic fields in cold and dense regions in other galaxies. One can reason as follows: It is generally easier to detect molecular lines than magnetic field strengths in external galaxies. The former should occur in regions analogous to the CNM of our Galaxy, where significant magnetic fields have been detected – as above. This may mean that where complex, "pre-biotic" molecules are detected, their CNM-like environment may well be commonly magnetised. See e.g. Salter *et al.* (2008) for Arp 220.

In general, the lines must not be blurred by turbulent–thermal Doppler or Stark broadening that exceeds $\sim eB/4\pi mc$ (Equation 2.5). In dense molecular cloud cores the fields can be at milli-G levels. Here conditions permit masering transitions where the very narrow line widths in small zones enable a splitting to be observed. Here the Zeeman splitting reveals the magnetic field sense, and in general produces satellite lines of either linear (Q, U) or circular polarisation (V). The Zeeman satellite lines can be interpreted to measure the total $|B|$, rather than just $B_{\parallel}$ (Crutcher *et al.* 1993). Zeeman probes of magnetic field strength and sign will continue to be of great value, especially as the sensitivities and resolution of radio and mm telescopes improve, and with VLBI observations that include large telescopes for added sensitivity. Zeeman splitting of maser lines in excited OH ($^2\Pi_{3/2}$, J = 5/2 and 7/2), and in H_2O maser line transitions (6_{16}–5_{23}) give $B_{\parallel}$ values from $\sim$1 to $\sim$50 mG, where the density régimes are above $\sim$$10^8$ cm^{-3} (e.g. Fiebig & Güsten 1989, Sarma *et al.* 2002, Caswell 2004).

The first extragalactic detection of a pre-biotic molecular species (methanimine) in a distant IR-luminous, dust-rich galaxy Arp 220 was made by Salter *et al.* (2008). The cold conditions that permit the formation and survival of such molecules will permit future Zeeman splitting measurements of local regions in external galaxies. Given steadily improving sensitivity and resolution of future generations of radio and millimetre telescopes, and the ability to resolve faint masering transitions by VLBI techniques, more direct Zeeman measurements of magnetic fields should be possible in external galaxies, e.g. in excited states of OH and H_2O, and C_2S and CH_4 masers. For more reading on the background and quantum physics relating to Zeeman splitting the reader could consult the textbooks by Gerhard Herzberg, *Atomic Spectra and Atomic Structure* and *Molecular Spectra and Molecular Structure*

2.4 Polarisation of optical starlight and dust radiation as a probe of interstellar fields

Optical polarisation observations have made great strides due to the recent availability of imaging CCD polarimeters. Optical polarisation is a tracer of the interstellar magnetic field direction, first shown by Davis & Greenstein (1951). In the "Davis–Greenstein Effect", non-spherical interstellar grains will align their angular momentum vector with the local magnetic field direction, thereby causing magnetic field direction-dependent absorption and scattering of the incident starlight. Their detailed interpretation requires some additional knowledge, or assumptions, about the size and scattering properties of interstellar dust and electrons. Interstellar optical polarisation can be caused by simple scattering, and also by dichroic extinction, i.e. differential transmission for X- and Y-polarised photons due to interstellar grains which are aligned with the local galactic magnetic field (Chapter 5 and Scarrott *et al.* 1986).

Observation of the wavelength dependence of the optical polarisation at a given location in the Galaxy can distinguish different "signature" $p(\%)$-λ curves (Zerkowski curves) due to these different effects. The grain size-range and composition need to be known or modelled. A notable advantage of optical polarimetry is the higher angular resolution than that normally possible with radio techniques. Consequently, optical polarisation data can detect small-scale fluctuations in galactic fields. The resolution is, potentially, as good as the optical seeing limit of 1–2″. This is considerably better than what is practically possible in Faraday rotation, although the latter will also steadily improve with better surface brightness sensitivity of radio telescopes and detectors. As with synchrotron radiation, optical (and infrared) polarimetry of dust emission, and absorption, do not reveal the field sign.

2.5 Radio telescope techniques for polarimetry

For wide-angle imaging of, say the (polarised) galactic interstellar synchrotron radiation, a small radio telescope with a correspondingly large field of view is used to scan the sky area of interest. Similar techniques apply to a higher resolution, large single-dish radio (e.g. 0.5′–10′ HPBW) for scanning the images of nearby galaxies (cf. Chapter 5). In both cases, the result is an image of some sky zone, convolved with the radio telescope's beam which is smaller than the angular size of the image. The instrumental polarisation can be thought of as a 2-D vector field, which must be deconvolved from the observed image (another 2-D array of vector quantities) to produce the true polarised sky image. The combination of telescope and antenna feed imperfections will cause the antenna to see a polarised point source as elliptically polarised, and by different amounts for different directions over the beam and side lobes. A simpler situation occurs when the radio source's angular size is much smaller than the single telescope beam – in which case the Stokes parameters are just integrated values over the entire source.

If a polarised source is imaged by scanning over two dimensions with a single-dish telescope, it is important to know the full specification of the instrumental response (telescope + feed) to a polarised source. The polarisation properties of a telescope can be completely described by a 9-element matrix, known as the Müller matrix, M_{ij}. If the Müller matrix is completely measured it is possible to recover the correct polarisation state at each given (x, y) position in the sky at a given wavelength (λ), i.e.

$$p(x, y, \lambda)^{\mathrm{obs}} = \|M_{\mathrm{ij}}\| . p(x, y, \lambda)^{\mathrm{true}}. \tag{2.6}$$

The construction of a Müller matrix for this purpose is described in Turło *et al.* (1985). Then, given a family of $p(x, y)^{\mathrm{true}}$ at 2 or more λ's, a Faraday *RM* value can be deduced at each (x, y) position over an extended radio source.

When the 2-D polarisation image is obtained with an interferometer, which samples the Fourier transform of $p(x, y, \lambda)$ (see below), corrections are needed for impurities in the polarisation response of each interferometer element. This information can be applied to the complex correlation coefficient, or visibility, of each antenna pair. This naturally involves many more correction terms than for the single antenna scenario above. The specification and removal of the instrumental polarisation for interferometers have been described in some detail by Roberts, Wardle, & Brown (1994), and by Sault & Cornwell (1999). As with single dish-scanned polarisation images, the instrumental effect-corrected images of Stokes parameters *I*, *Q*, and *U* at 2 or more frequencies enables 2-D $RM(x, y)$ images that can be used to probe the magnetic field structure of any polarised radio source.

Aperture synthesis radio interferometers build up spatial Fourier components of a source's image by correlating many antenna pairs. The Fourier components of the sky distributions of $I(\xi, \eta)$, $Q(\xi, \eta)$, $U(\xi, \eta)$ and $V(\xi, \eta)$ are thereby in effect sampled by each antenna pair. Since *V* is nearly always zero in the situations discussed here, it will be omitted in this discussion. We define ξ and η to be the source's Cartesian co-ordinates centred on some local, asymptotically planar patch on the sky. A common practice is to have each antenna receive left- and right-hand circular polarisation. When the output of a given antenna pair, say A and B, is correlating signals of the same handedness, e.g. "R" with "R", and "L" with "L" then, if $V = 0$, either of these combinations gives exactly one-half of the Fourier component of $I(\xi, \eta)$ for an unpolarised source. The combinations "RL" and "LR" can be shown to measure Fourier components of the distribution of only the linearly polarised component, i.e. $Q(\xi, \eta) + jU(\xi, \eta)$, a 2-D vector field. As in the single dish case above, the responses of antenna pairs A, B, etc. will in practice be contaminated by some small amount of spurious instrumental polarisation. This can be determined for each antenna by observing standard calibration sources (e.g. an unpolarised thermal source such as a planetary nebula and a "polarisation calibrator" source of independently measured *m* and χ). The theory of interferometer responses to extended polarised sources is described in more detail by Cohen (1958), Conway & Kronberg (1969), and Roberts *et al.* (1994).

Radio polarisation measurements using the above techniques can be designed to give either the calibrated integrated polarisation of a radio source at several wavelengths, or to measure the polarisation distribution over an extended radio source at one or more wavelengths.

2.6   Faraday rotation

2.6.1   Faraday rotation combined with independent thermal electron densities

In a more physically realistic application of Equations (2.1)–(2.3), we measure the Stokes parameters $I(\xi, \eta, \lambda_{\mathrm{n}})$, $Q(\xi, \eta, \lambda_{\mathrm{n}})$, and $U(\xi, \eta, \lambda_{\mathrm{n}})$ at wavelengths λ_1, λ_2, λ_3 … A simpler method is to measure their integrated values over (ξ, η) at each λ. Given that $\chi(\lambda_{\mathrm{n}})$ is an orientation angle, each measurement of $\chi(\lambda_{\mathrm{n}}^2)$ contains the inherent ambiguity $\chi = \chi \pm \mathrm{n} \cdot 180°$.

This means that, for the purpose of determining RM – i.e. the slope of the χ–λ^2 relation – coverage in λ^2 must be sufficiently complete that any $n\cdot 180°$ ambiguities can be resolved in the presence of noise and other uncertainties. Over the cm bands, usually 4 or more measurements are required, suitably distributed in χ^2 "space" (that is, at both small and larger $\Delta\chi^2$ intervals), to unambiguously determine the RM of the integrated radiation from a given radio source (see e.g. Simard-Normandin, Kronberg, & Button 1981).

Because we shall later discuss Faraday rotation probes of intervening systems at significant redshifts where cosmological scale factors are important, we can write a more general version of Equation (1.5) to describe the RM, when both the source and intervenor(s) can be at any redshift.

$$RM = \frac{\Delta\chi}{\Delta\lambda^2} = 8.12 \times 10^5 \int_0^{z_s} (1 + z)^{-2} n_e(z) B_{||}(z) dl(z). \tag{2.7}$$

As before, n_e (cm^{-3}) is the local density of non-relativistic electrons, $B_{||}$, the line-of-sight component of magnetic field (μgauss), and l the path length (parsecs). The RM is measured in the observer's rest frame at $z = 0$. The $(1 + z)^{-2}$ factor arises from the transformation $\lambda_{obs} = \lambda_{cloud}(1 + z)$, and $dl(z)$ varies according to the cosmological model adopted. Figure 2.1 illustrates a Faraday rotating system, in this case a cluster of galaxies, projected as an "intervenor" in front of some background, polarised radio source.

If the cloud in question co-expands with the Universe, the quantities $n_e(z)$ and $B_{||}(z)$ change with z due to pure cosmological scaling, e.g. $n_e(z)= n_e(0)(1+z)^3$, Alternatively, the exponent will be something other than 3, e.g. if the Faraday rotating system is already in a gravitationally bound system whose scaling is no longer determined by universal

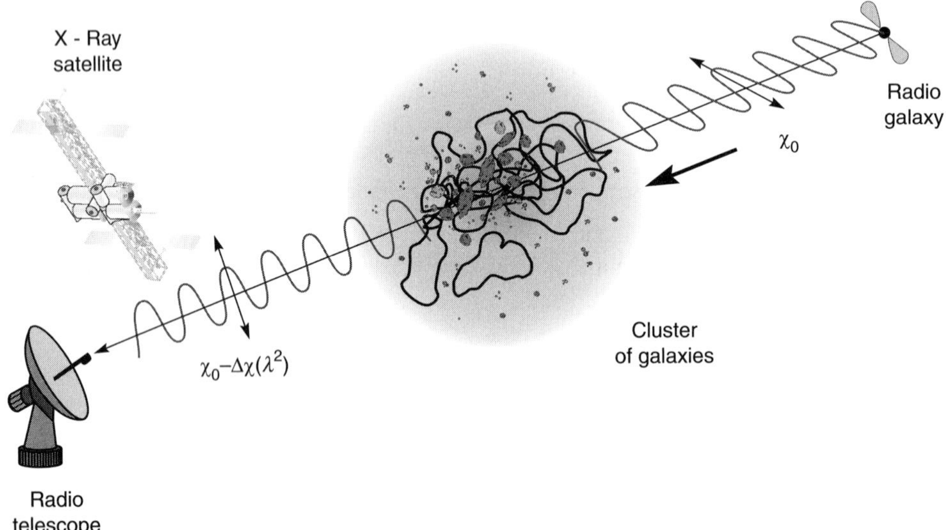

Figure 2.1 Illustration of the Faraday rotation of a background radio source at $z = z_s$ due to the intra-cluster magnetic field and hot 10^{6-7} K gas in a cluster of galaxies at z_{cl}. Multi-wavelength radio polarimetry to obtain RM requires complementary X-ray measurements to obtain n_e in the medium that is permeated by the magnetic field (Equation 2.7).

coexpansion. A similar statement can be made for $B_\parallel(z)$ where $B_\parallel(z) = B_\parallel(0)(1+z)^2$ for the case of adiabatic coexpansion. In general at large redshifts, $n_e(z)$ and $B_\parallel(z)$ can be influenced by a combination of cosmological scaling and evolutionary effects, since z is also a (cosmological model-dependent) function of proper time, $T(z)$.

The magnetic field strengths and baryonic matter densities in interstellar and intergalactic space are such that Faraday rotation is usually detectable only at radio wavelengths. To estimate the magnetic field strength from Faraday rotation it is obvious from Equations (1.5) and (2.7) that we require both an independent measurement of the free electron density *and* knowledge of its B-weighted distribution along the line of sight.

For Galactic pulsars, the latter is independently obtainable from a pulsar's Dispersion Measure (DM):

$$DM = c_{DM} \int_0^{l_p} n_e dl \; \text{cm}^{-3}\text{pc}. \tag{2.8}$$

DMs can be measured from the relative pulse delay vs. frequency relation. The ratio RM/DM (Equation 2.7 divided by Equation 2.8) has units of B and thus gives a weighted value of the line-of-sight component of field strength along the sight line to the pulsar. This type of magnetic field measurement was first performed by Lyne & Smith (1968). Unfortunately, most pulsars are too faint to observe in external galaxies, let alone at large z, so this method is limited to the interstellar medium (ISM) of the Milky Way until extragalactic pulsars are within observational reach.

The much improved sensitivity of X-ray satellites is making it possible to convert the X-ray count rate images and temperatures to thermal free electron *column* densities, N_e, especially in clusters of galaxies. If Faraday rotation can be measured in the same hot gas system then $n_e(l)$ can be determined, thus giving some line-of-sight weighted estimate of $|B|$. Some basic principles for doing this are described below.

N_e (cm^{-2}) or n_e (cm^{-3}) can be estimated from the overall bremsstrahlung emissivity, typically visible as thermal X-rays at 0.1–10 keV photon energies:

$$\frac{dE}{dt} \approx 1.43 \times 10^{-27} n_e n_i \sqrt{TZ^2} g \; \text{erg s}^{-1}\text{cm}^{-3}. \tag{2.9}$$

It results from charge acceleration in an ensemble of coulomb interactions between electrons and ions of atomic number Z in a plasma, which is assumed here to be optically thin.

Apart from Faraday rotation combined with independent electron density and field reversal scale estimates, possibilities are relatively few for *directly* measuring magnetic field strengths in the extragalactic universe. But they do exist, as we show in later chapters. However, even in the absence of companion data on n_e, detailed imaging of the polarised synchrotron emission (n_r and B) in and around external galaxies and galaxy clusters can give considerable information on the morphology and degree of magnetic field ordering. We will discuss later some less direct, but potentially powerful methods for estimating or limiting magnetic field strengths even in the absence of separate measurements of n_e.

Finally, separate parts of a multi-component radio source may have different RMs. We discuss this further below, but note here the simple case of a two-component source in which $RM_1 \neq RM_2$. Here it can be intuitively understood as two rotating vectors that "beat" against each other in λ^2 "space". This is a simple example of an integrated *RM* that is λ-dependent, in which case the observed *RM* has a sinusoidal behaviour in λ^2. When examined in detail with precisely measured $\chi(\lambda_n{}^2)$, many extragalactic radio sources show small non-linear "undulations" in an integrated plot of χ vs. λ^2. Over a large λ^2 interval, however, a single average slope (*RM*) in a χ vs. λ^2 plot can often be identified. In this sense a unique *RM* can be assigned to most sources over at least some large (wavelength)^2 range, and there are several such compilations of integrated *RM* in the literature. The preceding example of a two-*RM* component is a prelude to Section 2.7, where we describe more general and sophisticated *RM* probes of magnetic field structure that can apply to any class of intrinsically polarised radio source.

2.6.2   When is Faraday rotation negligible?

Faraday rotation, named after its discoverer Faraday (1844), is, microscopically, the consequence of the Lorentz force due to an electron's motion ($v_e \times B$), induced by a linearly polarised electromagnetic wave that propagates through a magnetised plasma containing a magnetic field at angle (ψ) to the propagation direction. A linearly, or more generally, elliptically polarised wave with initial major axis orientation χ_0, can be construed as a superposition of oppositely polarised ordinary "O" (LH) and extraordinary "E" (RH) circularly polarised propagation modes. When the EM wave's frequency $v \gg v_p$ (the plasma frequency), and $\gg v_B$ (the non-relativistic gyro frequency, $eB/2\pi m_e$), these modes have slightly different refractive indices (n) by

$$\Delta n = \frac{v_p{}^2 v_B}{v^3} \cos(\psi).$$
(2.10)

Consequently, the phases of the E and O modes, after propagating over a distance l at a given λ^2, differ by $\Delta\phi$. Because the superposition of the E (LH) and O (RH) modes produces a linearly polarised wave, it is easy to see that the orientation χ of the resultant linearly polarised wave, (or of the major axis of the resultant elliptically polarised wave if $|E| \neq |O|$), repeats over $180°$, not $360°$). Thus, $\Delta\phi = 2\Delta\chi$; that is, the measured polarisation is a pseudo-vector whose orientation angle needs to be multiplied by 2 in order for it to be represented mathematically as a vector quantity (Section 2.2.1). This leads to the Faraday rotation relation (2.7).

The constant in front of the integral equation in the Faraday rotation relation (2.7) contains the quantity $e^3\lambda^2/(2\pi m^2 c^4)$. This shows that $\Delta\chi$ is sensitive to the electron charge and mass by e^3/m^2. It is obvious that protons in the same ionised gas cloud would cause a rotation in the opposite sense, and *reduced* by $(m_e/m_p)^2$, i.e. $1/(1836)^2 = 3.37 \times 10^{-6}$. This is negligible relative to that caused by electrons (or positrons) in a non-relativistic plasma, and shows why the nuclei in an ionised gas cloud cause virtually no Faraday rotation in the presence of a magnetic field.

This discussion naturally leads to the question of whether *relativistic*, synchrotron-radiating electrons can cause Faraday rotation. The answer depends on the *effective* mass of a relativistic electron. Where $\gamma = E_e/mc^2$, the mass of a relativistic electron, γm, is comparable to a proton's rest mass when $\gamma = 1836$, and $mc^2 = 0.512$ MeV. As just discussed, the relativistic proton's Faraday rotation is negligible at this energy level. Even for electron γ's of order 10 ($E_e = 5.1$ MeV), the *RM* would be lower by a factor of 100 compared with a non-relativistic electron.

If we exclude the most compact extragalactic sources, galactic and extragalactic synchrotron emission observed above ~30 MHz requires electron γ's above ~500. As just noted, these relativistic electrons produce no significant contribution to Faraday rotation. For similar reasons, $\gamma \gtrsim 10$ ($E_e \gtrsim 5.1$ MeV) corresponds to plasma temperatures $\gtrsim 5.9 \times 10^{10}$ K. The hottest common temperatures in the thermal intra-cluster gas are ~10 keV (1.16×10^8 K), that is, more than two orders of magnitude below the threshold for reduced Faraday rotation. At these electron temperatures there clearly would be little relativistic reduction of their Faraday rotation.

In astrophysical plasmas, *RM*s are found to range from ~1 to $\gtrsim 10000$ rad/m^2. The lower limit of 1 is comparable to short term *RM* variations in the Earth's ionosphere, whose total *RM* can be up to ~10 rad/m^2 depending on time of day and the geomagnetic coordinates on the Earth. The λ^2 dependence of $\Delta\chi$ furthermore tells us that, even for electrons, $\Delta\chi$ is negligible below $\approx$ mm wavelengths unless the magnetic energy and electron densities are enormous. Faraday rotation in optical bands can occur in some laboratory plasmas, for example in some dense and magnetised laser-heated plasma. But this régime of parameter space does not apply to typical situations encountered in this book. Possible exceptions could be in environments close to magnetised neutron stars, or to super-massive black holes, where some Faraday rotation in mm bands might be seen. For the plasmas treated in this book, post-Recombination Faraday rotation is detectable mainly in cm wave and longer bands.

2.7 The concept of Faraday depth and magnetic field probes in the 3rd dimension

2.7.1 *Idealised models*

It is instructive to define the Faraday depth of a volume element emitting polarised radiation as the Faraday *RM* that its emission undergoes on the way to us. As Figure 2.2 illustrates, contributions to the Faraday depth (Φ_i) of the *i* th element can come from any $r < r_i$, either from somewhere within the emitting source's volume, or in an intervening, source-external magnetised ionised gas cloud.

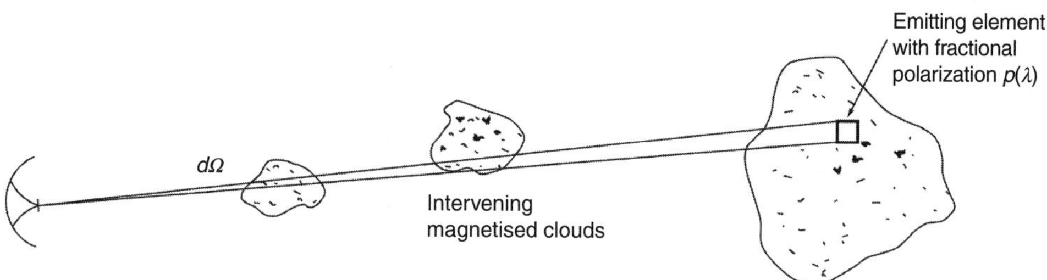

Figure 2.2 A polarised emitting element $d\Omega$ at position vector r_i whose Faraday depth is Φ_i within a radiating cloud of solid angle Ω. The intervening, Faraday rotating "clouds" might include the local interstellar medium of the Milky Way, and/or one or more extragalactic Faraday rotating interveners between us and the source cloud.

Here, $\Phi_i(r)$, expressed by Equation (1.5) for $z_{cloud} \approx 0$, can have several RM contributions, $\Phi_i(r)$. If any of the systems illustrated in Fig. 2.2 has a redshift significantly greater than zero, we need Equation (2.7) to describe $\Phi_i(r)$. The integrated (over Ω) fraction of polarisation, $P(\lambda^2)$, measured at wavelength λ by a radio telescope whose beam is larger than Ω, can be written as

$$\underline{P}\left(\lambda^2\right) = \frac{\iint\limits_{\Omega} \varepsilon_i(\underline{r}_i, \lambda)\Pi_i(\underline{r})e^{2i[\chi^0{}_i(\underline{r}_i, \lambda)+\phi_i(\underline{r}_i, \lambda)\lambda^2]}\, dld\Omega}{\iint\limits_{\Omega} \varepsilon_i(\underline{r}_i, \lambda)dld\Omega}. \tag{2.11}$$

P and Π denote the integrated and local linear polarisation fractions, so that the polarised emission of each i th element at vector location r_i is $\Pi_i\,\varepsilon_i$ at an intrinsic position angle $\chi^0{}_i$. $\varepsilon_i(\lambda)$ is the total emissivity (ergs s^{-1} cm^{-3}) of the i th element. The latter is Faraday rotated by $\Phi_i\lambda^2$, where Φ_i is the Faraday depth of the i th element, and $r = \mathbf{k}\cdot l$, $\mathbf{k}$ being a unit vector. We express the measured (complex) fraction $P(\lambda^2)$ as a function of λ^2 rather than λ because of the intrinsic λ^2 dependence of the Faraday depth (rotation) in the exponent.

Ideally we would invert the measured $P(\lambda^2)$ in Equation (2.11) to deduce Π_i, ε_i, and Φ_i at all r_i, after combining with independent 3-D information on n_e. The latter is included in Φ. With this information we would come closer to the goal of a 3-D magnetic field probe within the source. It would help, though not be entirely sufficient, if we significantly increased the telescope resolution ($\Omega_{beam} \ll \Omega_{source}$) so that $P(\lambda^2)$ could be separately measured for many independent, adjacent sightlines through the same radio source.

In the idealised situations discussed at this point, we assume that the spectral index of the radio emission, $\varepsilon_i(\lambda)$, is constant over the λ or v range under investigation. In some realistic situations a radio source can be the summation of different sub-regions having different spectral index ($\varepsilon \propto v^\alpha$) – e.g. a radio component with $\alpha \sim 0$ combined with a "normal" radio spectrum ($\alpha \sim -0.7$) associated with a highly polarised, optically thin synchrotron emitting source. This would require a full specification of $\varepsilon_i(r_i, \lambda)$ in Equation (2.11). We will ignore this refinement for the purpose of a didactic discussion. Further discussion of the incorporation of "spectral depolarisation" can be found in Burn (1966), Gardner & Whiteoak (1966), and Brentjens & de Bruyn (2005).

Following Burn (1966), Equation (2.11) can be reduced to one dimension by superimposing all cells with the same Faraday depth Φ, i.e. substituting r with Φ. Most, but not all, of Burn's nomenclature has been followed here. The integrated linearly polarised flux for this subset of cells at a single Faraday depth in Fig. 2.2 is the complex quantity $\boldsymbol{P(\Phi)}$, $= \Pi(\Phi)$ $E(\Phi)\,e^{2i\chi^0(\Phi)}$. Here, the integrated (scalar) total flux at Φ is $E(\Phi)$. $\int_\phi E(\phi)d\phi$ is also equal to the denominator in Equation (2.11). A Faraday dispersion function can be defined as $F(\Phi) = E(\Phi)P(\Phi)$, likewise a complex function. These substitutions permit Equation (2.11) to re-written in a simplified form as

$$P(\lambda^2) = \int\limits_{-\infty}^{\infty} F(\phi)e^{2i\phi\lambda^2}\, d(\phi). \tag{2.12}$$

The complex polarised intensity $P(\lambda^2)$, can be recognised as very similar to that for the Fourier transform $F(\Phi)$, which is

$$F(\phi) = \frac{1}{\pi} \int\limits_{-\infty}^{\infty} P(\lambda^2) e^{-2i\phi\lambda^2} d(\lambda^2).$$ (2.13)

The qualifying word "similar" in the previous sentence was used because $P(\lambda^2)$ is not defined at negative values of λ^2; hence, one cannot straightforwardly invert it mathematically to obtain a unique $F(\Phi)$. This is an important limitation. To make the observed $P(\lambda^2)$ invertible it is necessary to assume special conditions, for example taking $P(\lambda^2)$ to have Hermitian symmetry about $\lambda^2 = 0$. In that case $F(\Phi)$ must be a real function (see e.g. Burn 1966). This is satisfied, for example, either if $\chi_i(\Phi) \equiv 0$, or if it is constant; in both cases, $e^{2i\chi_i(\Phi)}$ can be taken out of the integral in Equation (2.13). Then direct Fourier inversion in order to obtain $F(\Phi)$ *is* possible.

Figure 2.3 shows an example of such an idealised source consisting of a single slab, containing a uniform magnetic field, and relativistic electrons uniformly mixed with the thermal electrons. This "special case" would also accommodate a constant foreground RM between the source and us, such as the Galactic ISM. By the Shift Theorem in Fourier transforms the foreground RM would simply be a shift of $F(\Phi)$ in Φ space by Φ_{fg}, leaving $|P(\lambda^2)|$ unchanged, and the model source in Fig. 2.3 would have an observed RM modified by Φ_{fg} rad/m^2. Figure 2.3(b) shows the case where $\Phi_{fg} = 0$.

This simple example shows how the observed λ-dependence of polarisation ($P(\lambda^2)$) can, in limited ways, be related to a 3-D distribution of emissivity and n_e. In this simplified example, $|F(\Phi)|$ is a top-hat function having width Φ_0 rad/m^2, where Φ_0 is the back-to-front RM through the slab. It is displaced from $\Phi = 0$ by $\Phi_0/2$ rad/m^2 (Fig. 2.3). Its Fourier transform, the measured $P(\lambda^2)$ is a $\sin\theta/\theta$ function whose width is $(\Phi_0)^{-1}$. The observed RM will be $\Phi_0/2$, corresponding to ½ the total Faraday depth of the slab. Provided n_e can be independently measured, the line-of-sight magnetic field component, $B_{||}$, within this slab source is directly determined. The measured $|P(\lambda^2)|$ in the upper plot is a sinc function, analogous to a diffraction image of light passing through a small rectangular slit:

$$\frac{P(\lambda^2)}{P(0)} = \frac{\sin\left(\lambda^2\varphi_0\right)}{\lambda^2\varphi_0}.$$ (2.14)

Note that $\chi(\lambda^2)$ abruptly flips at the first null in $|P(\lambda^2)|$ by 90° (180° in 2χ space). This happens at a λ for which the back-to-front $\Delta\chi$ within the "slab" is 180°. If we now were to randomise B and/or change the slab to a more realistic sphere, the $\chi(\lambda^2)$ variation will be linear at first, then deviate from a straight line (Fig. 2.3). That is, the "phase" of $P(\lambda^2)$ and RM cease to be constant in λ^2 when $P(\lambda^2)/P(0)$ falls to ≈ 0.7. For most straightforward models having the (magnetised) thermal and non-thermal emitting zones co-extensive, $RM\,(\lambda^2)$ becomes non-linear when $P\,(\lambda^2)/P(0)$ falls by a significant factor.

Special model source configurations that have other than $\Pi_i\,(r) = $ const (Fig. 2.3) can also be devised to make $P(\lambda^2)$ Fourier-invertible to $F(\Phi)$. An example is Gaussian B-fluctuations about a mean direction such that $<\Pi_i\,(\Phi)> = $ constant even though $\Pi_i\,(r)$ is not the same everywhere. Another possibility is that the RM is generated externally to the source in some foreground "Faraday screen" with different adjacent lines of sight varying their RM at

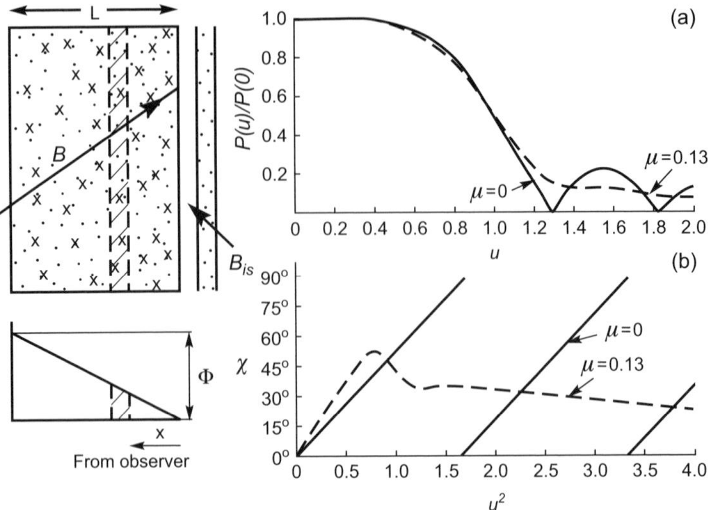

Figure 2.3 Idealised synchrotron emitting and Faraday rotation models. The solid lines in (a) and (b) show $P(\lambda^2)$ for a slab of uniformly mixed constant B, n_i, and n_e, $\mu = 0$ specifies a constant, uniformly aligned B in which $\varphi \propto L$, and without the effect of the thin RM-only "slab" to the right that represents an external foreground magneto-ionic medium φ_{fg}. The dashed lines show how $P(\lambda^2)$ would be altered if the magnetic field is partially randomised ($\mu = 0.13$). $\chi(\lambda^2)$ becomes non-linear in λ^2 as $P(\lambda^2)/P(0)$ begins to decline. If the external Faraday rotating medium is now "inserted", part (a) remains unchanged, but $d\chi/d\lambda^2$ is reduced everywhere – in this example – by φ_{fg}. The dashed $P(\lambda^2)$ are qualitatively similar in this case if the slab model is changed to a more realistic spherical model, not shown. (Adapted from Burn 1966).

random. Additional favourable, special cases are discussed by Burn (1966), Gardner & Whiteoak (1966), Tribble (1991), and Sokoloff *et al.* (1998, *erratum* 1999). In general, Fourier inversion of $P(\lambda^2)$ to obtain a unique 3-D "image" of the magneto-ionic medium in the radio galaxy or supernova remnant, etc. is not straightforward. The essence of the above is that we must create some informed model of $F(\Phi)$ and $F(r)$. However, with increasing angular resolution, we can generate a family of $F(\Phi, x_i, y_i)$, where (x_i, y_i) represent hundreds, or thousands of contiguous lines of sight over a well-resolved 2-D image of the source in question. The *lateral* constraints on $F(\Phi)$ and consequently $F(r)$ can bring us closer to a true 3-D "Faraday tomography" of a radio source. A more realistic example from actual observations is shown and discussed in Section 2.8.

2.7.2 *Faraday rotation in cosmic radio sources*

When we examine real RM observations of the integrated radiation of discrete radio sources over a wide range of λ^2, we often find that in reality the RM is approximately linear over a wide range, even to large λ's at which the source is almost depolarised! This is a clear discrepancy with the model radio sources in Fig. 2.3 in which a Faraday rotating plasma is mixed in with the source. Thus, the widespread linearity of observed RMs immediately rules out many models of 3-D magnetised thermal plasma that is *mixed uniformly* within cosmic radio sources.

That said, a mixed-in thermal plasma is very applicable to the ISM of our Galaxy, and by extension other late-type galaxy disks, and perhaps the stellar wind-driven galaxy halos discussed later. Multi-frequency "Faraday synthesis" of the kind illustrated very simply in Figs. 2.2 and 2.3 will be increasingly able to probe the Galactic ISM as suitable techniques become increasingly available. Some companion measurements include interstellar scintillation of pulsars and compact extragalactic radio sources, RMs and DMs of Galactic pulsars, interstellar γ-rays and CR nucleons below $\sim 10^{17}$ eV. Radio continuum measurements of the thermal gas distribution in the Milky Way, IR observations, and interstellar recombination lines in radio and optical transitions can all be important complements. For 3-D radio probing of the Milky Way's magnetic field, the addition of wideband $P(\lambda^2)$ measurements hold considerable future promise. They can involve multi-band polarimetry over a wide range of λ^2, from typically ~ 2 GHz to 100 MHz and below. Relevant instruments for the longer λ régime are the LOFAR array in the Netherlands and Germany, the Mileura Widefield (low-frequency) Array (MWA) in Australia, and the Long Wavelength Array (LWA) project in New Mexico, USA. The detailed 3-D B-probes discussed above have largely not been possible at the time of writing this book.

For further mathematical and theoretical descriptions of a mixed-in Faraday rotating plasma, the interested reader could also consult Cioffi & Jones (1980) and Spangler (1982, 1983). To get a recent picture of the scope of detailed magnetic field-related polarimetric observations in the Milky Way, see e.g. Wolleben & Reich, (2004), Wolleben (2007), and references therein. For the Large Magellanic Cloud, see Gaensler *et al.* 2005. Large scale magnetic fields in some nearby-universe spiral and irregular galaxies are discussed in Chapters 5 and 6.

2.8   The Crab Nebula as a 3-D Faraday synthesis model

The radio-bright Crab Nebula supernova remnant is an instructive example of a radio source having strong synchrotron radiation and an associated magnetised thermal plasma. In a seminal paper cited earlier, Burn (1966) also analysed observations of the integrated $P(\lambda^2)$ of the Crab Nebula at many wavelengths from optical ($\lambda^2 \approx 0$) to 1 GHz ($\lambda^2 = 900$ cm). Because $\chi(\lambda^2)$ was found, for example, to have departures from linearity (non-constant $RM(\lambda^2)$), Fourier inversion of $P(\lambda^2)$ for the Crab was in principle not possible to obtain $F(\Phi)$. However, the departures from linearity were small, and $|P(\lambda^2)|$ was well-defined, as shown in Fig. 2.4 below, adapted from Burn's paper. This suggests that the above-mentioned approach of "informed" models of $F(\Phi)$ can be successfully attempted. These, when Fourier transformed, can give an acceptable fit to $P(\lambda^2)$.

In the mid-1960s there were no detailed 2-D polarisation images of the Crab, except in the optical (Woltjer 1957, 1958). The optical polarisation and emission line data provided an initial template for the "informed" $F(\Phi)$ model by showing that the synchrotron radiating bulk of the Crab's volume is threaded/surrounded by a network of dense emission-line filaments.

The $F(\Phi)$ distribution function (Fig. 2.5) that produced the best fit to the observed $P(\lambda^2)$ had a large range of Faraday depths, including a substantial portion with low Faraday depth. This is consistent with partial coverage of the main body of the SNR, which has very low thermal gas content, by higher density foreground filaments which have radial velocities

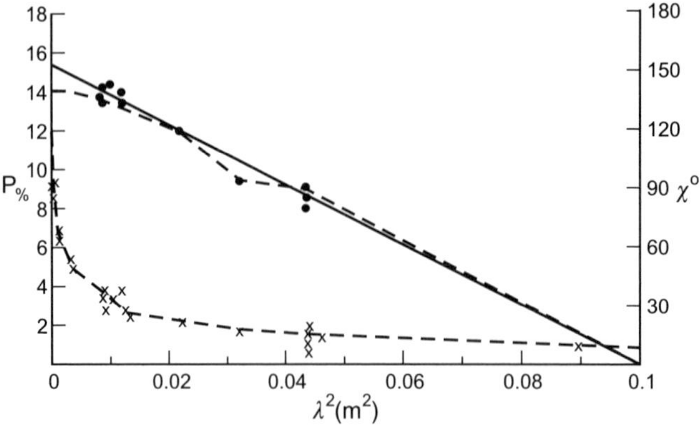

Figure 2.4 The measured $|P(\lambda^2)|$ and $\chi(\lambda^2)$ of the integrated Stokes' parameters Q and U of the Crab Nebula from optical wavelengths to 30 cm. Refer to the original in Burn (1966) for sources of the measurements.

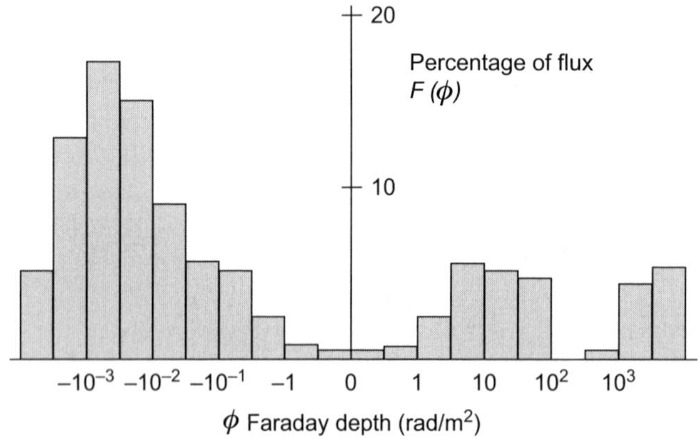

Figure 2.5 The model Faraday dispersion function $F(\Phi)$, fitted to the observations in Fig. 2.4 of $P(\lambda^2)$ of the integrated emission of the Crab Nebula from optical λs to 1 GHz radio (from Burn 1966).

toward us and magnetic field strengths of $\sim$0.3 mG, density $\sim$1 cm^{-3}, and areal coverage of $\sim$1.8 $\times$ 10^{37} cm^2) – numbers from Burn 1966. The large spread in $F(\Phi)$ is due to the high densities and strong magnetic fields in the Crab filaments. The corresponding estimate of filament mass, deduced from Fig. 2.5 and the optical morphology was $\sim$1.1 $M_\odot$. This agrees well with $\sim$0.46 $M_\odot$ independently deduced by O'Dell (1962) from optical H_β emission line strengths in the filaments. The foregoing results illustrate the potential power of Faraday synthesis probes of radio sources, and also the importance of ancillary data, for the construction of an informed physical model.

Subsequent to these analyses from the 1960s, multi-frequency 2-D radio images at $2''$ resolution in I, Q, and U were made with the VLA in the late 1980s by Bietenholz & Kronberg (1990, 1991). Figure 2.6(a), reproduced from Bietenholz & Kronberg (1991) shows the 2-D distribution of Faraday rotation with a resolution comparable to the optical polarisation image of Woltjer (1957). Figure 2.6(b) shows the distribution of Stokes parameters $(Q(\xi,\eta) + U(\xi,\eta))^{1/2}$ at $2''$ resolution, at $\lambda 21.3$ cm (1410 MHz) – the longest of 4 wavelengths used to construct the $RM(\xi,\eta)$ image in Fig. 6.2(a). This image shows strong (radio) depolarisation at the locations of the foreground (optical) filaments. It illustrates the astrophysical information that can be deduced from such multi-frequency RM and depolarisation images. It can be noted that the integrated $P(\lambda^2)$ behaviour of the Crab has some qualitative similarities with $P(\lambda^2)$ for many extended AGN/black hole-powered extragalactic radio sources. A key similarity is in the overall linearity of $\chi(\lambda^2)$, despite some wiggles, in the presence of a strong decline in $|P(\lambda^2)|$ toward longer λ (Fig. 2.4). As we saw above, this behaviour is generally incompatible with a thermal plasma that is uniformly mixed within the synchrotron-emitting volume.

The overall radio morphology of the Crab Nebula, a typical "plerion"-type Galactic SNR, is quite different from the double-lobe/jet structure of an extended FR I or FR II radio galaxy or quasar. However, what they have in common is the inflation and energisation of a "thermal plasma-poor" volume of synchrotron emitting CR electrons. Here, these are apparently accelerated or expelled by a coherent, dynamo-like, magnetic field tied to a rotating object. For the Crab Nebula this is the central pulsar, and for a radio galaxy it is the accretion disk or near-ergosphere of the central black hole. We might hypothesise that the similarity in $P(\lambda^2)$ in these morphologically diverse systems is fundamentally related to the

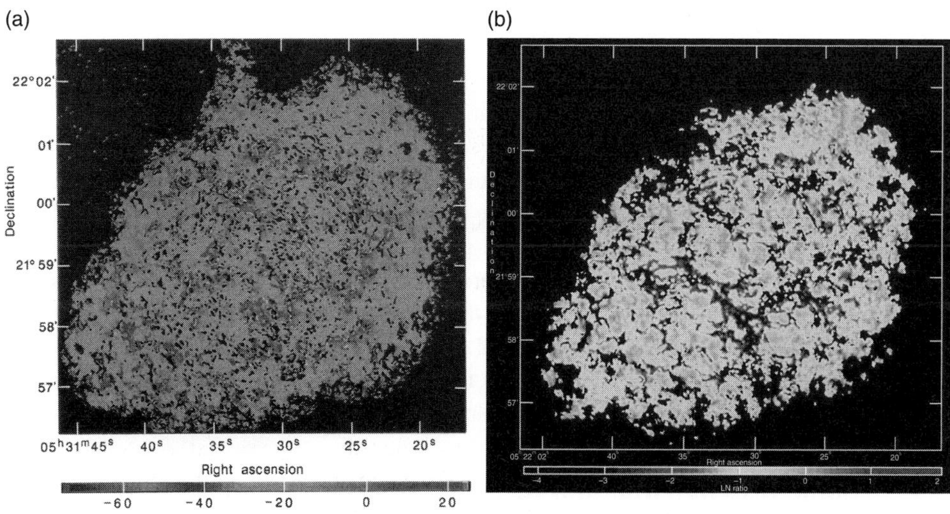

Figure 2.6 (a) At a resolution of $2''$, the 2-D image of the RM (x, y) over the Crab Nebula is produced here from four input (Q, U) images at $\lambda\lambda 6.1$, 6.4, 19.8, and 21.3 cm), made with the NRAO VLA (colour image from Bietenholz & Kronberg 1991). (b) Crab Nebula image at the same $2''$ resolution, showing the fractional polarisation at $\lambda 21.3$ cm, the longest wavelength used in the RM image in Fig. 2.6(a). It illustrates the dramatic depolarising effect of those (magnetised) filaments which are in front of significant synchrotron radiation from the interior of the Crab supernova remnant (reproduced from the original colour version by Bietenholz & Kronberg 1991).

CR acceleration mechanism, despite scale differences by a factor of $\gtrsim 10^6$. Related discussion can be found in Chapters 4 and 9.

Figure 2.6(b) in particular shows that the synchrotron radiation distribution is "overlaid" by a web-like, partially unresolved network of narrow, high *RM* zones (Fig. 2.6(a)). These zones exhibit sharp depolarisation at $2''$ resolution (Fig. 2.6(b)), and correspond well to the Crab's optically visible filaments that are expanding toward us.

In the *RM* image of Fig. 2.6(a) there is a nearly constant *RM* "plateau" at -21 rad/m^2, due to the foreground ISM. This corresponds to the "$B_{\text{interstellar}}$" slab in Fig. 2.3. Especially noteworthy is that this broad, low-*RM* component does not show significant *RM* modulation, even toward the edge of the nebula. Because the Crab synchrotron source has a network of ~ 0.1 pc ($\sim 11''$) magnetically coherent cells (Bietenholz & Kronberg 1990, Hickson & van den Bergh 1990), $B_{\parallel}$ should change by $\sim \mathcal{O} |B|$ over this scale, which is much larger than the $2''$ resolution. Thus it can be concluded from the "flatness" of the background *RM* (light orange or grey in Fig. 2.6(a)) that, within the *RM* errors, any thermal electrons mixed in with the main synchrotron source are not detected. The corresponding upper limit ($\lesssim 0.01$ cm^{-3}) is an order of magnitude below the average ambient interstellar ionised gas density. Future detection of the smaller *RM* variations shown in Fig. 2.3 will require a longer λ^2 baseline, which means extending such observations to lower frequencies, while maintaining a high angular resolution.

The *RM*s in the Crab's narrow filaments are, at $1.8''$ resolution, in the range 200 to >500 rad/m^2, have a size range from 0.1 to 5×10^{-3} pc, and magnetic field strengths ~ 0.5 mG (Bietenholz & Kronberg 1991). These numbers are consistent with Burn's original $F(\varphi)$ model in Fig. 2.5. They also reveal that the filament-internal magnetic field pressure is close to the thermal pressure estimated from the filaments' optical line emission (Péquignot & Dennefield 1983).

To perform similarly detailed Faraday rotation and optical diagnoses of the magnetised lobes and jets of extragalactic radio sources requires a combination of high surface brightness sensitivity *and* resolution that is beyond the capabilities of most current instruments. At this time of writing there are few extragalactic sources that can be studied in as much detail as the Crab Nebula. One of the first studied is the nearby radio galaxy Fornax A, in the Southern sky. Fornax A's surface brightness and angular size is high enough to also reveal a web of filaments (Fomalont *et al.* 1989). In a preliminary way this is consistent with the Faraday *RM* synthesis predictions discussed above, confirming that the thermal matter associated with Fornax A's lobes is, as with the Crab SNR, confined mostly to filaments and may not be part of the main radiating volume.

2.9   Some instrumental and measurement effects involved in Faraday rotation imaging

For the purpose of interpreting measurements of Faraday rotation images, we discuss some observational and instrumental effects that must be taken into account in deducing magnetic fields in radio sources. The following describes three examples:

(a)   Changes in polarisation angle within the beam area. One such effect is known as beamwidth depolarisation, in which the beam contains unresolved regions of size $\Delta\Omega \lesssim \Delta\Omega_{\text{beam}}$ having differing *RM*s. The different *RM*s might be associated with different depths in the 3rd dimension. Even in the absence of differential *RM*, unresolved sharp changes in the intrinsic χ within the beam distribution can also produce a sharp reduction in the observed polarisation degree, and mimic a depolarisation effect. To reveal and separate all of these effects requires higher angular resolution, and multiple λ's for the case of varying *RM*.

(b) Bandwidth depolarisation. Maximising sensitivity to faint emission requires the largest feasible bandwidth. But at some point differential *RM* within the bandwidth will reduce, or in the extreme, wash out the polarised signal. Differential *RM* over a finite bandwidth can be expressed by a change in polarisation angle, $\Delta\chi$ between opposite ends of a given frequency band, Δv, centred on v_0:

$$\Delta\chi_{\Delta v, v_0} = -2RMc^2 v_0^{-3}\Delta v \quad \text{radians.} \tag{2.15}$$

Here we express *RM* in units of rad/m^2, c in m/s^1, and v in Hz. As Δv increases, and/or if the centre (radio) frequency decreases for constant Δv, bandwidth polarisation becomes increasingly severe. It begins to set in when $\Delta\chi_{\Delta v, v_0}$ of order 1 rad.

The Crab Nebula images in Fig. 2.6 are affected by both effects (a) and (b) above. For the Δv = 25 MHz bandwidth used by Bietenholz & Kronberg (1991) of 1410 and 1515 MHz ($\lambda\lambda$21.3 and 19.8 cm), bandwidth depolarisation would begin to set in at their lowest frequency bands (Fig. 2.6(b) if the *RM* in the Crab filaments exceeded ~1000 rad/m^2. The observed Faraday depth (ϕ) range at their resolution of 1.8″ was $\Delta\phi$ ~300 rad/m^2, not far below this critical level. In fact, in this particular example, beam width depolarisation (a) at 1.8″ would compete with and possibly overwhelm the bandwidth depolarisation. Again using the Crab Nebula as an example, if the resolution at comparably high surface brightness sensitivity were ≪1.8″, possible smaller scale and larger $\Delta\chi$ variations might be bandwidth-smeared. This is an interesting issue for the coming generation of radio telescopes such as the planned Square Kilometer Array (SKA), which will have more (and adjustable) angular resolution, sensitivity, and (adjustable) bandwidth resolution than before.

A solution to this, especially important at lower radio frequencies, is to slice the total available bandwidth into N smaller segments $\Delta v' = \Delta v/N$. This would eliminate bandwidth depolarisation below ~$\Delta\phi = \phi/N$ rad/m^2.

(c) Spectral index (α) effects. Such effects, discussed briefly above, can be accounted for by independently imaging a source's I, (especially) and Q and U at a few suitably chosen v's to determine the α variations (see also Brentjens & de Bruyn 2005).

2.10 Faraday tomography to model magnetic structures in the 3rd dimension

The concepts outlined above and illustrated in Figs. 2.2–2.6 can be extended within a similar mathematical framework to provide quasi-3-D probes of the emission and *RM* structure of complex gaseous systems. Using Faraday synthesis techniques, de Bruyn & Brentjens (2005) have produced a polarisation "cube" giving the polarised flux as a function of Faraday depth (ϕ) for the intracluster medium of the Perseus cluster of galaxies in the format $|P(0.66–0.90 \text{ m}^2)|(\xi, \eta, \phi_n)$ where $n = 1, \ldots, N$ represent discrete "slices" in ϕ-space. The foreground *RM* distribution of the Milky Way, $\phi_{\text{fg}}(\xi, \eta)$ was separately determined from the RMs of discrete extragalactic radio sources in the sky field of the Perseus cluster. With the help of observations of this kind, it will be possible to infer the 3-D structure of B for a range of Faraday depth slices.

Given independent estimates or models of the 3-D distribution of n_e, e.g. from X-ray thermal bremsstrahlung emission, an ultimate goal will be to convert $P(\xi, \eta, \phi_n)$ to $P(\xi, \eta, z_n)$, where z_n is the line-of-sight dimension. This normally requires modelling, by incorporating additional information. Examples are the thermal gas density and temperature just mentioned, information from neighbouring discrete sources, and information on the transverse correlation scales of P, α where relevant, and ϕ. The latter can be gained by comparing data in adjacent contiguous lines of sight. A related technique can be used to apply RM smoothing in the (ξ, η) plane to resolve $n\pi$ ambiguities in RM. This was successfully applied to the RM imaging of the Hydra cluster of galaxies (Vogt, Dolag, & Enßlin, 2005).

Ideally we would like I, Q, and U to cover both small and large ranges of λ^2 with adequate sensitivity. Rapid "wiggles" in $P(\lambda^2)$ (associated with RM non-linearities) reveal information on $F(\phi)$ separated by large ϕ scales. Conversely, measurements of $P(\lambda^2)$ (including $\chi(\lambda^2)$) over a large and continuous range of λ^2 give a well-defined determination of $F(\phi)$ for all ϕ ranges at each (ξ, η) location.

The "inverse resolutions" in reciprocal coordinate space are exactly analogous to the relationship between beam width and maximum baseline of an interferometer, or time and frequency domain measurements of a time-varying signal $f(t)$. By analogy with beam pattern side lobes of a sparsely filled antenna aperture, or a comb filter in frequency-space in a circuit, gaps in the $P(\lambda^2)$ coverage will produce "diffraction lobes" in Faraday depth space.

Also, as depolarisation progresses for a given angular resolution to the point where $\chi(\lambda^2)$ deviates from linearity in λ^2, we obtain information on the interference between different parts of the $F(\phi)$ distribution along a line of sight. The value of λ^2 at which these non-linearities set in still depends on the electron densities, magnetic field strengths and physical scale of the system. As we saw for the Crab Nebula, non-linearities in $P(\lambda^2)$ develop at $\lambda \approx$ 4 cm and beyond. In the case of the Perseus cluster of galaxies, relevant $P(\lambda^2)$ set in at $\lambda \approx$ 80–90 cm. Here the relevant $|B|$ is $\sim$1–5 μG and $n_e \lesssim 10^{-3}$ cm^{-3}. Faraday synthesis probes of the latter systems require images at long λ, typically in the range $\lambda \gg 10$ cm up to $\sim \lambda 3$ m. Appealing to what we have learned from the Crab Nebula and Fornax A, the coverage in λ^2 should be not only wide, but as continuous as possible.

We are now equipped to apply the principles outlined above to set the specifications of a radio polarimeter to probe magnetic fields: The large overall λ^2 bandwidth should be finely segmented to detect regions with large $\Delta\phi$. The large maximum λ (or λ^2) sets an angular resolution requirement – λ/D so that D must be many 10's of km to achieve an arc second level beam at the longest desired wavelengths. This requires an interferometer, hence cross correlation of many (hundreds of) elements. Further, the segmented $\Delta\nu$ slices to cover inverse-λ^2 (large ϕ-) space require that each of the pairwise correlations be done separately for I, Q, and U (and V). Then for each pair, each $\Delta\nu$ sub-channel of an antenna must be correlated with the sub channels of all other antennas, and vice-versa. For N antennas, the number of correlated pairs for each Stokes parameter and frequency band is $N(N-1)/2$ – for the 27-antenna VLA this is 361 correlations per channel. Cross correlations against other antennas are not needed for a single dish, so the computational and electronic requirements are far fewer. But for most extragalactic (and Galactic) applications, we have seen that the simultaneous requirement of including long λ^2 and sub-arc second resolution demands a large multi-element interferometer of high sensitivity. At meter wavelengths the single-element beam widths are so large that Fourier transform methods must also deal with non-planar sky fields (in effect 3-D transform methods).

The correlators required to fulfil these specifications need supercomputer speeds and data throughput, in addition to very large and fast data storage technology. This capability is only becoming available in the 21st century. An example is the new WIDAR correlator, implemented in the 2009–2010 timeframe at the NRAO to create the Enhanced VLA (EVLA, aka the Jansky VLA). Similarly powerful electronics will extend capabilities of new low frequency interferometers such as LOFAR, and eventually the conceptual SKA and others.

2.11　Total energy and magnetic field estimates for synchrotron-radiating clouds

It is of interest to estimate the magnetic field energy density, (ε_B) in a synchrotron-emitting radio source that is observable over metre to millimetre wavelengths. This applies to e.g. supernova remnants and radio galaxies in any luminosity range. Unfortunately, we cannot normally just take the total synchrotron luminosity over a limited frequency interval $\nu_1 \rightarrow \nu_2$ as a complete proxy for the electron energy spectrum. For each source volume element that we detect, a relativistic electron energy density spectrum $\varepsilon_e(E)$ is weighted with the magnetic energy density, $B^2/8\pi$. In the absence of other measurements (sometimes possible) we cannot unscramble ε_e and B and must therefore invoke assumptions to infer B. Among these is the relativistic proton/electron energy ratio $K(E) = \varepsilon_p(E)/(\varepsilon_e(E)$ – usually unknown, since ε_p is not detected in the radio.

The integrated synchrotron luminosity of a radio galaxy, quasar, or SNR is

$$L = 4\pi D_L^2 \int_{\nu_1}^{\nu_2} l(\nu)d\nu \text{ erg/s,} \tag{2.16}$$

where $D_L(z)$ is the (cosmological model-dependent) luminosity distance of a radio source at a measured redshift z and, for generality, the frequencies are in the emitted reference frame, i.e. $\nu = \nu_{obs}(1 + z)$. Of course, where a source in question can be imaged, the luminosity measurement can be made through several (column) resolution elements over the source. For simplicity and illustrative purposes in most of the following, we refer to a measured integrated luminosity as in Equation (2.16) of an assumed homogeneous radio source whose magnetic field strength we wish to estimate.

To do this, we assume that the source is in an equilibrium state, in which case we calculate the total energy content (E_T), and then minimise it by differentiating it with respect to $|B|$. Other usual assumptions are that the electron pitch angle distribution is locally isotropic within the source volume, the magnetic field structure is random within the source, and that $K(E) = \varepsilon_p(E)/(\varepsilon_e(E)$ is a constant, k, independent of energy (and time).

With these simplifying assumptions, following e.g. Pacholczyk (1970), we can write a formula that relates the minimum total energy content (E_{tot}^{min}) to the directly measurable quantities of a radio source (in frequency space):

$$E_{tot}^{min} \approx 0.5 C^{4/7}(1+k)^{4/7}(\phi V)^{3/7} L^{4/7} \text{ ergs.} \tag{2.17}$$

The total energy is minimised when there is approximate equipartition in energy density, ε, between the relativistic ions and electrons (ε_i, ε_e), and that of the magnetic field ($B^2/8\pi$). We approximately follow the convenient formulation of Pacholczyk (1970), in which V is the overall source volume, Φ is the volume filling factor of the source's synchrotron radiating zones within that volume. The simplified parameter $k = \varepsilon_i/\varepsilon_e$ specifies the relativistic ion-to-electron energy ratio k, which is often not directly obtainable from radio observations. However, in the case of Galactic CRs that can be directly detected on Earth, it is ≈ 100. For distant central black hole (BH)-powered CRs in radio lobes with a different acceleration mechanism than the Galactic CRs, it has yet to be firmly specified. k may well be closer to 1 than 100 (see Chapter 7). C is a slowly varying function of v_1, v_2 (Hz), the observed lower and upper cut-off frequencies, and of the radio spectral index, α. In this book I use the convention $S \sim v^{\alpha}$, where S is the observed radio spectral flux density (erg cm^{-2}s^{-1} Hz^{-1}). For "typical" values of $\alpha = -0.7$, $v_1 = 10^7$ Hz, $v_2 = 10^{10}$ Hz, $C = 5.4 \times 10^7$, the analytical expression is

$$C = 1.056 \times 10^{12} \left(\frac{2\alpha + 2}{2\alpha + 1} \right) \times \left(\frac{v_1^{\left(\frac{1+2\alpha}{2}\right)} - v_2^{\left(\frac{1+2\alpha}{2}\right)}}{v_1^{(1+\alpha)} - v_2^{(1+\alpha)}} \right) \text{ (cgs).} \tag{2.18}$$

It comes from a substitution of CR particle energy (which is not directly observable) with the more directly observable frequency of the maximum spectral density of synchrotron radiation from a CR electron that has energy E_e. C is tabulated in Pacholczyk's book as "c_{12}" for some commonly used values of v_1, v_2, and α. The maximum spectral density occurs at about 0.3 of v_C, the so-called critical frequency at which v_C and E_e are related by

$$v_C = 6.3 \times 10^{18} (B \sin \vartheta) E_e^2 \text{ Hz,} \tag{2.19}$$

and ϑ is the electron velocity pitch angle relative to the local magnetic field, B.

Incorporating all of the above assumptions and simplifications, the magnetic field corresponding to the minimum total energy condition can be written as

$$B_{\mathrm{minE}} \approx 2.3(1 + k)^{2/7} C^{2/7} (\phi V)^{-2/7} L^{2/7} \text{ gauss,} \tag{2.20}$$

again approximately following the terminology of Pacholczyk (1970).

Ideally we want to independently know how the full energy spectrum of CR ions and electrons, i.e. $N(\varepsilon_i)$ and $N(\varepsilon_e)$ vary throughout the source, which would give $K(E)$. Also, the term C results from integrating over a fixed frequency interval. Inferring a range of integration in particle energy $E_1 \to E_2$ from an observed luminosity over frequencies $v_1 \to v_2$ has a dependence on the observed frequency spectral index, (α), though usually not severe. More detailed discussion and (still assumption-dependent) calculations of corrections to "equipartition" magnetic fields can be found in, among others, Brunetti *et al.* (1997), Pfrommer & Enßlin (2004), and Beck & Krause (2005) who include a graph of calculated corrections to B_{minE}. For a straight logarithmic spectrum (frequency-independent α) that is not too steep ($\alpha \gtrsim -1.2$), the correction factors to B_{minE} are modest, usually within a factor of ~ 2.

Much of the literature, and much of the above discussion on correcting estimates of the "equipartition" magnetic field or B_{minE}, omits a significant, possibly dominant correction factor. This is the volume filling factor ϕ in Equations (2.17) and (2.20) (in this case not Faraday depth). It is difficult to measure, and a consequence of limited instrumental surface brightness sensitivity and/or angular resolution. Thus, it is easy to miss the fact that synchrotron emitting "clouds" are, or may be, highly filamented, i.e. $\phi \ll 1$. A plausible physical understanding of this is that e.g. radiative cooling, will naturally lead to filamentation, thereby producing a lower total energy content of the magneto-plasma above, i.e. $E_{\text{tot}}^{\text{min}}$. A conservative estimate of $\phi = 0.1$ would raise the "equipartition" magnetic field by a factor of 5.7, and reduce $E_{\text{tot}}^{\text{min}}$ by 43% (from Equation 2.17). The few very bright, and sufficiently resolved radio sources observed so far do show significant filamentation, suggesting that the effective ϕ may commonly be less than ~0.1, and thus implying a non-trivial correction to the energy budget.

In a more rigorous way, given a full image having both high resolution *and* high brightness sensitivity, one would ideally perform a volume-weighted energy calculation that combines the lower volume (lower ϕ), brighter filament components with truly diffuse, low brightness (higher ϕ) emission. High surface brightness sources such as in the Crab Nebula (Fig. 4.2), and some high surface brightness extragalactic sources such as Virgo A and Cygnus A would lend themselves to this type of analysis of B_{minEtot} and E_{mintot}.

2.12　Prospects for magnetic field measurement in other energy bands

Up to this point, we have focused mostly on methods and instrumental considerations based on synchrotron radiation, Faraday rotation, and the Zeeman probes of magnetised gaseous systems. Here we have alluded only briefly to other methods, and details have been left to later chapters that discuss specific systems. These include extragalactic radio sources, the intracluster medium of galaxy clusters, the intergalactic medium, and magnetism in the pre-galactic era beyond $z \approx 50$.

2.12.1　*Far ultraviolet and X-ray observations*

Many X-ray satellites work in the ~0.1–10 keV region, and can detect diffuse emission in the intracluster medium (ICM) of galaxy clusters, outflow halos of galaxies, diffuse emission around galaxies and clusters, and the jets and lobes of active galactic nucleus (AGN/black hole-powered radio sources). The resolution of the latter varies from about an arcsecond (e.g. the Chandra satellite), to a few arc minutes. Compact AGN sources, quasars, blazars, BL Lac objects, supernovae and γ-ray bursters are also detectable sources of X-rays. In fact, X-ray catalogues of discrete extragalactic objects are dominated by the first four object types in this list.

There are three common types of luminous X-ray emission: (1) Thermal bremsstrahlung of a hot ($\gtrsim 10^5$ K) optically thin gas; (2) inverse Compton radiation when a relativistic electron (Thompson) scatters a photon, usually a CMB photon; and (3) X-ray synchrotron radiation. A unique signature of mechanism (1) is X-ray lines, usually in emission, whose excitation temperature is, correspondingly, $\gtrsim 10^5$ K. Synchrotron emission is a direct detector of a magnetic field at some level. Aside from X-ray lines, the X-ray continuum spectrum of bremsstrahlung is difficult to observationally distinguish from the quasi-power law spectra of inverse Compton and synchrotron radiation in the standard X-ray bands. Synchrotron radiation may be highly polarised, but X-ray polarisation up to now has been technically difficult to measure.

However, the *ratio* of inverse Compton to synchrotron radiation, e.g. in the ICM of galaxy clusters where all three processes can happen, offers an important way to constrain the magnetic field strength. Basically the ratio of the CMB energy density, ε_{CMB} to the magnetic energy density in the diffuse plasma, $B^2/8\pi$ gives the approximate ratio of IC to synchrotron emissivity, independent of X-ray energy. Alternatively, if the radio synchrotron photon density is very high, as in compact radio hotspots, $\varepsilon_{SYNCH} \gg \varepsilon_{CMB}$, the synchrotron emitting electrons may preferentially Compton up-scatter these photons. In first order Compton scattering of a photon of frequency v, the frequency of the scattered photon is $v' = \gamma^2 v$, where γmc^2 represents the relativistic electron energy. Thus, for a power law energy distribution of relativistic electrons, $N(\gamma) \sim \gamma^{-\Gamma}$, where Γ is typically -2.7, the IC-scattered X-ray spectrum might be identified by having the same logarithmic slope as the pre-scattered radio spectrum.

2.12.2   Extragalactic fields, high energy cosmic rays, and γ-rays

Gamma rays having $E\gamma \gtrsim 10$ TeV will have photon-photon reactions with the extragalactic background light (EBL). These both produce and interact with e^+e^- pairs. In Chapter 10, we discuss different observable effects of the trajectories of these e^+e^- pairs, whose Larmor radius is determined by an intergalactic magnetic field.

Cosmic ray nucleons are also deflected in magnetic fields, whether in our galaxy or in intergalactic space – for example, the trajectories of ultra-high energy CR nuclei, UHECR defined as having $E \gtrsim 10^{17}$ eV, can be bent over intergalactic dimensions. This raises the possibility of using UHECR observations of different types to infer the strength and structure of intergalactic magnetic fields, especially in the nearby, low $- z$ universe at distances $\lesssim 100$ Mpc where UHECR proton energies are not too strongly attenuated. These effects can be quite complex, for example, involving nucleon–photon (p–γ) interactions. Along with γ-rays, their relation to intergalactic magnetic fields is discussed in greater detail in Chapter 10.

References

Beck, R. & Krause, M. 2005, Revised Equipartition and Minimum Energy Formula for Magnetic Field Strength Estimates from Radio Synchrotron Observations, *Astron. Nachrichten*, 326, 414

Bietenholz, M. F. & Kronberg, P. P. 1990, The Magnetic Field of the Crab Nebula and the Nature of Its 'Jet', *Astrophys. J. Lett.*, 357, L13

Bietenholz, M. F. & Kronberg, P. P. 1991, Faraday Rotation and Physical Conditions in the Crab Nebula, *Astrophys. J.*, 368, 231

Bolton, J. G. & Wild, P. R. 1957, On the Possibility of Measuring Interstellar Magnetic Fields by 21-CM Zeeman Splitting., *Astrophys. J.*, 125, 296

Brentjens, M. & de Bruyn, G. 2005, Faraday Rotation Measure Synthesis, *Astron. Astrophys.*, 441, 1217

Brunetti, G., Setti, G., & Comastri, A. 1997, Inverse Compton X-rays from Strong FRII Radio-Galaxies., *Astron. Astrophys.*, 325, 898

Burn, B. J. 1966, On the Depolarization of Discrete Radio Sources by Faraday Dispersion, *MNRAS*, 133, 67

Caswell, J. L. 2004, New OH Masers at 13 441 MHz, *MNRAS*, 352, 101

Chandrasekhar, S. 1950, *Radiative Transfer* (London: Oxford University), 24

Cioffi, D. F. & Jones, T. W. 1980, Internal Faraday Rotation Effects in Transparent Synchrotron Sources, *Astron. J.*, 85, 368

Cohen, M. H. 1958, Radio Astronomy Polarization Measurements, *Proc. I.R.E.*, 46, 172

Conway, R. G. & Kronberg, P. P. 1969, Interferometric Measurement of Polarization Distribution in Radio Sources, *MNRAS*, 142, 11

Crutcher, R. M., Troland, T., Goodman, A. A., Heiles, C., Kazès, I., & Myers, P. 1993, OH Zeeman Observations of Dark Clouds, *Astrophys. J.*, 407, 175

Davies, R. D. 1968, Structure of the Galactic Magnetic Field, *Nature*, 218, 435

Davis, L. & Greenstein, J. L. 1951, The Polarization of Starlight by Aligned Dust Grains, *Astrophys. J.*, 114, 206 de Bruyn, G. & Brentjens, M. 2005, Diffuse Polarized Emission Associated with the Perseus Cluster, *Astron. Astrophys.*, 441, 931

Dolag, K, Vogt, C., & Enßlin, T. A. 2004, PACERMAN – I. A New Algorithm to Calculate Faraday Rotation Maps, *MNRAS*, 358, 726

Faraday, M. 1844, *Experimental Researches in Electricity* (London: R. Taylor)

Fomalont, E. B., Ebneter, K. A., van Breugel, W. J., M. & Ekers, R. D. 1989, Depolarization Silhouettes and the Filamentary Structure in the Radio Source Fornax A, *Astrophys. J.*, 346, L17

Fiebig, D. & Güsten, R. 1989, Strong Magnetic Fields in Interstellar H2O Maser Clumps, *Astron. Astrophys.*, 214, 333

Gaensler, B. M., Beck, R., & Ferretti, L. 2004, The Origin and Evolution of Cosmic Magnetism, *New Astronomy Reviews*, 28, 1003G

Gaensler, B. M., Haverkorn, M., Staveley-Smith, Dickey, J. M., McClure-Griffiths, N.M., Dickey, J. M., & Wolleben, M. 2005, The Magnetic Field of the Large Magellanic Cloud: A New Way of Studying Galactic Magnetism, in *The Magnetized Plasma in Galaxy Evolution*, ed. K. T. Chyży, K. Otmianowska-Mazur, M. Soida, & R.-J. Dettmar (Kraków: Jagiellonian University, Astronomical Observatory), 209

Gardner, F.F. & Whiteoak, J.B. 1966, The Polarization of Cosmic Radio Waves, *Ann. Rev. Astr. Astrophys.*, 4, 245

Heiles, C. & Crutcher, R. 2005, Magnetic Fields in Diffuse HI and Molecular Clouds, in *Cosmic Magnetic Fields*, eds. R. Wielebinski & R. Beck (Berlin: Springer), 137

Heiles, C. & Troland, T. H. 2005, The Millennium Arecibo 21 CM Absorption Line Survey IV. Statistics of Magnetic Field, Column Density and Turbulence, *Astrophys. J.*, 624, 773

Herzberg, G. 1939, *Molecular Spectra and Molecular Structure* (Toronto: Van Nostrand)

Herzberg, G. 1944, *Atomic Spectra and Atomic Structure*, 2nd ed. (New York: Dover Publications. Trans. by J.W.T. Spinks and G.H.) (1st ed. Published by Prentice-Hall 1937)

Hickson, P. & van den Bergh, S. 1990, CCD Observations of the Polarization of the Crab Nebula, *Astrophys. J.*, 365, 224

Lyne, A. G. & Smith, F. G. 1968, Linear Polarization in Pulsating Radio Sources, *Nature*, 218, 124

Mestel, L. & Landstreet, J. L. 2005, Linear Polarization in Pulsating Radio Sources, in *Cosmic Magnetic Fields*, ed. R. Wielebinski & R. Beck, (Berlin: Springer), 183

Minchin, R. & Momjian, E., eds. 2008, *The Evolution of Galaxies Through the Neutral Hydrogen Window: Arecibo Observatory, Puerto Rico, 1–3 February 2008* (Melville, NY: American Institute of Physics)

O'Dell, C. R. 1962, Photoelectric Observations of the Crab Nebula, *Astrophys. J.*, 136, 809

Pacholczyk, A. G. 1970, *Radio Astrophysics* (San Francisco: W. H. Freeman & Co.)

Péquignot, D. & Dennefield, M. 1983, *Astron. Astrophys.*, 120, 249

Pfrommer, C. & Enßlin, T. A. 2004, Estimating Galaxy Cluster Magnetic Fields by the Classical and Hadronic Minimum Energy Criterion, *MNRAS*, 352, 76

Roberts, D. H., Wardle, J. F. C., & Brown, L. F. 1994, Linear Polarization Radio Imaging at Milliarcsecond Resolution, *Astrophys. J.*, 427, 718

Salter, C. J., Ghosh, T., Catinella, B., Lebron, M., Lerner, M. S., Minchin, R., & Momjian, E. 2008, The Arecibo ARP220 Spectral Census. I. Discovery of the Pre-Biotic Molecule Methanimine and New cm-Wavelength Transitions of Other Molecules, *Astron. J.*, 136, 389

Sarma, A. P., Troland, T. H., Crutcher, R. M., & Roberts, D. A. 2002, Magnetic Fields in Shocked Regions: Very Large Array Observations of H₂O Masers, *Astrophys. J.*, 580, 928

Sault, R. & Cornwell, T. 1999, The Hamaker-Bregman-Sault Measurement Equation, in *Synthesis Imaging in Radio Astronomy II: A Collection of Lectures from the Sixth NRAO/NMIMT Synthesis Imaging Summer School, held at Socorro, New Mexico, USA, 17–23 June, 1998, Volume 2*, ed. G. B. Taylor, C. L. Carilli, R. A. Perley, ASP Conf. Ser. (San Francisco: Astronomical Society of the Pacific), 180, 657

Scarrott, S. M., Brosch, N., Ward-Thompson, D., & Warren-Smith, R. F. 1986, An Optical Polarization Study of Two Possible Bipolar Nebulae, *MNRAS*, 223, 505

Simard-Normandin, M., Kronberg, P. P., & Button, S. 1981, The Faraday Rotation Measures of Extragalactic Radio Sources, *Astrophys. J. Suppl.*, 45, 97

Sokoloff, D. D., Bykov, A. A., Shukurov, A., Berkhuijsen, E. M., Beck, R., & Poezd, A. D. 1998, Depolarization and Faraday Effects in Galaxies, *MNRAS*, 299, 189

Sokoloff, D. D., Bykov, A. A., Shukurov, A., Berkhuijsen, E. M., Beck, R., & Poezd, A. D. 1999, Erratum: Depolarization and Faraday Effects in Galaxies, *MNRAS*, 303, 207

Spangler, S. R. 1982, The Transport of Polarized Synchrotron Radiation in a Turbulent Medium, *Astrophys. J.*, 261, 310

Spangler, S. R. 1983, Determination of the Properties of Magnetic Turbulence in Radio Sources, *Astrophys. J. Lett.*, 271, L49

Stokes, G. 1852, On the Composition and Resolution of Streams of Polarized Light from Different Sources, *Trans. Cambridge Phil. Soc.*, 9, 399

Tribble, P. C. 1991, Depolarization of Extended Radio Sources by a Foreground Faraday Screen, *MNRAS*, 250, 726

Turło, Z., Forkert, T., Sieber, W., & Wilson, W. 1985, Calibration of the Instrumental Polarization of radio telescopes, *Astron. Astrophys.*, 142, 181

Verschuur, G. L. 1968, Positive Determination of an Interstellar Magnetic Field by Measurement of the Zeeman Splitting of the 21-cm Hydrogen Line, *Phys. Rev. Lett.*, 21, 775

Vogt, C., Dolag, K., & Enßlin, T. A. 2005, PACERMAN- II. Application and Statistical Characterization of Improved RM maps, *MNRAS*, 358, 732

Wolleben, M. 2007, A New Model for the Loop I (North Polar Spur) Region, *Astrophys. J.*, 664, 349

Wolleben, M. & Reich, W. 2004, Faraday Screens Associated with Local Molecular Clouds, *Astron. Astrophys.*, 427, 537

Woltjer, L. 1957, The Polarization and Intensity Distribution in the Crab Nebula Derived from Plates Taken with the 200-Inch Telescope by Dr. W. Baade, *Bull. Astr. Insts. Neth.*, 13, 301

Woltjer, L. 1958, The Crab Nebula, *Bull. Astr. Insts. Neth.*, 14, 39

3

Mechanisms for magnetic field generation and regeneration

3.1 Introduction

The Equations (3.1)–(3.12) – which govern the interactions between moving, conducting or partially conducting plasmas and magnetic fields – have wide application in astrophysics. They govern the evolution of magnetic fields in galaxies, and also influence the complex and non-linear processes that produce magnetic fields in stars, stellar outflows, supernovae, and jets and outflows from accretion disc systems. They affect self-gravitating magneto-plasmas on all scales, from the laboratory to intergalactic scale plasma flows in cosmological large scale structure (LSS). Also embedded in this very large subject is the evolution of magnetic fields on "intermediate" scales of galaxy clusters and groups.

Excepted from this book are the basic concepts of magneto-plasma dynamics in the presence of strong gravity, e.g. close to a black hole horizon when Schwarzschild metric terms are significant, or when a distorted space–time structure has significant rotation, in which case Kerr metric terms apply. The latter become important near the inner accretion disk around a black hole where $R \lesssim R_{\rm s} = 2GM/c^2$, $R_{\rm s}$ being the Schwarzschild radius.

The creation and the redistribution of magnetic flux in a plasma where magnetic fields and mass are coupled is governed by a basic set of equations. These apply to turbulent ionised plasmas on many scales, such as the Solar wind, and the interstellar and intergalactic medium. They can also be applied to the generation of magnetic energy and flux by a coherent rotator – such as a neutron star magnetosphere or a central galactic black hole.

They also govern the interaction between plasmas and magnetic fields in shearing flows, where two-stream interfaces develop between the Solar wind and the Earth's magneto-sphere, a differentially rotating galaxy disk, within galaxy outflow winds, and in galaxy clusters. In an intergalactic context, these basic equations are applicable to the large scale shearing flows and shocks that occur, as intergalactic plasma on megaparsec scales "falls into" large scale filaments and sheets of relative mass overdensity, and then into galaxy clusters and groups.

The amplification of astrophysical magnetic fields is commonly associated with a transfer of angular momentum. An overview of the physics of angular momentum transfer and magnetic fields, from laboratory to geophysical and astrophysical contexts, can be found in Ji *et al.* (2008).

In some of the above settings, magnetic energy is converted back into particle energy. This can happen through the phenomenon of *magnetic reconnection*, discussed in Section 3.5. Here, opposing magnetic lines of force come into sufficiently close proximity that they can annihilate, whereby their magnetic energy is converted into the kinetic energy of particles. Such situations might also produce cosmic ray acceleration. The microphysics of reconnection, as with angular momentum transfer in magneto-plasmas, involves non-linear processes in 3-D, and is a challenge to understand. But it is fundamentally important in many astrophysical settings (Section 3.5).

Computer simulations are increasingly indispensable for gaining insight into the non-linear, complex behaviour of magnetised plasmas. The basic equations below form a mathematical basis for understanding astrophysical magneto-plasmas on all scales.

3.2   Some basic equations and the magnetic induction equation

The equations governing the evolution of magnetic fields in a non-viscous, conducting plasma consist of Maxwell's equations, Ohm's law, the equation of continuity, and an equation of motion. The compact vector formulation of Maxwell's equations is attributed to Oliver Heaviside (see e.g. Hunt 2012 and earlier bibliography therein):

$$\frac{\partial B}{\partial t} = -c\nabla \times E \tag{3.1}$$

$$\nabla \cdot B = 0 \tag{3.2}$$

$$\nabla \cdot E = 4\pi\rho_c \tag{3.3}$$

$$\nabla \times B = \frac{4\pi}{c}j + \frac{1}{c}\frac{\partial E}{\partial t} \quad \text{Maxwell's equations} \tag{3.4}$$

$$J = \sigma\left(E + v \times \frac{B}{c}\right) \quad \text{Ohm's law} \tag{3.5}$$

$$\frac{\partial \rho}{\partial t} + \nabla \cdot (\rho v) = 0 \quad \text{The continuity equation} \tag{3.6}$$

and

$$\rho\left[\frac{\partial v}{\partial t} + (v\cdot\nabla)v\right] = -\nabla\Phi - \nabla P + j \times \frac{B}{c} \quad \text{Equation of motion.} \tag{3.7}$$

The electric and magnetic fields are E and B, respectively. ρ is the mass density, j the current density, σ the conductivity, ρ_c the net charge density, P the pressure, v the velocity, and Φ the gravitational potential.

The evolution of a magnetic field is described by solutions to the *magnetic induction equation* (cf. Parker 1979), which can be derived from Maxwell's equations and Ohm's law above:

$$\frac{\partial B}{\partial t} = \nabla \times (v \times B) - \nabla \times (\eta\nabla \times B). \tag{3.8}$$

The magnetic diffusivity, $\eta = c^2/4\pi\sigma$, has units of (length)^2/(time). Provided η is constant within the system, the induction equation can be re-written as

$$\frac{\partial B}{\partial t} = \nabla \times (v \times B) + \eta \nabla^2 B. \tag{3.9}$$

An important parameter for characterising magnetic turbulence is the (dimensionless) magnetic Reynolds number,

$$R_{\mathrm{m}} = \frac{V_0}{v_{\mathrm{d}}} = \frac{L_0 V_0}{\eta}, \tag{3.10}$$

where v_{d} is the speed of magnetic diffusion ($=\eta/L_0$), and V_0 and L_0, respectively, are the characteristic velocity and scale of the system. If R_{m} is finite, then in the absence of external pressure and density perturbations, magnetohydrodynamic (MHD) waves will propagate with the Alfvén speed, v_{A}. We note that the propagation speed of MHD waves (also called Alfvén or magnetosonic waves) can be straightforwardly understood by analogy with non-magnetic, mechanical waves propagating with velocity, v. The medium is generally characterised by a density, (ρ), and an elasticity or restoring force constant,

$$v = \sqrt{\frac{\text{stiffness parameter, or stored energy density}}{\text{density parameter}}}, \tag{3.11}$$

which explains the expression (1.3) for v_{A}, where the magnetic stiffness has units of B^2.

In many astrophysical settings, we are interested in the régime $R_{\mathrm{m}} \gg 1$. Other important dimensionless numbers are the Lundqvist number, S

$$S = \frac{v_{\mathrm{A}}}{v_{\mathrm{d}}}, \tag{3.12}$$

and the magnetic Prantl number,

$$P_{\mathrm{m}} = \frac{\nu}{\eta}. \tag{3.13}$$

Here, ν is the coefficient of kinematic shear, which is related to the dynamic viscosity, expressed by $\rho\nu$.

It is instructive to connect some of the above quantities to the magnetic induction equation (Equation 3.8/3.9). If the magnetic field is "frozen in" to the plasma, then the diffusivity η goes to zero (or $\sigma \to \infty$), in which case the term $\nabla \times (\eta \nabla \times B)$ vanishes, and only the first term on the right side of Equation (3.8) is important. This corresponds to $R_{\mathrm{m}} \gg 1$. Conversely, where $R_{\mathrm{m}} \lesssim 1$ the rate of change of the magnetic field reduces to the $\eta \nabla^2 B$ term (Equation 3.9). In very diffuse astrophysical plasmas, $R_{\mathrm{m}} \gg 1$. To illustrate with the Sun, consider the solar corona above an active flare region. Here $T \approx 10^6$ K, $\eta \approx 1$ m^2 s^{-1}, $L_0 \approx 10^5$ m, and $V_0 \approx 10^4$ ms^{-1}. This gives $R_{\mathrm{m}} \approx 10^9$ so that only the first term on the right side of Equation (3.8/3.9) is significant.

An important point to note is that neither the induction Equation (3.8/3.9), nor Maxwell's equations contain a *source term* for magnetic field. That is, they assume some *ab initio* field.

This underlines the need for a theoretical basis to account for the original source of magnetic fields. Examples of field seeding such as the Biermann battery described in the next section, or some other (e.g. cosmological) seed field, need to be identified. Having identified and estimated a seed field, the induction equation can describe how a magnetic field regeneration process is maintained, or grows against resistive and diffusive decay processes. The latter are represented by the second term on the right hand side of Equation (3.9).

Several articles and books give extensive discussions of basic physical processes in magnetoplasmas. Some references to consult are Parker (1979), Zel'dovich, Ruzmaikin & Sokoloff (1983), Priest (1985), and Asseo & Sol (1987). At this time of writing, the book by Priest and Forbes (2000) can be especially commended. However, advances in our understanding of magneto-plasma processes have been so rapid that the reader may also want to supplement this book with some other important literature. Some recent articles and useful reviews are mentioned throughout this book.

Large-scale, ordered motions are naturally caused by the gravitationally-driven dynamics of disk galaxies. In Chapter 4, we describe the "conventional" galactic dynamo mechanism, which was based on concepts developed above for the Sun and Earth (Parker 1955, Steenbeck *et al.* 1966, Steenbeck & Krause 1969 and others). It describes how ordered galactic motions can amplify, and at the same time impose an underlying order on the magnetic fields. An important primary energy source here is the gravitational energy "reservoir" in large spiral galaxies. The mean field galactic dynamo assumes that the induction equation can be applied to *ensemble averages* over small-scale variations in magnetic field, velocity, temperature, and density in a large spiral galaxy. The mean field applications of the induction equation have had good initial success in reproducing observed magnetic field patterns. However, more detailed observations of galaxy discs and galaxy clusters have revealed many small scale fluctuations of magnetic fields and variations of plasma velocity, temperature and density. By incorporating this more detailed information into ever more powerful computer simulations, it is increasingly possible to more fully simulate the complexities of astrophysical plasmas without the need to input mean parameters into solutions of the induction equation.

Applications of dynamo theory to global galaxy magnetic field evolution have mostly assumed that magnetic fields were orders of magnitude weaker in the earlier history of a galaxy. This is reflected in much of the pre-1990s literature on this subject. Now, many lines of evidence suggest that this widely-held hypothesis of initially weak early magnetic fields is not always the case. Another more recent appreciation is that the conventional galactic dynamo needs to incorporate the effect of cosmic ray energy input (Parker 1992, Hanasz & Lesch 2003). This input would add an additional pressure component to the equations of motion, as discussed later. Additionally, we can now better specify and describe the production of strong magnetic fields from the accretion disks of rotating collapsed objects such as massive black holes. These topics will re-emerge later in the book.

3.3   Battery processes and seed fields

3.3.1   General comments

At some initial time we will assume that $B \equiv 0$ within what we shall call, hypothetically, the initial "cosmic fluid" (The case of non-zero initial fields will appear in Chapter 13). A battery describes whatever effect is able to produce an EMF (voltage) and some initial current, which generates a "seed" field, however weak. Once a seed field is produced, a number of processes, including the dynamo, can amplify or regenerate the field via Maxwell's equations, and specifically the magnetic induction equation (Equation 3.8).

So we now wish to ask: At what cosmological epoch and by what process(es) was an initial seed field produced? This leads us to the dark beginnings of physical and plasma states in the primordial Universe, which are incompletely understood. The question can be applied separately to two broad epochs: (i) before the time, at approximately $z = 1500$, when recombination into the first atoms occurred; and (ii) to the post-recombination period, closer to our direct observational grasp, when the first stars and galaxies began to form, and subsequently to the present epoch.

Epoch (i), discussed in Chapter 13, covers many different phases and important transitions, beginning with the "plasma epoch" and going all the way back to the physics of the primordial Universe when important physical processes happened at or near the Planck scale. Various scenarios have been proposed for seed field generation between the QCD phase transition (when the cosmic temperature was ~200 MeV), and the plasma epoch just before recombination. The generation of seed fields before $z \sim 10^{10}$, corresponding to Universe scale size $R \sim 1$ cm, has many links to current research in fundamental particle physics and field theory. These include a proposed link to gradients in the vacuum expectation value of the Higgs field (Vachaspati 1991), the magnetic moments and masses of different neutrino flavours (Enqvist *et al.* 1993), and the generation of primordial magnetic fields during the proposed inflation epoch of cosmology (e.g. Turner and Widrow 1988, Quashnock *et al.* 1989). Reviews of many of these ideas can be found in Kronberg (1994), Grasso & Rubenstein (2001), Widrow (2002), and Widrow *et al.* (2008).

The scope of the related physics is too extensive to include in this book. This is partly because of the wide variety of physical settings, and also because it is generally difficult to connect field generation at these earliest epochs to a unique observational imprint. It has also been suggested that cosmic strings (e.g. Kibble 1976) and magnetic monopoles may have played a role in seed field generation. These are also not covered in this book.

If significant, post-Recombination seed fields were generated at these primordial times, they must have survived certain key transitions in order to have "emerged" at a calculable level in the present, post-Recombination Universe. Chapter 13 discusses possible seed field origins at the time of QCD and electroweak transitions, an epoch still poorly understood. Magnetogenesis at key phase transitions, as well as its later effects, adds another layer of uncertainty to questions about the strength and turbulence scale of any seed fields in the post-Recombination Universe that originated in the primordial Universe. In spite of this, we will show in Chapter 13 that considerable progress is being made in linking pre-Recombination magnetogenesis to gravitational lensing, baryogenesis, and gravity at the inflation epoch. Our discussion of seed fields therefore deals mostly with events in the epoch range (ii), that is, after Recombination. However, at $z \sim 1500$, some of the following basic plasma physics concepts may also apply through the inflation and plasma epochs covering Hubble times of $\approx 10^{-36}$ s to ≈ 1 s.

3.3.2   *Gas dynamics and the Biermann battery effect*

We now examine how a seed magnetic field can be generated in an ionised gas even if no previous magnetic field existed. A mechanism for this was proposed by Ludwig Biermann (1950) in the context of stellar interiors. The free electrons in a stellar or interstellar plasma, if it is not an e^-e^+ pair plasma, have a charge-to-mass ratio 1836.1 times that of protons. This large ratio, m_p/m_e, is a fact of our Universe that has the "collateral" effect of complicating the simulation and understanding of magneto-plasma physics in many situations, both in the laboratory and in space. It especially complicates computer simulations of plasma processes in astrophysics, and, by further extension, the design and engineering of magnetic fusion energy devices.

Either radiation or thermal pressure, or the force of an electric or magnetic field, will cause an acceleration a_e and a_i of the electrons and ions respectively in a gas whose thermal pressure is $p_G = p_e + p_i = 2nkT$. We assume the gas has zero net charge, i.e. $n = n_e = n_i$ cm^{-3}. The macroscopic velocities of the electrons and ions are v_e and v_i, respectively.

From basic gas theory (Jeans 1925 Ch. XIII, Chapman & Cowling 1939), we can write an expression for the diffusion speed, $d = v_e - v_i$, of the electrons relative to the ions:

$$d = \frac{2D_{ei}}{kT}\left\{-\frac{1}{2N}\nabla P_G + m_e(P_{re} - P_{ri})\right\} + D_T\frac{\partial \ln T}{\partial r} \tag{3.14}$$

(see also L. Biermann 1950), where D_{ei} is the coefficient of diffusion between the electrons and the ions. D_{ei} is related to the conductivity, σ, by

$$\sigma = \frac{2e^2}{kT}\cdot D_{ei}\ \left(s^{-1}\right). \tag{3.15}$$

Equation 3.15 is sufficiently general to extend to any two fluid components, not just electrons and ions. The radiation pressure on the electrons, P_{re}, affects almost only a_e, since the *accelerations* are in inverse proportion to the ion/electron mass ratio. In consequence, the external force of radiation pressure produces an electric current density given by

$$j = end.\left((\text{e.s.u.})\,\text{cm}^{-2}\text{s}^{-1}\right). \tag{3.16}$$

Ignoring for the moment the radiation pressure term in Equation (3.14), the force on an electron due to gas pressure and thermal gradients can be expressed as (Biermann 1950)

$$-eE^e = 1/(2n)\cdot\nabla P_G - \frac{1}{2}(D_T/D_{ei})\,k\nabla T. \tag{3.17}$$

If $1/(n)\cdot\nabla P_G$ is a potential in a star assumed to be in solid body rotation under gravity, the isobaric and isothermal surfaces will coincide. The radiation force also has a potential, so that a current-free electrostatic field can exist. Consequently, we can simply replace ∇P_G in Equation (3.17) with ∇P_{rad}:

$$-eE^e = 1/(2n)\cdot\nabla P_{rad} - \frac{1}{2}(D_T/D_{ei})\,k\nabla T. \tag{3.18}$$

In the radiative zone of a sun-like star, $\nabla P_G = \nabla P_{rad}$. Thus, the net electric field vanishes, even for very small space charge separation. Ludwig Biermann pointed out that if eE, a force field, has any component with non-zero curl, then the compensating effect of the electrostatic field does not occur, and currents will flow, thereby producing a magnetic field. This is known as the *Biermann battery effect*.

In his 1950 paper, Biermann also proposed that this battery effect is general: It will occur in interiors of Sun-like Main Sequence stars, in rapidly rotating stars, and in the much more rarefied conditions of the interstellar medium where the gravitational potential is negligible relative to gas dynamical forces. Early estimates of these battery-generated fields in a Sun-like Main Sequence star were based on $\sigma \approx 10^{17}$ s^{-1}, and $\nabla \times E \approx 10^{-10.5}$ cm^{-1}. These values led Biermann to deduce a stellar radiative zone seed field $B \approx 10^3$ G. Dynamo activity, especially at the Solar tachocline, the interface between the radiative and convective zones (see Chapter 4), amplifies the Sun's internally generated field, eventually to the much higher values at the photosphere, which we measure in sunspots. In the latter, magnetic fluxes are found to be 10^{20}–10^{22} G cm^2. In the massive and rapidly rotating 78 Virginis-type stars the comparable number is 10^{25} G cm^2. This value, observationally verified, can be attained over a timescale of 3×10^8 years. Over this time, the rate of flux generation is $(\partial B/\partial t) \times$ (area) $\approx 10^9$ G cm^2 s^{-1} for a circulation velocity of 30 kms^{-1}. These numbers are roughly consistent with values actually observed in stars (see Section 4.3).

Virtually all stellar objects transport magnetic flux into the interstellar medium via their outflow winds, as do more energetic objects such as novae, supernovae, pulsars, and stellar and protostellar jets. This can naturally explain how the interstellar medium becomes infused with magnetic fields that are seeded in stars. It shows how the interstellar medium *can* be seeded with a magnetic field even if none existed *ab initio* from some prior or larger scale event. Even for a Sun-like star, observed sunspot magnetic fluxes of 10^{20}–10^{22} G cm^2, if carried by the solar wind into interstellar space, would naturally magnetise the ISM. If we make the simple assumption of flux freezing, solar type fields alone, when expanded into the ISM, would magnetise the latter at a level of $\approx 10^{-16}$ G. It remains to be understood just how seed fields from individual stars in our Galaxy can produce *ordered* magnetic fields on kpc scales. This will be discussed elsewhere; however, we can make some general physical arguments at this stage.

The high conductivity of the interstellar medium implies that any level of magnetic flux, once produced, has a long lifetime against ohmic losses. The decay time of a magnetic field is given by

$$\tau \approx l^2 \frac{\sigma}{c^2} \text{ s,} \qquad (3.19)$$

where l is the scale size of an interstellar magnetic "cell". Taking $l = 100$ pc (3×10^{20} cm), $\sigma_{\text{ISM}} = 10^{11}$ s^{-1}, these numbers give an ohmic loss time $\tau = 10^{31}$ s or 3.2×10^{23} years – orders of magnitude longer than the Hubble time! Over long timescales of Galactic differential rotation, small stellar wind bubble–sized magnetic cells will plausibly be sheared and then merged into much larger scales.

Because there are other phenomena that can shorten the decay time in Equation (3.19), we can assume that a galaxy disc lifetime is generously long to allow for processes that can bring the interstellar magnetic energy density ($B^2/8\pi$) into closer equilibrium with that of the interstellar turbulence ($\approx 10^{-12}$ erg cm^{-3}) and/or of cosmic ray particles in the Galactic disc. The latter has also been established to be $\approx 10^{-12}$ erg cm^{-3} (close to ≈ 1 eV cm^{-3}).

If stars were the original source of the interstellar seed field, then to ask what level, i.e. energy density, was seeded into the ISM through stellar outflow would require a calculation involving all stellar types and phases integrated over their lifetimes. This is obviously difficult, but it is clear that massive and rapidly rotating stars that are magnetically seeded would produce strong outflow winds, in addition to pulsars, novae, and supernova. Shell-type supernovae such as Cassiopeia A are also known to inject and amplify magnetic fields through shocks and in associated turbulence internally within the supernova ejecta. By isolating fast-varying X-ray synchrotron (rather than thermal or IC) emission in shocked regions of Cassiopeia A's ejecta, Uchiyama and Aharonian deduced local magnetic field strengths at ~1 mG-level (2008). Extrapolation and extension of such fields over time and space in the Milky Way disk clearly demonstrate the potential effectiveness with which SNR shells alone can magnetise the interstellar medium within only a fraction of a Hubble time. The SNR story is in addition to pulsar winds, and stellar and pre-SN winds.

Another example of stellar field regeneration is the class of pulsar wind-powered supernova remnants. The prototypical Crab Nebula shows no strong interaction with the ISM. It has regenerated the magnetic flux of the original star/neutron star to $\approx 10^{47}$ ergs within the nebula's 4.5 pc^3 volume. This magnetic energy is $\approx 10^{-4}$ times the entire energy released by the supernova, and it was generated within 950 years by a strongly magnetised pulsar rotating (currently) at 30 Hz. The largest magnetic coherence scale within the Crab Nebula has been found to be ≈ 0.1 pc (see Fig. 4.3), inferred from polarisation images of the nebula's synchrotron radiation in the optical (Woltjer 1957), and in the radio (Bietenholz & Kronberg 1991).

Interstellar medium elements heavier than hydrogen and helium were put into the ISM by the stars and supernovae. Since the same supernovae and their stellar cousins also inject magnetic flux into the interstellar (and eventually intergalactic) medium, it is clear that metallicity can be taken to some degree as an indicator or "proxy" for magnetic flux permeation. Metallicity is a useful proxy for magnetic fields because quantitative spectroscopic measurements of metallicity are often easier to make than direct measurements of magnetic field strength. Later, we similarly discuss intergalactic "pollution" by the heavier elements in connection with galactic magnetic fields, e.g. in galaxy clusters.

The foregoing has shown how *stars* can inject seed magnetic fields in galaxies. But this does not prove that stellar battery effects are the *only* originators of galactic magnetic fields. It could still be argued that the very first stars formed out of gas that was cosmologically pre-seeded with a magnetic field. Seeding could have happened before the Recombination era, before formation of the first stars, or at some time after recombination. Stars of approximately the Sun's mass and greater were born out of interstellar gas that was enriched by previous generations of stars, pulsars and supernovae, etc. Metals came predominantly from supernovae and other high mass stars that have short lifetimes. Hence, one could argue that magnetic fields in most stars were "polluted" by the violent outflows associated with high mass stars, which pre-magnetised the interstellar gas.

In an instructive *gedanken experiment*, let us now assume that stars were the *only* original providers of seed fields in the universe. In this idealised scenario, the first magnetic fields will have been generated by battery effects within the very earliest generations of stars, after they have collapsed. These would be stars with no pre-collapse non-hydrogenic atoms, except for primordial helium. If this were so, then the very first stars would have formed out of magnetic field-free collapsing gas clouds. That now raises a paradox: How did the first stars form if, as believed, a pre-existing magnetic field was required in order to shed enough angular momentum for the initial stellar collapse? It raises the question of other battery-like effects that can seed fields in an ionised or partly ionised gas. Some examples of field-seeding effects on large scales and early times are described next.

Given the appropriate conditions, primeval galaxies could contain Biermann battery-like effects in their turbulent, conducting plasma. One situation is the early primeval galaxy universe after recombination, where localised volumes close to sources of energetic photons could have reversed the recombination process. That is, reionisation would have occurred, with consequent chaotic heating and injection of turbulence. The basic ingredients for a Biermann battery-like process occur when a conducting fluid has non-coinciding gradients of pressure and density ($\nabla P \neq \nabla \rho$). This combines with non-zero vorticity ($\nabla \times \boldsymbol{v} \neq 0$) to create fresh ionisation fronts, propagating through density and temperature anomalies. Subramanian *et al.* (1995), (1996) and Gnedin *et al.* (2000) have produced calculations and simulations to demonstrate how a Biermann battery process in these conditions will naturally generate seed magnetic fields. An example is illustrated in Fig. 3.1, where $B_{\mathbf{seed}}$, the cross product $\nabla \rho \times \nabla T$, is oriented perpendicular to the page if $\nabla \rho$ and ∇T are in the plane of the page in Fig. 3.1.

At a time near Recombination, protons and electrons were tightly coupled to the radiation field, but differentially so, due to the large $m_{\mathrm{p}}/m_{\mathrm{e}}$ ratio and the consequently large difference in particle-photon (Compton) coupling due to the different effectiveness of Thompson scattering of protons and electrons (Harrison 1970). *If* this fluid had a *non-zero* helicity, then the co-moving eddies in a coupled lepton–photon "soup" would have had a lower increase in angular velocity compared with hadrons as the cosmic scale factor $R(t)$ increased ($\omega^{\,\mathrm{e-,hv}} \propto R^{-1}$, $\omega^{\,\mathrm{p,hv}} \propto R^{-2}$). The difference between these components could have generated a Biermann battery-like current and a corresponding seed field. As implied

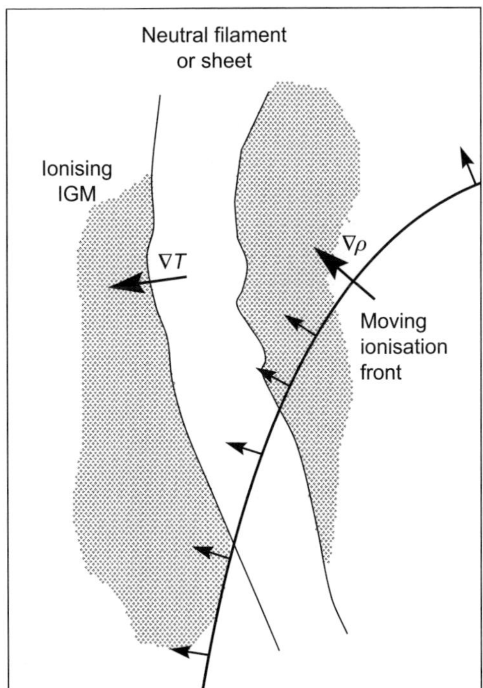

Figure 3.1 Schematic illustration of how a temperature or pressure gradient (∇P) arising from the photon flux from a group of newly formed stars in a protogalaxy will generally differ from other density gradients ($\nabla \rho$), e.g. due to some moving ionisation front in the ambient intergalactic plasma. (Adapted with permission from Gnedin, Ferrara, & Zweibel 2000).

above, the effectiveness of Harrison's battery mechanism depends on vorticity in the primordial perturbations at this epoch. This would need to be reconciled with the expected predominance of *irrotational* density perturbations that arise from curvature fluctuations associated with the formation of the first galaxies (Rees 1987).

Having discussed how seed fields can be generated and injected into the interstellar and intergalactic medium, we are ready to examine some mechanisms for amplifying galactic fields from initial, hypothetically weak seed fields up to the observed microgauss levels seen in Milky Way-like galaxies. Do conditions in the ISM favour field regeneration via Equation (3.8) up to currently observed μG levels? Below, we discuss some mechanisms relevant to such questions.

3.4   The role of cosmic ray pressure in galactic magnetic fields

Supernovae appear to be a primary source of cosmic ray gas in the stellar disks of galaxies. On a galaxy scale, their injection of CRs and magnetic fields into the ISM is a stochastic process, whose average rate is approximately known. The smoothed-out interstellar CR energy density, $\langle \varepsilon_{CR} \rangle$, is ≈ 1 eV cm^{-3}. It was noted previously that ε_{CR} is comparable to $\langle \varepsilon_{gas} \rangle$ and $\langle \varepsilon_{mag} \rangle$ in the Galactic disc. Locally it can be much greater, since the ε_{CR} is very non-uniform, having, for example, strong peaks in the vicinity of recent supernovae and concentrations of intense star formation. Figure 3.2 below shows a Parker instability, driven by a local excess of ε_{CR}, and computed in a more recent, realistic simulation.

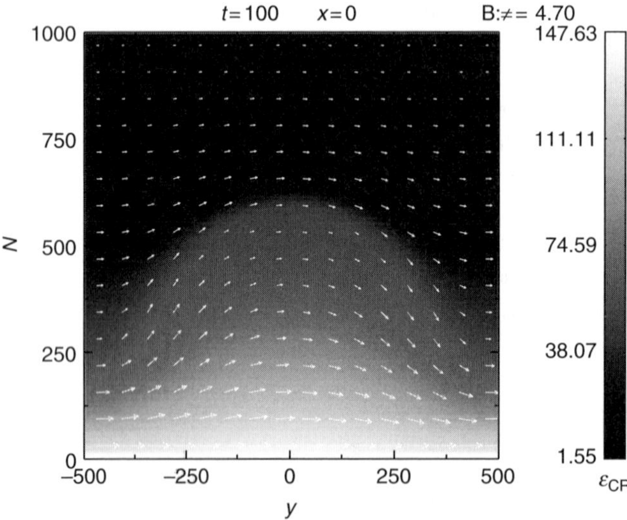

Figure 3.2 A time step for a computed "Parker loop" at 10^8 year in a galaxy disc's $(y - z)$ plane after an explosion at $(x, y, z) = (0, 0, 0)$. Arrows show the perturbation of an initially uniform disc magnetic field $\boldsymbol{B}(0, y, 0) = 4.7$ μG, and ε_{CR} (right: grey scale), which also traces ε_B, is shown at the right. A ZEUS 3-D code was used, modified by the addition of a CR pressure term (Hanasz & Lesch 2003).

Over time, dispersion and diffusion spread the cosmic rays and magnetic energy into the wider ambient medium. The diffusion is anisotropic, being directed primarily along local preferred magnetic directions in the Galactic disc (Chapter 5). The local CR overpressures also power local outflows and produce corresponding vertical perturbations of the disk magnetic field, as Fig. 3.2 illustrates. The Canadian Galactic Plane Survey (CGPS) image in Fig. 5.5 conveys a good sense of this inhomogeneity, and of the stochastic and chaotic nature of both CR and turbulent energy injection into the interstellar medium.

Supernova-driven outflow field loops were first proposed by Parker (1992) to modify the Galactic dynamo and greatly shorten the magnetic field growth time, from $\geq 10^9$ to ~10^8 year or ~1 galactic rotation at the Sun's radius. In short, a realistic model of galactic magnetic field evolution must incorporate the cosmic ray component, which adds large local energy injections into a galaxy disk, and on a timescale much smaller than a galaxy disk lifetime. Based on the Parker instability concept, a considerable literature exists on the subject of CR-driven galactic dynamos and their ability to provide efficient magnetic field amplification (Blitz & Shu 1980, Berezinsky *et al.* 1990, Hanasz & Lesch 2003, Hanasz *et al.* 2004, Otmianowski-Mazur *et al.* 2007). Much of the CR-driven modelling has used the Milky Way ISM model of Ferrière (1998a, b). See also Chapter 5.

There is some evidence, mentioned above, for local milligauss-level fields in young SNR shells. The convection and diffusion of such strongly magnetised plasma ($\approx$three orders of magnitude greater than the average interstellar field) suggest that the diffused SNR ejecta can carry non-trivial magnetic fields into the ISM and, by extension, into galaxy disk outflows. Some examples are given in Chapter 6.

3.5   Magnetic reconnection

3.5.1   *Introduction and early insights*

The process of magnetic reconnection is required to explain several phenomena in astrophysics that otherwise cannot be explained by kinetic gas theory, nor by conventional MHD theory. This is because reconnection depends on the magnetic field's geometrical *topology* in 3-D. It is also a highly non-linear process, and its essential nature and significance are well summarised in the words of Priest and Forbes (2000):

> ... reconnection is essentially a topological restructuring of a magnetic field caused by a change in the connectivity of its field lines. This change allows the release of stored magnetic energy, which in many situations is the dominant source of free energy in a plasma.

By processes not yet fully understood, magnetic reconnection has the remarkable ability to couple magneto-plasma dynamics on large scales to dynamical processes on microscopic scales. The "action" mostly occurs on these microscopic scales and, when fully understood, may have many implications for astrophysical and laboratory-scale magnetic fields.

The relevance of magnetic reconnection to astrophysical phenomena was first realised by Giovanelli (1946) and Hoyle (1949). It provides a means whereby stored magnetic energy is released in various Solar and Sun-Earth phenomena, including those that generate particles in aurorae. This kind of release occurs near a neutral X-point, where two oppositely directed magnetic fields are "pressed together". Here, a narrow current sheet forms at a location of local collapse of the surrounding magnetic field structure. This narrow sheet has been commonly called the *diffusion region*. It was shown by Dungey (1953) that, at the X-line location (e.g. Fig. 3.3) magnetic lines of force can be broken, then rejoined or reconnected. The unexpected narrowness of the current sheet had been predicted by Cowling (1953). A variant of this configuration can involve the presence of a *guide field*. In this case, **B** rotates by $<180°$ across the diffusion layer, and does not go to zero at the X-line. Reconnection can also occur if there is a guide field, i.e. when B does not go to zero.

Figure 3.3(a) (left panel) shows a reconnection diagram as proposed by Sweet (1958) and Parker (1957, 1963). The right hand panel (b) shows a modified version by H. Petschek (1964) in which the diffusion zone is much shorter (in y). In both cases plasma and magnetic field are fed in along the x-axis direction from both sides, and are ejected along the y-axis with higher energy, gained in the diffusion zone.

These pioneering efforts to explain the mysterious workings of magnetic reconnection have been followed by many subsequent advances. E.N. Parker pointed out at an early stage that advanced computational simulations would be necessary, combined with further insights into the physics – particularly the microphysics of magnetised plasmas. Fifty to sixty years later, much of this has happened, helped by the crucial computer simulations. The basic Sweet–Parker reconnection model depicted in Fig. 3.3(a) is a useful starting point for many purposes because it has since been tested and confirmed in numerical simulations within a limited parameter space for low Lundquist numbers ($S < 10^4$).

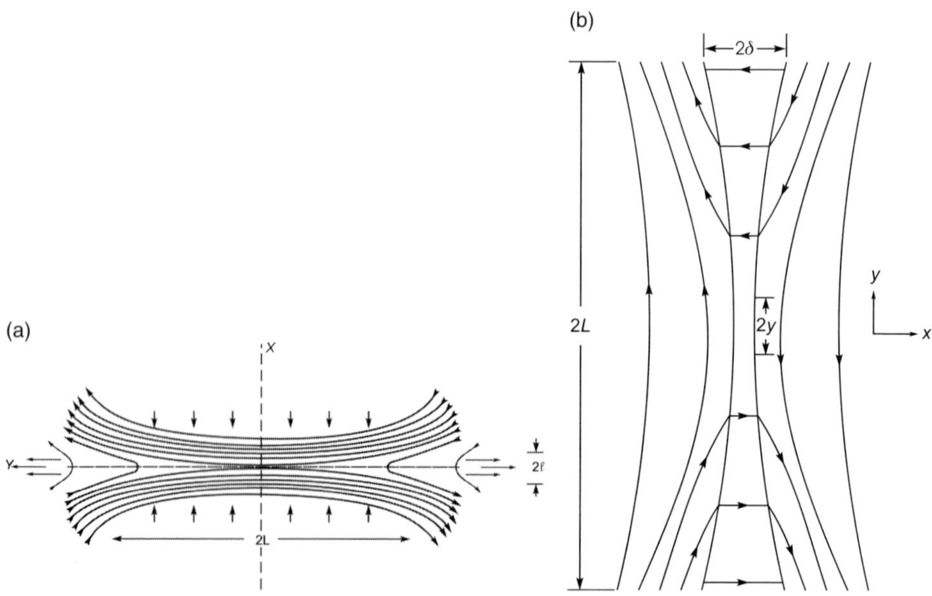

Figure 3.3 Reproduction of early magnetic reconnection configurations proposed by (a) Sweet and Parker (reproduced from Parker 1963), and (b) Petschek (1964), where $2y$ is the dimension of the so-called "diffusion zone". (*Note:* in this figure, the authors used y to denote the axis of a 2-D reconnection layer; In later discussions in this book, we *sometimes use z* to denote distance along the reconnection layer).

As discussed by Parker (1963), the basic Sweet–Parker reconnection model is much too slow to explain observed rates of conversion of magnetic to kinetic energy and heat in Solar flares. There, conversion timescales are observed to be as short as $\mathcal{O}$ 1 minute! The reconnection timescale calculation in the Sweet–Parker model includes the assumption that the Spitzer (1962) resistivity applies in Solar flare reconnection zones. Also, the reconnection (diffusion) layer in Fig. 3.3(a) needs to be unrealistically thin and straight and it is known not to be stable above a certain maximum Lundqvist number. Two important parameters for understanding magnetic reconnection are the dimensionless Lundqvist number, S ($=v_A/v_d$ – Equation 3.12), and a dimensionless system scale, λ, normalised by the ion gyroradius, λ_i.

H. Petschek's alternative reconnection layer model, proposed in 1964, is shown in Fig. 3.3(b). Petschek modified the current sheet geometry so that the diffusion zone – where reconnection occurs – is much shorter. This modification had the advantage of producing a fast enough reconnection rate ($\sim\log(S)$) to agree with some observations. However, Petschek's model does not naturally produce a solution, unless the resistivity is artificially enhanced near the X-line. This contrasts with the Sweet–Parker model, which does not require such an *ad hoc* adjustment to produce a credible numerical solution in Solar System situations. Whereas the Sweet–Parker model has been verified both computationally and experimentally in some parts of S–λ parameter space, some aspects of the Petschek model have been difficult to confirm numerically in laboratory plasmas. In particular, as noted above, the diffusion zone (Fig. 3.3(b)) needs to contain added localised resistivity.

Both laboratory and near-space experiments, confirmed by computer simulations, have furthermore demonstrated the existence of an out-of-plane, or Hall magnetic field term (below), which adds to the electric field in the generalised version of Ohm's law in a fully ionised plasma. This is:

$$\frac{-j \times B}{n_e}. \tag{3.20}$$

Hall fields play an important role in the magnetic reconnection process. They develop because of electron-ion *de*coupling in the reconnection plane, which has consequences for the current distributions.

The electron skin depth, $\lambda_e = c/\omega_{pe}$ (ω_{pe} is the electron plasma frequency), specifies the maximum penetration depth of an EM wave into a conducting plasma. A condition for reconnection to occur is that the current sheet be sufficiently thin, in fact order of λ_e.

We can also note that for $\omega < \omega_{pe}$ the wave cannot propagate, and the attenuation scale is the order of d_e. Conversely, in collisionless plasmas, if $\omega > \omega_p$, EM waves can propagate indefinitely. Reconnection can also have onset in thicker current sheets; i.e. thick on the order of the ion gyroradius. When this happens they are unstable to tearing. In this case, the non-linear evolution will produce current sheets on the scale of λ_e.

When turbulent fluctuations of the electric field occur within the thin reconnection layer, they can greatly reduce electron mobility, and consequently increase the resistivity in the thin current sheet. This effect, known as *anomalous resistivity*, has improved our insight into the reconnection process. Turbulence in the E-field has the effect of dissipating the magnetic energy. Additionally it has been proposed, though not conclusively verified in laboratory plasmas, that reconnection can be induced, or driven, if the ion skin depth, λ_i ($=c/\omega_{pi}$), exceeds the thickness of the Sweet–Parker reconnection layer (Yamada *et al.* 1997).

The Lundqvist number, S, can be also written in a form that is relevant to reconnection in a current layer with half-length L_{CS} (Ji & Daughton 2011).

$$S = \frac{\mu_0 L_{CS} v_A}{\eta}. \tag{3.21}$$

This is another formulation of Equation (3.12) in which η (m^2/s) is the plasma resistivity due to Coulomb collisions, v_A is the Alfvén speed that is relevant to the reconnecting magnetic field component, and L_{CS} is some fraction of L, the total plasma system size; $L_{CS} \leq 0.5\,L$. (As before, S remains dimensionless.)

Reconnection can be either collisional or collisionless. Generally, collisional reconnection will occur in plasmas where the resistive layer is larger than the ion kinetic scale (see e.g. Ji & Daughton 2011). In the majority of astrophysical situations discussed in this book, reconnection happens in the collisionless régime. Collisional reconnection occurs in astrophysical environments such as the Solar chromosphere, the Solar tachocline (at ~70% of the photospheric radius), in molecular clouds, in disks around supermassive black holes, and in protostellar disks. Astrophysical examples of collisionless reconnection are in the Solar corona and flares, and Solar wind, the flare in Sagittarius A*, and possibly the Crab Nebula flare, in extragalactic radio source jets and lobes, and the ICM of galaxy clusters. In general the boundary between collisional and collisionless reconnection occurs at a point where the (dimensionless) system scale λ is given by

$$\lambda \equiv \frac{L}{\rho_s}. \tag{3.22}$$

L, above, is the physical plasma system size, and ρ_s is the ion sound gyroradius:

$$\rho_s \equiv \frac{\sqrt{m_i(T_i + T_e)}}{q_i B_T}. \tag{3.23}$$

The ion mass and charge are m_i and q_i, the ion and electron temperatures are T_i and T_e, and B_T is the total magnetic field strength, including the reconnection and guide field components (Ji & Daughton 2011, and references therein).

Astrophysical reconnection mechanisms have been reasonably confirmed in the above magneto-plasma systems, which lie within the Solar System. Evidence for reconnection in systems beyond the Solar System comes with varying degrees of certainty. With the improvement occurring in instrumental and computational capability, the role of reconnection will be increasingly clarified (see e.g. Fig. 3.5). One of many challenges is to understand the connection between *macro* scales which might, in the extreme, extend to kiloparsecs and more, and "micro" scales of the order of ρ_s. Computer simulations will continue to play a role in solving such puzzles.

In more advanced 2-D and 3-D simulations, reconnection becomes progressively more complex, as the following sections will illustrate. At sufficiently high Lundqvist numbers, resistive MHD simulations have shown that highly elongated Sweet–Parker reconnection layers can become unstable and break up into island-like instabilities called plasmoids. These develop along the X-line, giving the appearance of beads along a string. The plasmoids or "islands" are connected by smaller X-line segments; that is to say, the original X-line breaks up into multiple segments that connect the plasmoids. This process can be iterative, so that sub-levels of multiple plasmoids/X-lines can develop from each original X-line segment. This hierarchical process can continue its downward progression until the Lundqvist number, S, of these higher order "inter-plasmoid" current sheets falls below the critical S value for plasmoid stability. Alternatively, it appears to be terminated if the newly produced current sheets approach the scale of ρ_S. For the interested reader, these processes are described in more detail by Ji & Daughton (2011).

3.5.2 *Particle acceleration in reconnection configurations*

An important companion process to reconnection, not discussed above, is particle acceleration in the process of reconnection. The following sections give a brief, necessarily incomplete account of developments in the understanding of particle acceleration in reconnection ("diffusion") zones. This has become a major subject in plasma studies, in both space and laboratory contexts.

In astrophysical settings reconnection rates can be very fast, as in Solar flares, or orders of magnitude slower, as in some Solar wind and geosphere contexts. Often reconnection is termed "slow" if it scales inversely with resistivity or system size. In such cases the reconnection times are longer than the global Alfvén time ($=L/\upsilon_A$). One cause for slow reconnection rates appears to lie in the anomalous resistivity, which is correspondingly higher in slow reconnection situations.

Fast reconnection, on the other hand, is insensitive to resistivity or system size. It occurs on a time scale which is some fraction of the global Alfvén time. Plasma scientists often use the term "fast Alfvénic" reconnection to emphasise this point. In laboratory magnetic configurations, including Tokomaks, magnetic energy is first built up slowly (stored), then destroyed in a catastrophic (fast) reconnection release of magnetic energy in a flare-like process. For high power fusion devices that rely on magnetic confinement this is an unwanted, even dangerous event that needs to be designed out. Analogous sawtooth-like, reconnection-driven processes occur in the Earth's outer magnetosphere. Magnetic reconnection is inferred to play an important role in various explosive flaring events in stars. These occur all the way down in mass to brown dwarfs. We remark here that the fast-growing subject of stellar magnetic fields is covered only briefly in this book. The interested reader could begin with a good review by Mestel & Landstreet (2005).

Similar or analogous processes may occur as violent bursts of energy from the magnetised accretion disk of a collapsed object, although here details are more difficult to verify. Magnetic reconnection has been proposed to explain the sudden flare-like energy release from accretion disks (Galeev, Rosner, & Viana 1979). This very interesting possibility may also explain the flare-like energy release from Sagittarius A*, the central black hole system in our Milky Way, and, on much larger scales, release of energy by γ-ray bursters, and into the jets of radio galaxies and quasars.

In reality, reconnection is a 3-D phenomenon. However, until recently, 3-D reconnection has presented a computational challenge to simulate with sufficient cell resolution while also incorporating all of the plasma physics – see below, and in particular Section 3.5.5. Estimates and measurements of reconnection rates have been made in the laboratory, in Solar corona flares and prominences, and around the Earth's magnetopause and magnetotail. Detailed reconnection processes have not been directly measured outside the Solar System, or on interstellar and intergalactic scales. In the latter contexts, there are obvious "real-time" scaling difficulties. For galactic and extragalactic distances, the required linear resolution and sensitivity are lacking.

3.5.3   *Reconnection studies in the Earth's near-space*

Reconnection is known to occur in the following zones. (1) The bow shock region where the Solar wind magnetic field lines meet the oppositely directed field structure on the sunward side of the Earth's magnetic field; (2) Another nearby space reconnection "laboratory" is the geomagnetic tail at 10–30 Earth radii. Here again, oppositely directed magnetic lines of force are pressed together, creating a collisionless, 3-D reconnection zone. After years of uncertainty, it now seems well established that the accelerated particles in aurora events originate from magnetic reconnection, sometimes explosive in nature, that happens in zones within $\lesssim 30\ R_{\oplus}$ on the sunward side of the magnetosphere. The consequent aurorae are observed much closer to the Earth's surface, i.e. not where the original heating/acceleration takes place.

Great strides have been made by placing formations of satellites "in situ" into these largest directly probe-able reconnection zones. Groups (or "Clusters") of four or more satellites can directly measure quantities such as particle densities, energies, atomic species, magnetic and electric fields, currents, etc. – all with good spatial and time resolution. The very successful ESA 4-spacecraft *Cluster II* mission was succeeded by an ambitious NASA *Magnetospheric Multiscale* (MMS) 4-spacecraft system – launched in 2015. The MMS probes with greatly improved time resolution, and in 3-D, within the magnetospheric reconnection zones mentioned above.

3.5.4   *Ion and electron acceleration in reconnection zones*

The foregoing discussion has not explored another important aspect of magnetic reconnection – namely, the conversion of magnetic energy to thermal and kinetic energy of particles. This happens in the y-directed outflow (Fig. 3.3) at either end of the reconnection layer. At the Solar-Earth magnetopause, the conversion efficiency of magnetic field-to-particle energy is estimated to be substantial, of order 10%.

For Solar flares, observational analyses and simulations confirm that electrons and ions can indeed be accelerated out of the reconnection zones. The acceleration mechanism(s) are of great interest beyond the Sun where they have been best studied so far. If reconnection explains at least part of the energy conversion in accretion disks, as has been proposed for Sagittarius A at the nucleus or the Milky Way, then similar reconnection-driven phenomena might at least partially scale to the enormously massive and energetic supermassive black hole (SMBH) accretion disks in the wider Universe. In these latter systems, magnetic and CR particle energy are converted from gravitational infall energy with globally measured efficiencies of $\gtrsim 10\%$ (e.g. Kronberg *et al.* 2001). Another open question is whether some magnetic reconnection process in a supermassive BH-jet system might explain very high hadronic cosmic ray energies, E_{CR}, up to 10^{21} eV (Chapter 11).

Returning to the Solar flare context, mechanisms have been suggested for acceleration of ions in the reconnection layer direction, where the charge-to-mass ratio, $Z_i/m_{i,}$ is modestly low. An investigation by Drake *et al.* in 2009 (and included references) shows how slowly moving heavy ions that drift into the reconnection layer can be more effectively "picked up" in the X-line exhaust flow in the limiting case of a strong guide field. The criterion for

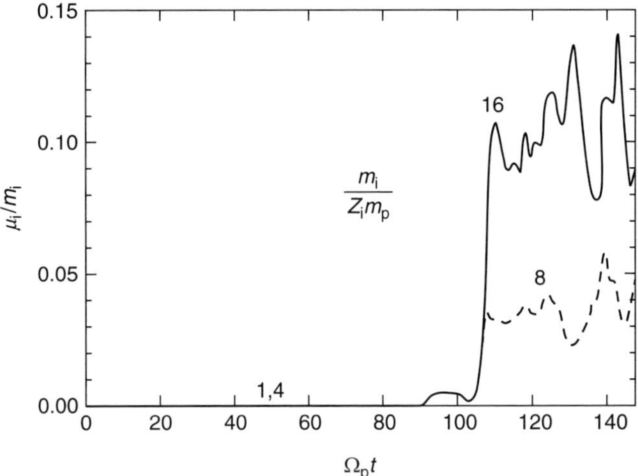

Figure 3.4 Time dependence of ion magnetic moment/mass ratio, μ_i/m_i, for ions having $m_i = 16$, 8, 4, and 1 m_p in a Hall reconnection simulation in which the guide field $B_{0z} = 5\,B_{0x}$. The protons and helium nuclei, $m_i = 1, 4$, show no jump in μ_i as they cross the reconnection exhaust flow, whereas higher mass particles satisfying the pickup threshold criterion in Equation (3.23) experience a strong jump in μ_i/m_i as they cross the exhaust boundary (adapted from Drake *et al.* 2009).

effective pickup, illustrated in Fig. 3.4, is a sufficiently large ratio of ion mass to charge (Drake *et al.* 2009):

$$\frac{m_i}{m_p Z_i} > \frac{10}{\pi}\,\frac{(\textbf{proton sound speed})}{(\textbf{proton Alfvén speed})}. \tag{3.24}$$

Figure 3.4 illustrates the sharpness of this threshold for increased ion pickup beyond Helium, i.e. $m_i > 4$, above which the ion's energy no longer behaves adiabatically.

The magnetic moments in Fig. 3.4 are given by

$$\mu_i = \frac{m_i \partial v^2_{\perp}}{2B}, \tag{3.25}$$

where $\delta v^2_{\perp}$ is the ion perpendicular velocity after subtracting the $\boldsymbol{E} \times \boldsymbol{B}$ drift. These experimental results illustrate how heavier ions may be preferentially accelerated in reconnection events. This appears to be confirmed in the Solar corona, and may be significant in larger astrophysical contexts where heavier nuclei are accelerated.

3.5.5 *Reconnection in 3-D and the role of advanced computer simulations*

As the foregoing discussion indicates, magnetic reconnection requires further insights into the physics of magneto-plasmas and, in particular, more advanced supercomputer simulations. Reconnection in the "diffusion zones" is a much more complex process than was envisaged in the pioneering concepts shown in Fig. 3.3. Within reconnection zones, the current sheet develops into a succession of individual magnetic "islands". In 3-D simulations these have the appearance of magnetic loops, and their formation can be understood as a combination of the tearing mode instability, and local reconnection. The loop (island) dimensions are of order the ion gyroradius. Destruction of the smooth current sheet by the tearing mode instability has the effect of transforming the "system" toward a minimum energy state, as often happens in Nature. The island length is determined by the most unstable tearing mode, and this is usually in the range $k \cdot L_{1/2} \approx 0.3$–$0.5$, where $L_{1/2}$ is the half-thickness of the layer. This turns out to be in the same range for sheets, from the d_i scale all the way down to the d_e scale (Daughton 2012).

Kinetic simulations, such as the one illustrated in 3-D below, normally start with a layer thickness that is comparable to the ion skin depth. What can develop are *multiple* reconnection sites within the reconnection zone. In the context of Section 3.5.4, the acceleration of both ions and electrons in multiple island reconnection layers proceeds to higher energies than it would in a single reconnection layer. Simulations relevant to the Earth's magnetosphere have produced reconnection-accelerated protons up to ~ 0.2 MeV, and up to ~100 keV for electrons – still in the non-relativistic régime. Nearer to the Sun, E_i and E_e are observed to reach relativistic energies, presumably through acceleration by similar processes, suitably scaled.

As they transform into lower energy states, the loops coalesce into ever larger islands, while reconnection-accelerated particles reach progressively higher energies. The acceleration process involves a combination of ion currents that are parallel to local E-fields, i.e. direct acceleration in an EMF, mixed with multiple reflections off magnetic field lines – a Fermi-acceleration-like process (Drake *et al.* 2006). This complex mix of processes happens in 3-D, and the interactions occur among ions, electrons, and local E- and B-fields. As E_i and E_e reach Mev and higher energies, relativistic electrodynamics will apply.

Full simulation of a 3-D reconnection layer is computationally easier in a relative sense, if the interacting particles consist of an $e^+ e^-$ pair plasma. As the mass ratio m/m_e increases from 1 for a pair plasma toward Nature's m_p/m_e ratio of 1836.1, the computational time increases dramatically. To illustrate this, computational times increase as approximately $(m/m_e)^2$ in a 2-D particle-in-cell (PIC) simulation, and as $(m/m_e)^{2.5}$ in a 3-D simulation – that is, in this case 1.4×10^8 times longer for a proton–electron fluid than for a pair plasma! Over the first decade of the 21st century, simulations have typically managed mass ratios in the range up to ~20, and at this time of writing some 2-D PIC simulations have actually been able to incorporate m/m_e up to 1836. Simulations of reconnection zones in 3-D with fast supercomputers such as the "Roadrunner" at Los Alamos National Laboratory and the NSF "Kraken" computer have incorporated m/m_e of 200–400 up to this time of writing. A sample result of a simulated 3-D reconnection zone is shown below for a mass ratio $m/m_e = 100$.

3-D reconnection layer simulations of the diffusion zone such as that in Fig. 3.5, show what appear as reconnection islands in 2-D. In 3-D, they appear as complex configurations of magnetic flux *ropes* generated by the tearing instability. Figure 3.5 graphically illustrates the result of complex interacting processes, some of which were neither previously recognised nor anticipated in 2-D simulations. A "collapsed" 2-D integration is shown in projection on one side (x–z) of the 3-D simulation "box". Electromagnetic waves can cause a kinking of the electron layer, with consequent secondary reconnection instabilities. These are what typically give rise to the formation of flux ropes that are seen Fig. 3.5. The flux ropes give new and unanticipated physical insights into the internal structure of a

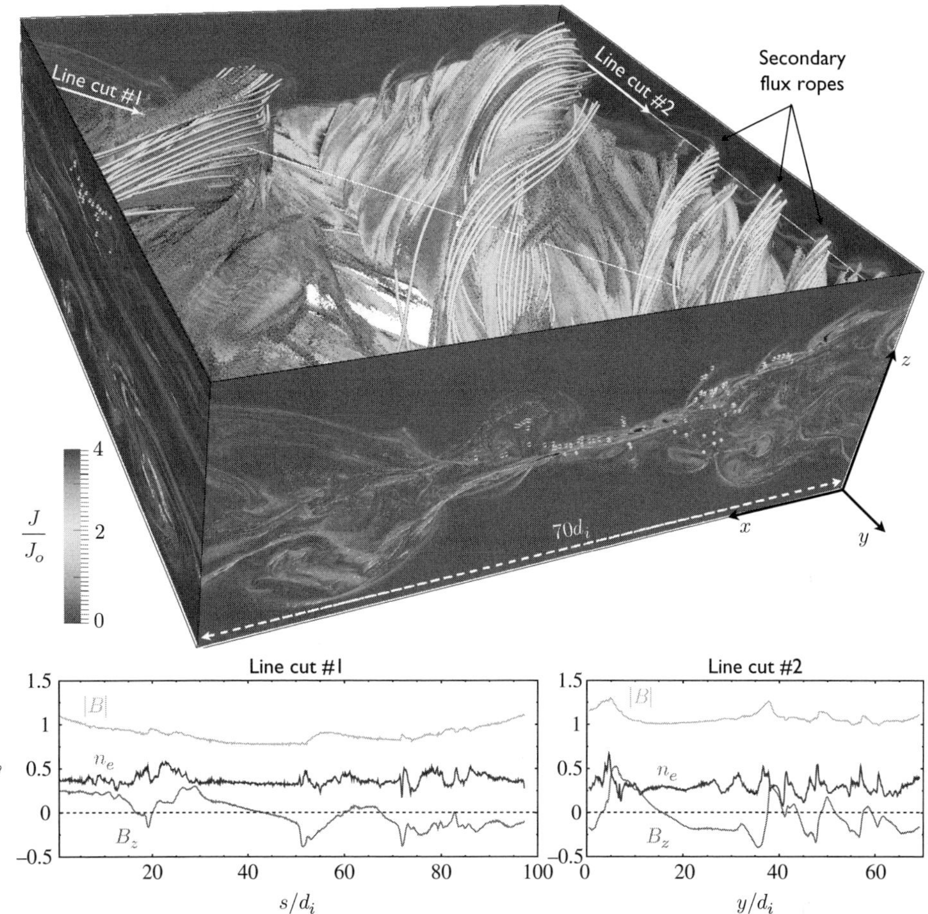

Figure 3.5 Three-dimensional view of a multi-layer, 3-D magnetic reconnection (diffusion) region simulation for a proton–electron plasma with $m/m_e = 100$. This kinetic particle-in-cell simulation with the NSF Kraken supercomputer used ~3.3×10^9 cells, ~1.3×10^{12} particles, and a guide field (Ji & Daughton 2011, Daughton 2011). (Courtesy of William Daughton, Los Alamos National Laboratory.)

reconnection diffusion region. This new "micro"-view of EM wave-induced kink waves and flux ropes within a reconnection zone has potentially far-reaching implications for processes on macro scales. The connection between the orders of magnitude larger macro scales and these micro scales appears to happen, but by interactions that are not yet well understood. The process includes the rate at which magneto-plasma can be fed from above and below into the reconnection (diffusion) layer. Complex as they are, these simulations still do not yet fully reveal processes of particle acceleration to high, energies, possibly even relativistic, within the diffusion zone.

In the next phase of reconnection and associated particle acceleration simulations, it will be important to incorporate relativistic electrodynamics. Systems such as BH-accretion disk-powered outflows, jets, and lobes represent higher order scalings of the systems discussed here. In the Solar context, evidence is already clear that magnetic reconnection zones can accelerate ions and electrons to at least transrelativistic energies. It seems entirely conceivable that ion acceleration to TeV and even EeV levels may also happen in some magnetic reconnection zones, on different scales. Reconnection acceleration may play an important role in energetic processes that produce TeV γ-ray bursts, and in SMBH/accretion disk-driven extragalactic radio and X-ray sources (Chapters 7 and 8). In the latter objects it is yet to be confirmed whether magnetic reconnection plays a role in CR acceleration. However, if it does, and if it is an important process in extragalactic source jets and radio lobes, the very high efficiencies that are implied for SMBH-powered lobes (discussed later in Section 8.5 and Fig. 8.1) might involve reconnection acceleration. If so, they might ultimately provide useful efficiency constraints.

Processes akin to reconnection acceleration could be one cause of the ultra high energy cosmic ray particles (UHECRs) discussed in Chapter 11, but it should be emphasised that this is not confirmed. Reconnection seems a more easily identifiable mechanism in the case of γ-ray bursts. Magnetic reconnection is discussed in various contexts in the books by Priest & Forbes (2000), Birn & Priest (2007), in some informative reviews (e.g. Zweibel & Yamada 2009), and other recent articles on the subject (e.g. Burch & Drake 2009, Mozer & Pritchett 2010, Daughton 2011, Ji & Daughton 2011).

3.6   Effects of neutral gas

In galactic discs, stellar winds, and galactic outflows, the available mechanical energy in the form of shearing and turbulence can, in the right circumstances, be transformed into magnetic energy if there is an initial imbalance in favour of the former. A substantial fraction of the interstellar gas may be neutral, which raises the question of whether or not the neutral gas might be substantially decoupled from MHD processes.

At the relatively low electron temperatures of 300–3000 K, the effective action radius for Coulomb forces is large, and the interactions between charged particles overwhelm those between charged and neutral particles. The cross-section for the former is of order 10^{-10} cm^2, compared with ca. 10^{-15} cm^2 for the charged-neutral interactions. This means that even regions that are mostly neutral, such as HI regions, can still behave like a plasma, and can maintain some coupling to the magnetic field. The neutral-ionised diffusion speed (ambipolar diffusion) is small. Because of the relatively low temperatures the conductivity, σ, is high, in the range 10^{10}–10^{12} s^{-1}, and the magnetic field decay time (Equation 3.19) is correspondingly long: For a magnetic cell size (L) of 100 pc, the decay time, τ, can exceed 10^{23} years, $\approx 10^{13}$ times longer than the age of the Universe! This illustrates that, given the very high conductivity of interstellar (and intergalactic) space, a "neutral" gas can also behave as a plasma, and its associated magnetic fields can be very long lived.

The question of the role of neutral gas is of interest, given observational evidence indicating that galactic outflow winds, which have a significant neutral component are strongly magnetised. Good examples of this include M82 (Figs. 6.2 and 6.3 in Chapter 6, Reuter *et al.* 1994) and NGC4569 (Chyży *et al.* 2006). The outflows contain significant amounts of neutral and molecular gas and dust (see e.g. Weiß *et al.* 2000). Birk *et al.* (2000) examined the effect of neutral gas entrained in fast galactic outflows, e.g. from starburst galaxies. They quantitatively investigate the effect of mixing neutral gas into Kelvin–Helmholtz instabilities, which can rapidly regenerate the magnetic field along two-stream instabilities. When they used a neutral/ionised mixture in a three-fluid gas code, and included compressibility, their numerical simulations produced magnetic field amplification for the neutral/ionised two-fluid mix similar to that of a purely ionised gas.

References

Asseo, E. & Sol, H. 1987, Extragalactic Magnetic Fields, *Phys. Rep.*, 148, 307

Berezinsky, V. S., Bulanov, S. V., Dogiel, V. A., & Ptuskin, V. S. 1990, *Astrophysics of Cosmic Rays*, ed. V. I. Ginzburg (Amsterdam: North-Holland)

Biermann, L. 1950, Über den Ursprung der Magnetfelder auf Sternen und im interstellaren Raum *Zeitschrift für Naturforschung A*, 5, 65

Bietenholz, M. F. & Kronberg, P. P. 1991, Faraday Rotation and Physical Conditions in the Crab Nebula, *Astrophys. J.*, 368, 231

Birk, G. T., Wiechen, H., Lesch, H., & Kronberg, P. P. 2000, The Role of Kelvin-Helmholtz Modes in Superwinds of Primeval Galaxies for the Magnetization of the Intergalactic Medium, *Astron. Astrophys.*, 353, 108

Birn, J. & Priest, E. R. 2007, *Reconnection of Magnetic Fields* (Cambridge: Cambridge University Press)

Blitz, L. & Shu, F. H. 1980, The Origin and Lifetime of Giant Molecular Cloud Complexes, *Astrophys. J.*, 238, 148

Burch, J. L. & Drake, J. F. 2009, Reconnecting Magnetic Fields, *Am. Sci.*, 97, 392

Chapman, S. & Cowling, T. G. 1939, *The Mathematical Theory of Non-uniform Gases*, (Cambridge: Cambridge University Press)

Chyzy, K., Soida, M., Bomans, D. J., Vollmer, B., Balkowski, Ch., Beck, R., & Urbanik, M. 2006, Large-Scale Magnetized Outflows from the Virgo Cluster Spiral NGC 4569: A Galactic Wind in a Ram Pressure Wind, *Astron. Astrophys.*, 447, 465

Cowling, T. G. 1953, Solar Electrodynamics, in *The Sun*, ed. G. Kuiper (Chicago: University of Chicago Press), 532

Daughton, W. 2012, *Private Communication*

Drake, J. F., Cassak, P. A., Shay, M. A., Swisdak, M., & Quatert, E. 2009, A Magnetic Reconnection Mechanism for Ion Acceleration and Abundance Enhancements in Impulsive Flares, *Astrophys. J. Lett.*, 700, L16

Dungey, J. W. 1953, Conditions for the occurrence of electrical discharges in astrophysical systems, *Phil. Mag.*, 44, 725

Enqvist, K., Semikoz, V., Shukurov, A., & Sokoloff, D. D. 1993, Neutrino Mass and the Origin of Galactic Magnetic Fields, *Phys. Rev. D.*, 48, 4557

Ferrière, K. 1998a, Global Model of the Interstellar Medium in Our Galaxy with New Constraints on the Hot Gas Component, *Astron. Astrophys.*, 497, 759

Ferrière, K. 1998b, Alpha-Tensor and Diffusivity Tensor Due to Supernovae and Superbubbles in the Galactic Disk, *Astron. Astrophys.*, 335, 488

Galeev, A. A., Rosner, R., & Vaiana, G. S. 1979, Structured Coronae of Accretion Disks, *Astrophys. J.*, 229, 318

Giovanelli, R. G. 1946, A Theory of Chromospheric Flares, *Nature*, 158, 81

Gnedin, Y., Ferrara, A., & Zweibel, E. N. 2000, Generation of the Primordial Magnetic Fields during Cosmological Reionization, *Astrophys. J.*, 539, 505

Grasso, D. & Rubinstein, H. R. 2001, Magnetic Fields in the Early Universe, *Phys. Rep.*, 348, 163

Hanasz, M., Kowal, G., Otmianowska-Mazur, K., & Lesch, H. 2004, Amplification of Galactic Magnetic Fields by the Cosmic-Ray-Driven Dynamo, *Astrophys. J.*, 605, L33

Hanasz, M. & Lesch, H. 2003, Incorporation of Cosmic Ray Transport into the ZEUS MHD Code. Application for Studies of Parker Instability in the ISM, *Astron. Astrophys.*, 412, 331

Harrison, E. R. 1970, Generation of Magnetic Fields in the Radiation ERA, *MNRAS*, 147, 279

Hoyle, F. 1949, *Some Recent Researches in Solar Physics* (Cambridge: Cambridge University Press)

Hunt, B. J. 2012, Oliver Heaviside, a First-Rate Oddity, *Phys Today*, 65, No. 11, 48

Jeans, J. H. 1925, *Dynamical Theory of Gases*, (Cambridge: Cambridge University Press), Ch. XIII

Ji, H. & Daughton, W.D. 2011, Phase diagram for magnetic reconnection in heliophysical, astrophysical, and laboratory plasmas, *Phys. Plas.*, 18, 111207

Ji, H., Kronberg, P. P., Prager, S. C., & Uzdensky, D. A. 2008, Mini-Conference on Angular Momentum Transport in Laboratory and Nature, *Phys. Plas.*, 15, 058302

Kibble, T. W. B. 1976, Topology of Cosmic Domains and Strings, *J. Phys. A Math. Gen.*, 9, 1387

Kronberg, P. P. 1994, Extragalactic Magnetic Fields, *Rep. Prog. Physics*, 57, 325

Kronberg, P.P., Dufton, Q.W., Li, H., & Colgate, S.A. 2001, Magnetic Energy of the Intergalactic Medium from Galactic Black Holes, *Astrophys. J.*, 560, 178

Lesch, H. & Hanasz, M. 2003, Strong Magnetic Fields and Cosmic Rays in Very Young Galaxies, *Astron. Astrophys.*, 401, 809

Mestel, L. & Landstreet, J. 2005, and Stellar Magnetic Fields, in *Cosmic Magnetic Fields*, ed. R. Wielebinski & R. Beck, Lecture Notes in Physics (Berlin: Springer), 664, 183

Mozer, F. S. & Pritchett, P. L. 2010, Magnetic Field Reconnection: A First-Principles Perspective, *Phys. Today*, 63, No. 6, 34

Parker, E. N. 1955, Hydromagnetic Dynamo Models, *Astrophys. J.*, 122, 293

Parker, E. N. 1957, Sweet's Mechanism for Merging Magnetic Fields in Conducting Fluids, *J. Geo. Res.*, 62, 509

Parker, E. N. 1963, The Solar-Flare Phenomenon and the Theory of Reconnection and Annihilation of Magnetic Fields, *Astrophys. J. Suppl.*, 8, 177

Parker, E. N. 1979, *Cosmical Magnetic Fields* (Oxford: Clarendon)

Parker, E. N. 1992, Fast Dynamos, Cosmic Rays, and the Galactic Magnetic Field, *Astrophys. J.*, 401, 137

Petschek, H. E. 1964, in *AAS-NASA Symposium on the Physics of Solar Flares: Proceedings of a Symposium held at the Goddard Space Flight Center, Greenbelt, Maryland, Oct.28–30, 1963* / ed. W. N. Hess (Washington, DC: National Aeronautics and Space Administration), 425

Priest, E. R. 1985, The Magnetohydrodynamics of S, Rep. *Prog. Phys.*, 48, 955

Priest, E. R. & Forbes, T. 2000, *Magnetic Reconnection: MHD Theory and Applications* (Cambridge: Cambridge University Press)

Quashnock, J. M., Loeb, A., & Spergel, D. N. 1989, Magnetic Field Generation During the Cosmological QCD Phase Transition, *Astrophys. J.*, 344, L49

Rees, M. J. 1987, The Origin and Cosmogonic Implications of Seed Magnetic Fields, *Q. J. R. Astr. Soc.*, 28,197

Reuter, H.-P., Klein, U., Lesch, H., Wielebinski, R., & Kronberg, P. P. 1994, The Magnetic Field in the Halo of M82. Polarized Radio Emission at Lambda-Lambda(6.2) and 3.6 CM, *Astron. Astrophys.*, 282, 724

Ryu, D. S., Kang, H. S., Cho, J. Y., & Das, S. 2008, Turbulence and Magnetic Fields in the Large-Scale Structure of the Universe, *Science*, 320, 909

Spitzer, L. 1962, *Physics of Fully Ionized Gases* (New York: Interscience)

Steenbeck, M. & Krause, F. 1969, On the Dynamo Theory of Stellar and Planetary Magnetic Fields. I. AC Dynamos of Solar Type, *Astron. Nachrichten*, 291, 49

Steenbeck, M., Krause, F., & Rädler, K.-H. Z .1966, *Zeitschrift für Naturforschung*, 21, 369

Subramanian, K. 1995, The Origin of Large Scale Magnetic Fields in Galaxies, *Bull. Astr. Soc. India*, 23, 481

Subramanian, K. 1996, The Origin of Large Scale Galactic Magnetic Fields, *J. Korean Astr. Soc.*, 29, 155

Sweet, P. A. 1958, The Neutral Point Theory of Solar Flares in *IAU Symp.* 6, ed. B. Lehnert (London: Cambridge University Press), 123

Turner, M. S. & Widrow, L. M. 1988, Inflation-Produced, Large-Scale Magnetic Fields, *Phys. Rev. D*, 37, 2743

Uchiyama, Y. & Aharonian, F. A. 2008, Fast Variability of Nonthermal X-Ray Emission in Cassiopeia A: Probing Electron Acceleration in Reverse-Shocked Ejecta, *Astrophys. J.*, 677, L105

Vachaspati, T. 1991, Magnetic Fields from Cosmological Phase Transitions, *Phys. Lett.*, 265, 258

Weiß, A., Kovács, A., Güsten, R., Menten, K. M., Schuller, F., Siringo, G., & Kreysa, E. 2008, *Astron. Astrophys.*, 490, 77

Widrow, L. 2002, Origin of Galactic and Extragalactic Magnetic Fields, *Rev. Mod. Phys.*, 74, 775

Woltjer, L. 1957, The Polarization and Intensity Distribution in the Crab Nebula Derived from Plates Taken with the 200-Inch Telescope by Dr. W. Baade, *B.A.N.*, 13, 301

Yamada, M., Ji, H., Hsu, S., Carter, T., Kulsrud, R. M., Bretz, M., Jobes, F., Ono, Y., & Perkins, F. W. 1997, Study of Driven Magnetic Reconnection in a Laboratory Plasma, *Phys. Plas.*, 4, 1936

Zel'dovich, Ya., Ruzmaikin, A. A., & Sokoloff, D. D. 1983, *Magnetic Fields in Astrophysics* (New York: Gordon and Breach)

Zweibel, E. G. & Yamada, M. 2009, Magnetic Reconnection in Astrophysical and Laboratory Plasmas, *Ann. Rev. Astr. Astrophys.*, 47, 291

4

Galactic "microcosms" of extragalactic magnetic systems

4.1 Introductory comments

Many phenomena in our Milky Way demonstrate interactions between magnetic fields and plasmas. Also, on laboratory scales (cm and m), insights on magnetised plasmas such as jets and magnetic plasma heating processes are increasingly available to detailed study. The significance for astrophysical magnetic fields lies in the enormous scalability of some MHD processes and gas turbulence. Research into magnetic confinement fusion, e.g. in Tokomaks, has provided a strong motivation for laboratory experiments of plasma beam formation, collimation, magnetic reconnection, dynamos, and other phenomena. It is possible to understand physically inaccessible systems on kpc to Mpc scales via controlled laboratory experiments.

Working laboratory jet experiments are described by Hsu & Bellan (2002). Ampleford *et al.* (2007) have likewise achieved a "working" laboratory model of a magnetically confined astrophysical jet. A Z-pinch can be produced in the laboratory using a twisted conical wire array. This configuration produces both a radial current and an axial magnetic field, which achieves the important conservation of both angular and axial momentum. In the latter experiment, the ratio of azimuthal to axial velocity is measured, and is found to be of order 10%. Interestingly, this is similar to the same ratio actually measured in a stellar jet – that associated with the Herbig–Haro (HH) object DGTauri. For these, the scale is ~ 100 pc and the velocities are a few kms^{-1}, as measured with the HST by Bacciotti *et al.* (2002). The observations use the reasonable assumption that measured side-to-side velocity differences are due to rotation and not to some peculiar shearing effect. For further reading on laboratory jets, spheromaks, *etc*, the reader is commended to Bellan (2006, 2008).

Ions have not directly been detected in astrophysical jets, and ion viscosity has therefore been difficult or impossible to directly measure in space. The ion viscosity is nonetheless an important parameter for clarifying the physics of astrophysical jets and other magnetoplasmas, including some terrestrial applications. Usually it is inferred via a theoretical formulation (Braginskii 1965). However a more recent laboratory plasma beam experiment at the Los Alamos National Laboratory succeeded in making an unambiguous measurement of the ion viscosity in a controlled magnetised plasma beam (Intrator *et al.* 2007). By directly measuring density, temperature, velocity, and magnetic field components at different distances from the beam axis and along the beam, Intrator et al. were able to relate the observed radial shearing directly to the ion viscosity. Besides astrophysical jets, such laboratory measurements are also of interest for the development of magneto-dynamic devices such as spheromaks, and for fusion plasmas in general. The following sections of this chapter briefly describe some key magnetic phenomena known in the Solar system and in the Milky Way.

4.2   The Sun

Because this book gives primary emphasis to galactic and extragalactic magnetic fields, it discusses only a few Solar highlights, which, because of their proximity, also give some physical insight into magnetic processes on galactic and extragalactic scales.

4.2.1   Magnetic fields in the Solar tachocline

Important magnetic dynamo processes in the Sun appear to operate in the tachocline, a remarkably thin layer at $r \sim 0.7$ R$_\odot$ and $\Delta R \lesssim 4\%$ of R$_\odot$. This sharp discontinuity marks the boundary between the inner solar radiation zone and the convection zone. The tachocline has a relatively high temperature ($T \sim 2 \times 10^6$ K) and density ($\rho \sim 0.2$ g cm^{-3}), corresponding to a pressure of $\sim 6 \times 10^{12}$ Pa. The magnetic field strength is of order 10^5–10^6 G, and the plasma β is $\sim 10^5$. The tachocline contains a well-stratified shearing zone that is believed to create the turbulent dynamo that effectively regenerates the magnetic field in the convection zone. Although not completely understood, dynamo processes at and near the tachocline appear to be very important in establishing the properties of the photosphere, and, by extension, the solar wind and the physics of the interplanetary wind. Below the tachocline, Solar rotation is solid body-like, and the magnetic field is much weaker. In the model of Gough & McIntyre (1998), the differential rotation above the tachocline serves both to drive meridional circulation in the Sun and to confine the magnetic field that is embedded below. The tachocline also appears to influence sunspot cycle behaviour.

Models of important angular momentum transport mechanisms are described and summarised by Tobias *et al.* (2007). In all of these processes, magnetic turbulence and dynamo processes are closely coupled with angular momentum transport. A complete understanding of this crucially important zone at the bottom of the convection layer will ultimately illuminate our future understanding of stellar magnetic fields and stellar interior processes in general. For further reading the reader is referred to a book devoted to the Solar tachocline by Hughes, Rosner, & Weiss (2007).

4.2.2   Magnetic phenomena above the photosphere

Photospheric spicules, sunspots, Solar flares; associated Solar storms, prominences, coronal mass ejections; and other phenomena related to the Solar wind are either intimately connected with, or driven by, magnetic fields. Our understanding of these phenomena has been unfolding over the past 60 years. A vast literature and many ground- and space-based instruments have been devoted to the Solar photosphere, chromosphere, and corona.

Magnetic loops in the corona, a well known phenomenon, show that magnetic energy can dominate the coronal hot gas motions. The increase in coronal temperature outward from the chromosphere is likely due to magnetic effects. More specifically, the sudden acceleration of protons and electrons in Solar flares and in coronal mass ejections appears largely due to conversion from magnetic to thermal and kinetic energy. One of the first suggestions of the importance of magnetic reconnection as an important process in flares came from Carmichael (1964).

There is evidence to suggest that the bulk acceleration in coronal mass ejections is efficiently powered by magnetic fields. Coronal magnetic fields are of direct interest for the questions of how coherent, low-β magnetic structures can form, and also how they transfer momentum – e.g. in shearing flows, and across a strongly magnetised accretion disc. It appears that flares are, in effect, large reconnection events in the Solar corona.

Based on these and other related facts, the corona has been modelled as an ensemble of magnetically dominated loops (Uzdensky & Goodman 2008). They are initially tied to footprints at or near the photosphere. Structured as a kind of magnetic "foam", the loop boundaries collide, and interface with neighbouring loops to form sites of magnetic reconnection. Uzdensky & Goodman's model claims to self-consistently reproduce the coronal loop sizes and orientations, their magnetic torque, and the dependence of $B^2/8\pi$ on height in the corona. With the advantage of actual measurements of ensembles of coronal loops, this work lays the basis for extending to magnetic loops that are formed above an accretion disc. The latter is so far inaccessible to detailed observation.

4.2.3   *Magnetic processes in the outer Solar wind*

Test beds even closer to Earth are the Solar wind perturbation zones due to "obstruction" by the Earth's magnetosphere. In the magnetopause on the sunward side, one can study a zone of shearing and turbulence in which magnetic reconnection and particle acceleration occur.

The Earth's geomagnetic tail, the "wake" of the Earth's obstruction of the Solar wind, is a particularly interesting test bed. It contains zones of magneto-ionic turbulence, ordered field lines, and a neutral zone. The neutral zone is approximately aligned with the Sun–Earth axis where oppositely directed fields come into close proximity. Here, the ill-understood process of magnetic reconnection can be studied (Chapter 3). The first attempt to do this was with "Cluster", a group of four satellites that could simultaneously probe reconnection events in three dimensions in the geomagnetic tail when they happen. Reconnection can happen in any system when a magnetic field structure suddenly reconfigures itself. A consequence is the release of energy by rapid and directed diffusion of high-speed ionised gas. Cluster has succeeded in verifying reconnection events in this type of field configuration (e.g. Mozer *et al.* 2005). The aurorae are the only systems in which the naked eye can witness the direct terrestrial consequence of magnetic reconnection in space (Chapter 3). However the energetic particles that cause visible aurora appear to be accelerated much higher up in the Earth's magnetosphere, near the zone of Solar wind–magnetosphere interaction.

4.3   Magnetic fields in other stars

4.3.1   *Brief overview of stellar magnetic field measurements*

The first magnetic field measurement in a star beyond the Sun was made by Babcock (1947) for a "peculiar A" (Ap) star, a sub-classification of main sequence (M-S) class A stars, which are in the mass range $\sim$2–3 $M_\odot$. Magnetic fields near the photosphere of M-S stars range from a few gauss to kilogauss, the strongest being of order 20–40 kilogauss. This can apply to dwarf Main Sequence (M-S) stars that are much smaller than the Sun. Post M-S stars have even higher fields, extending to $\sim 10^6$ G in the case of white dwarfs. Magnetic fields in neutron stars (pulsars) can be $\gtrsim 10^{12}$ G. A subclass of millisecond period pulsars appear to possess field strengths up to 10^{15} G. Such high-field strengths had been thought impossible, and they raise some interesting basic physics questions. In contrast to the relatively restricted applicability of Zeeman splitting in interstellar magnetic fields, variations of the Zeeman effect can detect stellar magnetic fields over a range of mG to 10^8 G, a range of at least 10^{11} in $|B|$.

A related question arises here: What are the scale and geometry of magnetic field structures above the photospheres of stars? We cannot yet conduct image-based magnetic field measurements on stars in the way that is possible for the Sun. However, one can take advantage of a star's rotation to try to conform a modelled magnetic field structure to observations: these are the integrated (over the star) flux of circular polarisation and the Zeeman effect. If, for example, an Ap or Bp star has a significantly inclined dipole magnetic field (a model) the longitudinal Zeeman effect provides an estimate of the average of the systematically varying $\langle B_\| \rangle$. The star's total magnetospheric field, $<|B|>$, can be independently measured if $|B| \gtrsim 10^3$ G, in which case the Zeeman splitting is large enough to overwhelm the Doppler smearing effects of stellar rotation. If $\langle B_\| \rangle$ does not average to a vanishingly small number, the star likely has a relatively simple magnetospheric structure. Further cases are often found for Ap and Bp stars in which $<|B|>$ is only slightly larger than $\langle B_\| \rangle$. This would confirm a relatively simple stellar magnetic structure which often can be successfully modelled, e.g. as an oblique rotator dipole field.

Very small circular polarisation, down to a few parts in 10^5, can be measured in the line wings, where the circular polarisation is detectable in the two σ (+ and −) components of the circularly polarised Zeeman-split lines. Such measurements can detect $|B|$ as weak as a few gauss.

In the high field, megagauss régime in white dwarfs, the degree of optical circular polarisation can be significant, and the high-field Zeeman effect enables $|B|$ determinations in the $\sim 10^6$ G range. In addition to the high-field Zeeman effect, the consequent +/− handedness of the electrons' orbital spin in ordered megagauss fields causes a preferential absorption of one component of Stokes V in the star's continuum radiation (Landstreet 1967, Kemp 1970). Linear polarisation can be detected at the $\sim 10^8$ G level in the stellar continuum radiation. For further reading, and references on stellar magnetic fields over more stellar types, the reader is referred to an authoritative review by Mestel & Landstreet (2005).

4.3.2   *Dynamical and magnetic effects in stars compared*

The discovery of kilogauss-level magnetospheres in main sequence dwarf stars leads naturally to the question of whether such strong stellar fields can be dynamically significant. In pre-collapse protostars and molecular clouds, the magnetic energy density, ε_B, is a non-negligible fraction of the thermal and turbulent energy density in the gas, which is threaded with magnetic fields. This means that ε_B may work against the self-gravity ε_G of, say, the pre-stellar molecular cloud. The obvious question is how the dynamical significance of these energy densities ε_B and ε_G compare in a star which is the end point of collapse from a magnetised molecular, and then protostellar, cloud.

A useful parameter for comparison with the self-gravity in this context is the (dimensionless) virial fraction, η_B, of magnetic flux, Φ, which is coupled to the collapsed stellar interior. This fraction can be expressed as

$$\eta_B = \frac{\Phi}{M_* \sqrt{G}} \tag{4.1}$$

(Mestel & Landstreet 2005), where M_* is the star's mass. Whereas η_B can be substantial in pre-collapse systems, it is vanishingly small in the star, even when the star's magnetospheric field strength is $\sim 10^4$ G. This remains true even if we extrapolate the magnetospheric $<|B|>$ in Section 4.3.1 into the stellar interior (a factor of ~ 100 in $|B|$), in which case η_B would increase from $\sim 10^{-5}$ to $\sim 10^{-3}$. The essential conclusion here is that magnetic (Lorentz) forces, in contrast to the internal thermal energy, play a negligible role in counteracting the star's self-gravity.

This transition of η_B from somewhere near unity to a vanishingly small number in the star means that magnetic flux must be very efficiently removed in the course of stellar collapse. The precise explanation of this process is closely related to other plasma physics questions. Basic magnetohydrodynamic considerations suggest that magnetic flux will float to the star's surface in a time that is related to the Kelvin–Helmholtz time, τ_{KH}:

$$\tau_{\text{flux loss}} \simeq \frac{\tau_{KH}}{\eta_B^2}. \tag{4.2}$$

4.3.3   *Magnetic flux removal by stellar jets*

Apart from the slow buoyant transfer of *B*-flux just mentioned, magnetised proto-stellar and stellar jets may be important in removing magnetic flux in the course of stellar collapse. For stellar collapse to occur to the "end point" of slowly rotating M-S stars, both angular momentum and magnetic flux need to be removed. It is likely that the magnetised stellar jets seen in collapsing stars (next section) are a principal mechanism for removing both magnetic flux and angular momentum. These are physically connected phenomena, and even brown dwarf stars are now known to have magnetised outflow jets – see Section 4.4 below.

It is reasonable to speculate, even hypothesise, that removal of angular momentum and magnetic flux in the formation of central collapsed cores of *galaxies* might have some analogies to the processes just discussed. We show evidence in Chapter 8, Section 8.7–8.10, that a significant fraction of the gravitational self-energy of supermassive black holes is released externally in the form of magnetic flux.

The removal of magnetic flux in gravitational collapse happens on a wide range of mass and size scales – i.e. it is a fundamentally important phenomenon of the Universe affecting planet formation, brown dwarf stars, galactic central black holes, and possibly even galaxy formation.

4.4   Jets from Galactic stars

It was noticed by Richard Schwartz (1975) that the emission line spectra of HH objects suggest a post-shock cooling process, rather than photo-ionisation. This was an early clue, supported by subsequent measurements, for several hundred kms^{-1} outflow speeds from HH objects, and spectral evidence for low velocity shocks. Outflow winds from many star forming regions were verified by the early 1980s from CO J = 1 – 0 observations and also higher order rotational transitions (e.g. Lada 1985). Their typical outflow velocities were a few 10's of kms^{-1} and they are poorly collimated. Their sizes ranged up to ~ 1 pc.

More spectacularly, *some* HH objects are known to have jet-like outflows at several hundred kms^{-1} – exceeding the Galactic escape velocity! Outwardly moving knots in HH jets are also seen, mostly in the infrared. These outflow velocities are somewhat lower than those cited above, which may mean that energy transfer is occurring from the faster moving jet to condensations of presumed ambient shocked gas. Jet Mach numbers are in the range of 10–30. A notable fact is that the widths of these stellar jets appear not very dependent on distance from the source. This suggests that they are highly collimated, i.e. with low opening angles. This is one of several lines of evidence pointing to magnetic field domination of the jet outflows. The proximity of some of these jet systems to us enables measurement of the jet width and proper motions in some cases, as well as the temperatures of regions where the jets interact with the ISM, etc.

The latter are in the range of 10^4 K. Radiative cooling plays an important role in these stellar jets, and this is one aspect that differs from black hole-powered AGN jets in external galaxies. Figure 4.1 gives a basic illustration of how the magnetic field is integral to the energy outflow process in a stellar jet. The rotational kinetic energy at the stellar accretion disc ($z = 0$) exceeds the magnetic energy, so that a field line rooted at $r = r_0$, $z = 0$ rotates at $\omega(r_0)$. Up to the Alfvén distance along the jet, z_A, the B-field forces the outflow plasma to follow the Keplerian motion of the field line footprints in the disc below. In this régime, magnetocentrifugal acceleration occurs when the angle $(\boldsymbol{B} \hat{\ } z) \lesssim 60°$ (Tsinganos 2007). At $z > z_A$ (Figure 4.1), B_z becomes weaker than the force of the z-directed bulk motion, so that the B-field now follows the plasma. The net current of the plasma flow creates a strong $B\varphi$. It is $B\varphi$ that provides the collimation of the jet.

Milder forms of outflow jets, at a few kms^{-1}, have been detected even in brown dwarf stars (Whelan *et al.* 2007). Generally, stellar outflow jets are accompanied by an accretion disc, which appears to be integral to the jet's formation and collimation. This property is analogous to galactic scale jets (Section 4.6) all the way in energy and mass up to the most powerful galaxies and quasar jets, which we discuss later.

These facts tell us that an accretion disc-jet system is a natural and basic process associated with gravitational collapse. Magnetic fields play an integral role in these systems, which, in effect, release energy out of the collapsing or collapsed system through transfer of angular momentum. The all-important transfer of angular momentum can thus be regarded as a fundamentally important process in Nature. The main theatre of action is a differentially rotating and magnetised accretion disc approximately perpendicular to the angular

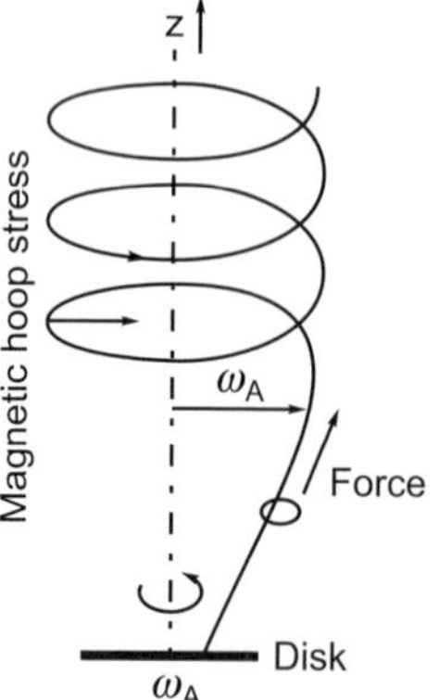

Figure 4.1 A simple model of magnetic acceleration and confinement in a stellar accretion disc-jet system such as in an HH object. (Adapted from Tsinganos 2007).

momentum axis of a collapsing system. It is apparent that an accretion disc is required for stars and planets – and galaxies – to form. The current available evidence shows that in these Galactic collapsing systems, e.g. stellar and protostellar discs and jets, magnetic fields play a key role, especially within the accretion disc. By further extension, it can be argued that our cosmogenic existence depended on initially diffuse magnetic fields in order that the angular momentum transfer, essential for gravitational collapse of planets, stars, and galaxies, could occur!

The subject of Galactic stellar jets and their progenitor variations is too extensive to cover beyond the cursory introduction given in this chapter. The reader interested in this area could begin with the book *Jets from Young Stars: Models and Constraints*, edited by Ferreira *et al.* (2007). It contains extensive references to previous work.

4.5 Molecular clouds and the role of magnetic fields in star formation

4.5.1 *Observations and numbers*

Molecular clouds in the Milky Way are regions where stars are currently forming. The protostellar scales have been difficult to probe in detail because (1) they are shrouded in dust, making optical observations impossible and (2) the angular sizes of protostellar cores in the clouds puts them below the resolution of most contemporary instruments. There are exceptions, however, such as VLBI radio observations of maser lines in the protostellar cores of molecular clouds, and Zeeman splitting of OH in methanol maser clouds around forming high-mass stars. Zeeman splitting measurements in the main OH lines (Green *et al.* 2012) reveal magnetic field strengths within these regions in the range of -1.3 to 3.8 mG. A further interesting outcome of this study in the Galactic Carina-Sagittarius spiral arm tangent is that the field sense is, curiously, coherent for 11 objects whose locations stretch over ~ 5.3 kpc along the arm tangent.

Small molecular clouds (SMCs) are $\lesssim$ a few $\times 10$ pc in size, and have a mass of at most a few $\times 10^2$ $M_\odot$, which includes their gas, molecules, and dust. They are relatively cool, ≈ 10–20 K, and have no massive OB stars. An example is the well-studied Ophiuchus Cloud. Giant molecular clouds (GMCs), in contrast, have dimensions of order 100 pc, masses of 10^4–10^6 $M_\odot$ and cloud-internal ambient temperatures of ≈ 50–100 K. According to recent counts, GMCs in the Milky Way number about 4000, and there are many more thousands of SMCs. On a cosmic time scale, molecular clouds are relatively transient events, so that if we extrapolate these numbers over $\sim 5 \times 10^9$ years of the Galaxy's lifetime their total number per galaxy is huge. This illustrates the importance of molecular clouds as key building blocks of the baryonic Universe.

4.5.2 *Summary*

Magnetic fields in molecular clouds are integral to the process of stellar collapse within the clouds. Because of the limited observational accessibility mentioned above, detailed dynamical and magneto-plasma processes are not yet completely understood. Magnetic fields almost certainly play an important role in the dynamics and chemical physics of protostellar clouds. The successive processes of gravitational collapse from the interstellar medium, to cloud, then stellar dimensions, require that there is a very effective transfer of angular momentum outwards at each stage of collapse. The very low angular momenta of stars like the Sun testify to the effectiveness of angular momentum removal in star formation. This also applies to galaxy formation, and to the formation of supermassive black holes, as mentioned earlier. There is no known purely hydrodynamical process that can achieve this with the required high efficiency. Magnetic fields are necessary, and the basic MHD processes involved are becoming increasingly better understood, though many challenges remain.

4.6   The generation and regeneration of magnetic fields in supernova remnants

Supernova remnants are capable of filling a substantial volume of interstellar space with magnetic flux in the aftermath of their explosions. The large energy they release into the surrounding interstellar medium propagates cosmic rays (CR) and magnetic fields. The collective energy release of hot, massive stars and their associated supernovae not only magnetises the interstellar medium of the host galaxy, but can also be released via a galactic wind into the surrounding galaxy halo, and ultimately the nearby intergalactic medium in the case of a strong starburst (Chapter 6). Even the polar regions of the quiescent Sun have outflow speeds of $\gtrsim 400$ kms^{-1}, although the energy flux of this outflow is small.

4.6.1   The Crab Nebula and other plerion-type supernova remnants

A 5 GHz radio image of the Crab Nebula is shown in Fig. 4.2. The nebula is powered by a rotating pulsar magnetosphere with a period of ~33 ms. The magnetic flux and CRs producing the observed synchrotron radiation appear to have come directly from a rotating object, the Crab pulsar – as distinct from a random turbulent process. Quasi-circular, hoop-like acceleration zones were identified in VLA radio images (Bietenholz & Kronberg 1992; Kronberg, Lesch, & Bietenholz 1993; Bietenholz, Frail, & Hester 2001), then in an HST optical image, and in X-rays by the Chandra Observatory (Weisskopf *et al.* 2000). The Chandra X-ray image also reveals a faint, vertically propagating jet, which can be seen in addition to the "rings" shown in the radio inset from Bietenholz *et al.* in Fig. 4.2. Some parts of these features have been referred to as "wisps", following Scargle's (1969) first discovery of the brightest inner structures.

The Crab Nebula's proximity and young age make it an enduring high-energy laboratory for the production of magnetic fields and CRs, and for understanding the role of pulsars. The "rib cage" of intensity and/or spectral index perturbations and wisps (Fig. 4.2(b)), and their variability, appear to be associated with the acceleration of synchrotron-emitting electrons. These are detected all the way from the radio to γ-rays. Particle acceleration models and the energetics of the Crab have been discussed by many authors (e.g. Rees & Gunn 1974, Kennel & Coroniti 1984, Gallant & Arons 1994, Melatos & Melrose 1996, Atoyan 1999).

Most of the models discussed involve shock acceleration to produce the CRs. A successful model must explain particle acceleration over a very broad range of CR γ–energies ($\gamma = E_e/mc^2$). In contrast to Cassiopeia A, discussed next, the Crab Nebula shows no evidence for an outer shock zone, where any significant component of the CRs could be energised. The Crab Nebula's rate of energy supply, $\approx 5 \times 10^{38}$ erg s^{-1}, appears to come from the spin-down energy of its central $P = 33$ ms central pulsar. This energy of the high-energy particles is thus fundamentally gravitational, in that the magnetic and CR energy derive from the transfer of angular momentum of the collapsed pulsar. The pulsar input power can be compared with $\approx 5 \times 10^{37}$ erg s^{-1}, the total electromagnetic loss rate of the

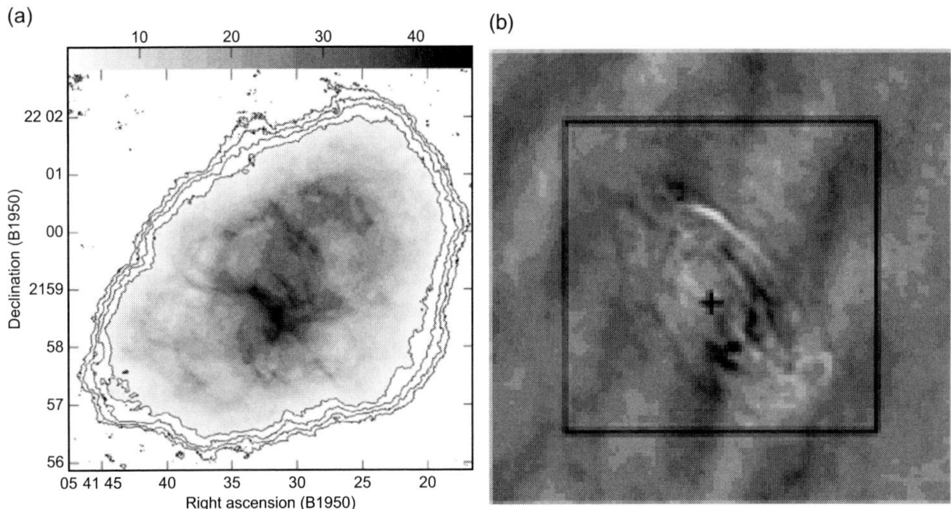

Figure 4.2 (a) Image of the Crab Nebula (Taurus A) showing its synchrotron radio emission at 5 GHz (Bietenholz, Frail, & Hester 2001). (b) The inner zone around the pulsar, showing the radio-luminous rings (Bietenholz & Kronberg 1992) that have also been seen in X-rays.

Crab Nebula. Most of this energy is emitted between optical and TeV bands, where τ is relatively short. Other forms of energy output are PdV work on the pre-existing medium (probably small for the Crab), and the kinetic energy of the denser expanding filaments, which may derive from the thermonuclear energy of the SN explosion.

The magnetic field amplification and particle acceleration are largely thought to have originated in relativistic shocks in the wisp regions. One key test for the origin of CR and magnetic field energy is the overall conversion efficiency from the spindown energy. This can be expressed as

$$\frac{\frac{dE}{dt}\,(\text{losses})}{\frac{dE}{dt}\,(\text{spindown})}. \tag{4.3}$$

The prominence of the Crab pulsar makes it relatively easy to quantify the input energy as opposed to, for example, shell-type supernovae, which have a less uniquely identifiable energy source. The efficiency of energy transfer in supermassive black hole-driven, galaxy scale systems is discussed in Chapter 8.

In the Crab Nebula, most of the pulsar rotational energy is converted directly and electro-dynamically via its magnetosphere into a magnetised relativistic wind. The first phase is a Poynting flux power flow from the rotating magnetosphere. This is thought to be converted to kinetic energy via a flow of electrons and positrons having $\gamma \approx 10^5$–10^6. In this scenario, these leptons (e^+, e^-), and possibly nuclei, are accelerated directly by local E-fields in the rotating magnetic structure. The ensuing relativistic wind encounters a shock zone, most visible as radiation. This occurs in the inner wisp/ring structure (Fig. 4.2). In this scenario the shocked relativistic plasma is more a consequence of the accelerated leptons and hadrons than the basic cause of their acceleration. The general question of the relative importance of particle acceleration by magnetic field annihilation and/or direct electrodynamic acceleration, as distinct from Fermi-type shocks, will be encountered later in connection with black hole-powered jets and radio lobes in galaxies. What is common to the Crab and supermassive black holes is that, in both systems the energy initially comes from a single, coherently rotating object. Nonetheless, many details of the Crab's relativistic particle acceleration and magnetic field production not yet fully clarified. In particular, the synchrotron electrons with $\gamma \sim 10^3$–10^4 radiating at cm and metre wavelengths have implied ages of a few hundred years. In contrast, the freshly accelerated X-ray emitting electrons, with γ's 10^5–10^6, radiate in X-ray bands up to ~10 TeV, and have lifetimes of only a few years.

At the outer periphery of the Crab Nebula, we see no evidence of a global, outer shock due to interaction with the ambient interstellar medium. This is in contrast to shell-type SNRs, such as Cas A discussed below, where there is evidence suggesting shock acceleration at the SNR periphery.

This absence of an obvious outer shock can be understood from independent estimates that exist for both the thermal gas density within the bulk of the Crab Nebula and the local ambient ISM gas. Two-frequency VLA images by Bietenholz & Kronberg (1991) produced a detailed Faraday rotation map of the Crab at ~2″ resolution – see Fig. 2.6 – showing that the ionised gas density within the bulk of the nebula has an upper limit of only ~0.01 cm^{-3}. This is about an order of magnitude below the average ISM ionised gas density, $n_e \sim 0.1$ cm^{-3}. However, within a ~120 pc zone around the Crab the ISM density has also been found to be anomalously low, suggesting that it may lie in an evacuated HI bubble (Wallace *et al.* 1999). A recent deeper X-ray search in Chandra data by Seward, Gorenstein & Smith (2006) confirms the absence of any outer X-ray emitting shell beyond the photons reflected by interstellar dust. Both of these results are consistent with an unusually low ambient ISM density around the Crab; of order 0.01 cm^{-3}, they approach Bietenholz & Kronberg's upper limit for nebula-internal thermal gas density (excluding the filaments). Although it is not yet known whether all plerion-type SNRs occur in such "evacuated", low density interstellar "holes", the bulk of the Crab Nebula volume is a relativistic plasma, largely empty of thermal gas.

4.6.2   *Magnetic fields and CR energisation in shell-type supernova remnants*

A second, more commonly observed type of supernova remnant is the shell SNR. Here the energised gas appears to derive its energy more directly from a thermonuclear source – the supernova explosion. In shell-type supernova remnants, shock acceleration and other related mechanisms, not always clearly separable, appear to play a role in magnetic field amplification and CR acceleration. At only 3.4 kpc distance and ~330 years old, Cassiopeia A is a prime such object for detailed examination. Figure 4.3 shows a composite multiband X-ray image from the Chandra Observatory (Allen, Stage, & Houck 2007, Stage *et al.* 2006). The spectral information from three X-ray bands can nicely distinguish the outer, hotter forward shock from the inner, clumpy remnants (lighter grey) of the ejecta that have been heated by a slower, reverse shock wave. The outer shock front is remarkably thin, with some consequences that are discussed below.

The thermonuclear blast wave energy of the supernova remnant is converted into photons, high-energy particles and a magnetic field. The photons are produced by a mixture of thermal and relativistic bremsstrahlung and synchrotron radiation. At the highest energies, photons up to γ-ray bands could also be produced by π^0 decay, possibly added both to inverse-Compton scattering in local dense infrared (IR) radiation field zones of the dense

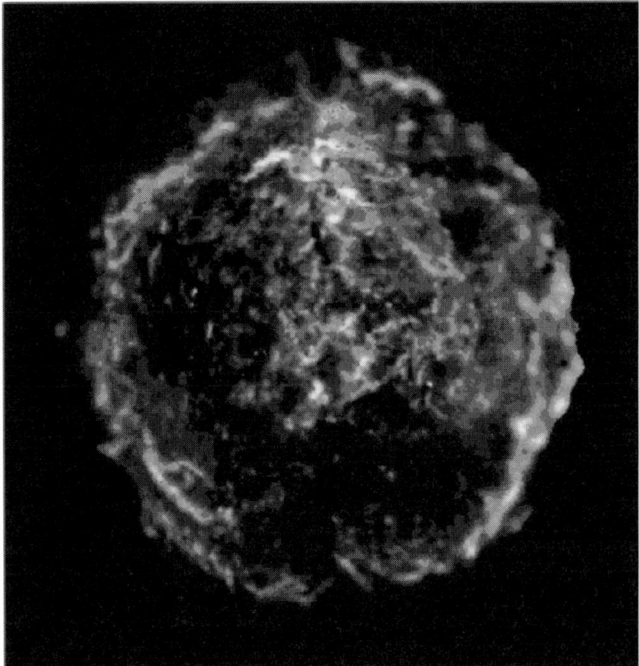

Figure 4.3 Cassiopeia A in X-rays, imaged by the Chandra X-Ray Observatory (Stage *et al.* NASA photo). The false colour scale is coded by temperature using a combination of the 0.5–1 keV, 1.5–2.5 keV, and 4–6 keV bands, respectively. (Allen *et al.* 2007). Color image exists and it is mentioned in section 4.6.2

knots, and to relativistic bremsstrahlung. The field regeneration is part of the conversion process from the mechanical and CR energy of the outbound shock front. The details of this process are not yet clear. Berezhko & Völk (2004) argued that the X-ray morphology and luminosity of Cas A are consistent with efficient acceleration of CR nuclei. Their acceleration model is consistent with a relatively strong SNR-interior magnetic field ($\sim$0.5 mG), and requires that the γ-ray emission be produced largely by π^0 decay. Detailed discussion of the complex mix of plasma processes operating in Cassiopeia A can be found in Cowsik & Sarkar (1984), Aharonian *et al.* (2001), Laming (2001), Berezhko, Pühlhofer & Völk (2003), Vink & Laming (2003), and Bell (2004). It is possible that the different mechanisms above operate in different sub-regions of the supernova remnant, and we discuss some of them briefly below. Magnetoplasma processes in Cassiopeia A, which can now be probed in considerable detail from radio to γ-ray bands, may help to explain extragalactic systems that are too distant to diagnose with a comparable degree of detail.

The local magnetic field strengths and the IR photon and particle densities within and around the SNR are obviously important in discriminating interactions among magnetic fields, particles, and photons. It is likely that the Cas A supernova ejecta have encountered an ambient magnetic field well above the ambient pre-SN interstellar value of $\sim$3 μG. Biermann & Cassinelli (1993) argued that the pre-existing hot stellar wind outflow from Cas A's progenitor provides this, and came from a Wolf–Rayet (W–R) star. The W–R wind could have been magnetised at a level of ~10× the ambient interstellar field strength. However, Krause *et al.* (2008) were more recently able to identify the echo spectrum near peak light of the SN explosion, and conclude that the progenitor was a Type IIb supernova. This makes it easier to assume that near-milligauss-level magnetic fields (Atoyan *et al.* 2000) are produced within the Cas A remnant. It is difficult to explain the TeV emission up to ~10^{15} eV without these high-magnetic fields – see also Uchiyama & Aharonian (2008). The high (mG-level) fields also limit the inverse Compton component of emission, which appears insufficient to explain the fluxes observed at the highest energies. X-ray synchrotron radiation, another important mechanism, appears to occur in the very thin outer shell in the 4–6 keV X-ray band but is not seen in the radio (Vink & Laming 2003). The thin outer X-ray shell's measured width of $1.5''$–$4''$, combined with an estimate of the plasma velocity behind the shock front of ~1400 kms^{-1}, enables a direct estimate of the synchrotron loss time, τ_{loss}. Vink & Laming (2003) obtain numbers in the range 18–50 years. This means that strong synchrotron radiation at the very highest energies predominates in a thin shell, as observed.

This radiation can now be incorporated into the expressions for the peak of the photon spectral density, v_{p} (E_{e}, B), and the synchrotron loss time, τ_{loss} (E_{e}, B) for electrons. The photon spectral density can be represented as follows:

$$v_{\mathrm{p}} = 1.8 \times 10^{18} B \sin \varphi E_{\mathrm{e}}^2 \quad \mathrm{Hz}$$

where φ is the relativistic electron's average pitch angle ($<\sin \varphi > ~0.5$) with respect to the local field direction (Ginzburg & Syrovatskii 1965), and v_{p} corresponds in this case to 5 keV.

The synchrotron loss time, τ_{loss} (E_{e}, B) expressed as a half-life is

$$\tau_{1/2} = \frac{637}{B^2 E_{\mathrm{e}}} \, \mathrm{s} \tag{4.4}$$

(Reynolds 1998). The small linear dimension of the synchrotron-radiating, thin outer shock front sets an upper limit on τ, by comparing the transport and (very short) loss times of the freshly accelerated electrons in the thin shell. Equations 4.4 and 4.5 can be combined with post-shock velocity estimates to estimate the magnetic field strength immediately behind the thin outer shell. The result is ~0.1 mG in this zone (Vink & Laming 2003).

A more direct estimate of the magnetic field strength in Cas A's knots comes from the observed *variability* in X-rays of the knots and filaments. This was achieved by Uchiyama & Aharonian (2008), who compared Chandra telescope images from 2000, 2002, and 2004 and found variability in the featureless power law-like spectrum of several knots and filaments within the Cas A nebula. The synchrotron loss time (Equation 4.4) of ~1 year leads to a magnetic field strength of 1 to a few mG in the filaments and knots.

Given that high-energy photon detectors have ever better sensitivity and spectral and spatial resolution, one can place new constraints on the SNR-internal magnetic field using high-energy photons. Recently the spectral shape and intensity of Cas A's γ-ray spectrum are measurable from ~1 keV all the way up to 10 TeV and more. Relevant instruments include BeppoSAX/OSSE, CGRO/COMPTEL, ASCA, EGRET, HEGRA, VERITAS and HESS. By verifying the relative absence of synchrotron radiation in the X-ray bands compared with the radio, limits can be placed on the electron γ's that are much lower than those in the outer thin synchrotron-emitting shell discussed above. The synchrotron brightening in the radio knots (Anderson & Rudnick 1996) could arise *prima facie* either from a rapid B amplification or from the local production of more relativistic electrons.

We next examine whether the TeV photons might be inverse Compton radiation. That seems unlikely for CMB photon scattering when we combine the CMB photon flux with a ~1 mG magnetic field in the radio luminous parts of Cas A. At the higher field strengths, likely $\gtrsim 1$ mG, the correspondingly reduced relativistic particle density, n_r, would not produce enough Compton-scattered high-energy photons. A higher photon density than the CMB might conceivably come from an IR photon density, n_{IR}, generated within SNR-internal knots. Vink & Laming (2003) conclude that n_e–n_{IR} photon (IC) scattering would require pushing B to improbably low values in the Cas A knots (to increase n_e), and/or alternatively having a ρ_{IR} photon density $\gtrsim 150$ cm^{-3}. Straightforward estimates of ρ_{IR} can be made from IRAS flux measurements (Braun 1987). Assuming a spherical knot size consistent with observations, Vink & Laming (2003) obtain $\rho_{IR} \approx 72$ cm^{-3}. While this is smaller than the ρ_{IR} above, we should keep in mind that IC emission, like any classical scattering is direction/anisotropy dependent, and of course n_{IR} scales inversely as the cube of the dimension of the region generating the IR photons.

To summarise, the observed extension of Cas A's spectrum is significantly higher at photon energies above ~10 keV than the predicted photon fluxes from synchrotron, IC or bremsstrahlung mechanisms in the optical and keV X-ray bands. At least some of the excess radiation appears to come from the "knot" concentrations seen in Fig. 4.3 and in high-resolution optical images. Virtually all physical arguments suggest that B in these knots is relatively high, though not sufficient to explain all of the TeV excess as synchrotron radiation. This, in turn, suggests that if there is a missing radiation component from the Cas A knots, it probably involves some other magneto-plasma process and significant local magnetic fields. This leads us to examine how the bremsstrahlung-producing thermal electrons might be accelerated beyond their normal highest energies to produce a relativistic bremsstrahlung "tail" of the highest energies.

4.6.3 Electron acceleration by lower hybrid waves

Lower hybrid waves (LWH) are known in other plasma situations. They can anisotropically accelerate electrons and ions, preferentially along magnetic field lines. This may be happening in the Earth's magnetotail and in other astrophysical settings where the two stream instability develops. Production of electrons accelerated in this way has been modelled analytically for comet X-ray emission by Krasnosel'skikh *et al.* (1985), and these accelerated electrons have been detected, for example, in Halley's comet by Klimov *et al.* (1986). The LHW arise from electrostatic ion oscillations. The LHW frequency ω_{LHW} is the geometric mean of the ion and electron gyrofrequencies, ω_i and ω_e.

$$\omega_{LHW} = \sqrt{\omega_i \omega_e}. \qquad (4.5)$$

The oscillations are in the $B_\perp$ direction, and they produce a resonance with ions moving across the magnetic field lines and electrons moving along them. In most astrophysical situations of interest, like the SNR knots of Cas A, the electron acceleration time is much shorter than the electron–electron collision time. In this situation, a non-Maxwellian distribution of electron momenta can develop in the direction of the magnetic field. The electrons are beamed along the field lines into an angle given by ~$2\omega_p/\omega_i$. In the Cas A knots, this gives a relativistic bremsstrahlung component of the continuum emission as high as ~100 keV, and it can qualitatively explain the observed high-energy photon spectrum of Cas A.

This mechanism is also interconnected with the Ta Kα and Kβ lines, since their broadening and excitation is also integral to the suprathermal plasma "tail". A good starting background on LHW acceleration can be found in Vink & Laming (2003). LHWs might also be an important mechanism in extragalactic sources that are discussed later. Unfortunately, in these latter situations we are usually missing the detailed plasma diagnostic information that is more easily available for the Sun, comets, and nearby supernovae, etc.

Independent of the actual acceleration mechanism for the relativistic particles, it is sometimes helpful to apply the Hillas criterion, as we do later in Chapter 11, to a volume of dimension L containing energetic particles and a coherent magnetic field. The product BL sets an upper limit in energy to which some acceleration mechanism can possibly accelerate a charged particle. The Hillas criterion is often useful as necessary, but not always sufficient, to identify a particle acceleration mechanism. If we consider some of Cassiopeia A's magnetised, quasi-linear filaments to have a coherent magnetic field structure (B_{FIL}) over their length (L_{FIL}) we can write

$$\frac{B_{\mathrm{FIL}}}{1mG} \cdot \frac{L_{\mathrm{FIL}}}{1\mathrm{pc}} \leq 3 \times 10^{15}\ \mathrm{eV}. \tag{4.6}$$

By scaling L to the observed lengths of the filamentary features (up to ≈ 1pc for Cas A) and to the mG levels discussed above, we can set an upper limit to proton energies of order 10^{15} eV. This is interestingly close to the energy of the "knee" in the CR spectrum, which has been thought to originate in supernovae.

4.7   A magnetised jet in the Galactic centre

The Milky Way is known to have a $\sim 4.8 \pm 0.3 \times 10^6\ \mathrm{M}_\odot$ black hole at its centre (Ghez *et al.* 2007). This mass is modest compared with typical values in the range of 10^8–$10^9\ \mathrm{M}_\odot$ inferred for the elliptical galaxy progenitors of powerful extragalactic radio galaxies and quasars. It is, however, notable that the mass of the Milky Way's central black hole has been directly measured in real time using close-in, orbiting stars as "test particles" in the BH's gravitational potential.

The Galactic centre is known to possess a circumnuclear disc (e.g. Güsten *et al.* 1987, Christopher *et al.* 2005). Protruding away from the central BH is a remarkable twin helix jet-like structure, discovered with 24 μm observations by Morris *et al.* (2006). This apparent twin helix structure extends about 20 pc vertically away from the Galactic centre. The transverse dimension of the two helical strands is about 3.5 pc, and the entire structure has no known counterpart on the opposite side of the Milky Way.

This transverse dimension is, interestingly, comparable to that of the circumnuclear disc of the Galaxy's central black hole. Its rotational velocity is independently known to be $\sim 100\ \mathrm{kms}^{-1}$, corresponding to a period of $10^4 R$ year, R being the radius of the circumnuclear disc. With the plausible assumption that the double helix jet is an outwardly propagating and magnetically dominated structure, and connected to the circumnuclear disc (Fig. 4.4), the above numbers permit a direct estimate of the Alfvén speed (Equation 1.3) in the outflowing jet (Morris *et al.* 2006). This is $v_{\mathrm{A}} \approx 1000\ \mathrm{kms}^{-1}$, and it follows from a direct measurement of the propagating helical structures seen in Fig. 4.4. Combining the estimate of v_{A} with the proton density, $\rho = n_{\mathrm{p}}m_{\mathrm{p}}$, the magnetic field strength in this torsional wave can be estimated. Morris *et al.* used the emissivity of the thermal bremsstrahlung X-ray emission to estimate

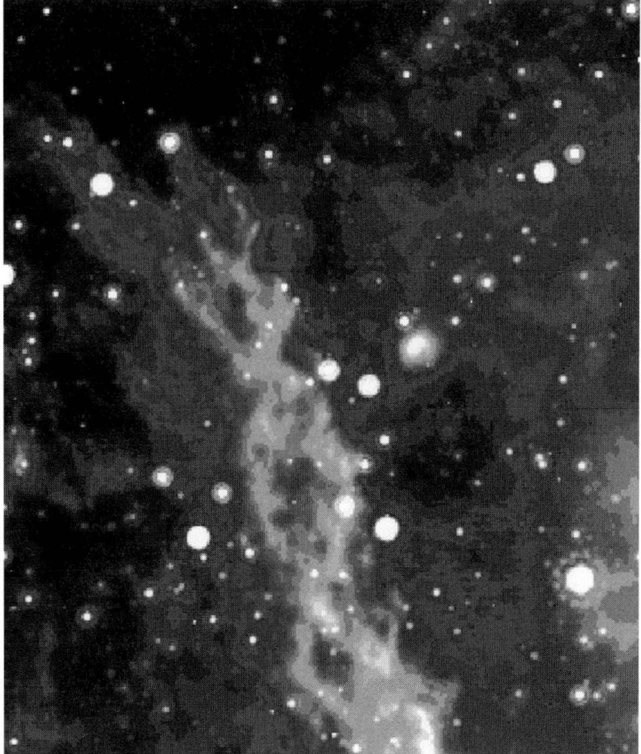

Figure 4.4 An Alfvénic torsional wave outflow jet emanating from the Milky Way's central black hole/accretion disc (Morris *et al.* 2006). The height of the figure panel is 14 arcmin, or 32 pc, at the distance of the Galactic centre. (Figure courtesy of M. Morris, private communication.)

n_p, where $B \approx 0.5\, n_p^{1/2}$ mG. Plausible values of n_p put B at ≈ 1 mG, although the uncertainty could be up to a factor of 10. These results make this jet one of the best diagnosed of all jets, and it may serve as a useful "template" for investigating the magnetic properties of other, extragalactic jets. In connection with galaxy outflows discussed later, we note that an Alfvén speed of 1000 kms^{-1} is well above the escape speed of the Milky Way, so that the energy carried away is likely removed at least into the Milky Way's halo.

References

Aharonian, F. *et al.* 2001, Evidence for TeV Gamma Ray Emission from Cassiopeia A, *Astron. Astron. Astrophys.*, 370, 112

Allen, G. E., Stage, M. D., & Houck, J. C. 2008, Nonthermal Brensstrahlung v. Synchrotron Radiation: The Nature of the Nonthermal X-ray Emission from Cas A, in *Proceedings of the 30th Annual International Cosmic Ray Conference, July 3–11, 2007, Mérida, Yucatán, Mexico*, ed. R. Caballero *et al.* (Mexico City: Universidad Nacional Autónoma de México), 2, 839

Ampleford, D. J., Lebedev, S. V., Ciardi, A., Bland, S. N., Bott, S. C., Hall, G. N., Naz, N., Jennings, C. A., Sherlocke, M., Chittenden, J. P., Frank, A., & Blackman, E. 2007, Laboratory Modeling of Standing Shocks and Radioactively Cooled Jets with Angular Momentum, High Energy Density Lab. *Astrophys.*, 307, 51

Anderson, M. C. & Rudnick, L. 1996, Sites of Relativistic Particle Acceleration in Supernova Remnant Cassiopeia A, *Astrophys. J.*, 456, 234

Atoyan, A. M. 1999, Radio Spectrum of the Crab Nebula as an Evidence for the Fast Spin of Its Pulsar, *Astron. & Astrophys.*, 355, 211

Atoyan, A. M., Aharonian F. A., Tuffs, R. J., & Völk, H. J. 2000, On the Gamma-Ray Fluxes Expected from Cassiopeia A, *Astron. & Astrophys.*, 355, 211

Babcock, H. W. 1947, Zeeman Effect in Stellar Spectra, *Astrophys. J.*, 105, 105

Bacciotti, F., Ray, T. P., Mundt, R., Eislöffel, J., & Solf, J., 2002, Hubble Space Telescope/STIS Spectroscopy of the Optical Outflow from DG Tauri: Indications for Rotation in the Initial Jet Channel, *Astrophys. J.*, 576, 222

Bell, A. R. 2004, Turbulent Amplification of Magnetic Field and Diffusive Shock Acceleration of Cosmic Rays, *MNRAS*, 353, 550

Bellan, P.M. 2006, 2008 *"Fundamentals of Plasma Physics"* (Cambridge: Cambridge University Press)

Berezhko, E. C., Pühlhofer, G., & Völk, H. J. 2003, Gamma-Ray Emission from Cassiopeia A Produced by Accelerated Cosmic Rays, *Astron. Astrophys.*, 400, 971

Berezhko, E. & Völk, H. J. 2004, The Theory of Synchrotron Emission from Supernova Remnants, *Astron. Astrophys.*, 427, 525

Biermann, P. L. & Cassinelli, J. P, 1993, Cosmic Rays. II. Evidence for a Magnetic Rotator Wolf-Rayet Star Origin, *Astron. Astrophys.*, 277, 691

Bietenholz, M. F., Frail, D. A., & Hester, J. J. 2001, The Crab Nebula's Moving Wisps in Radio, *Astrophys. J.*, 560, 254

Bietenholz, M. F. & Kronberg, P.P. 1991, Faraday Rotation and Physical Conditions in the Crab Nebula, *Astrophys. J.*, 368, 231

Bietenholz, M. F. & Kronberg, P. P. 1992, Activity and Radio Spectral Index Variations near the Center of the Crab Nebula, *Astrophys. J.*, 393, 206

Braginskii, S. I. 1965, Transport Processes in Plasma, *Rev. Plasma Phys.*, ed. M.A. Leontovich (New York: Consultants Bureau), 1, 205

Braun, R., 1987, The Structure and Dynamics of Young Supernova Remnants – New Constraints from Observations of Shock-Heated Dust, *Astron. Astrophys.*, 171, 233

Carmichael, H. 1964, A Process for Flares, *in Proceedings of the AAS-NASA Symposium held 28–30 October, 1963 at the Goddard Space Flight Center, Greenbelt, MD*, ed. W. N. Hess (Washington, DC: National Aeronautics and Space Administration, Science and Technical Information Division), 451.

Christopher, M. H., Scoville, N. Z., Stolovy, S. R., & Yun, M. S. 2005, HCN and HCO+ Observations of the Galactic Circumnuclear Disc, *Astrophys. J.*, 622, 346

Cowsik, R. & Sarkar, S. 1984, The Evolution of Supernova Remnants as Radio Sources, *MNRAS*, 207, 745

Ferreira, J., Dougados, C., & Whelan, E., eds. 2007, *Jets from Young Stars: Models and Constraints, Lecture Notes in Physics 723* (Berlin: Springer-Verlag)

Gallant, Y. A. & Arons, J. 1994, Structure of Relativistic Shocks in Pulsar Winds: A Model of the Wisps in the Crab Nebula, *Astrophys. J.*, 435, 230

Ghez, A. M., Salim, S., Weinberg, N., Lu, J., Do, T., Dunn, J. K., Matthews, K., Morris, M. Yelda, S., & Becklin, E. E. 2007, Probing the Properties of the Milky Way's Central Supermassive Black Hole with Stellar Orbits, in *Proc. IAU Symp. S248* (Cambridge: Cambridge University Press) 3, 52

Ginzburg, V. L. & Syrovatskii, S. I. 1965, Cosmic Magnetobremsstrahlung (Synchrotron Radiation), *Ann. Rev. Astron. Astrophys.*, 3, 297

Gough, D. & McIntyre, M. E. 1998, Inevitability of a Magnetic Field in the Sun's Radiative Interior, *Nature*, 394, 755

Green, J. A., McClure-Griffiths, N. M., Caswell, J. L., Robishaw, T., & Harvey-Smith, L. 2012, MAGMO: Coherent Magnetic Fields in the Star-Forming Regions of the Carina-Sagittarius Spiral Arm Tangent, *MNRAS*, 425, 2530

Güsten, R., *et al.* 1987, Aperture Synthesis Observations of the Circumnuclear Ring in the Galactic Center, *Astrophys. J.*, 318, 124

Hughes, D. W. Rosner, R., & Weiss, N. O., eds. 2007, *The Solar Tachocline* (Cambridge: Cambridge University Press)

Hsu, S. C., & Bellan, P. M. 2002, A Laboratory Plasma Experiment for Studying Magnetic Dynamics of Accretion Disks and Jets, *Mon. Not. Roy. Astr. Soc.* 334, 257.

Intrator, T. P., Furno, I., Ryutov, D. D., Lapenta, G. L., Dorf, L., & Sun, X. 2007, Long-Lifetime Current-Driven Rotating Kink Modes in a Non-Line-Tied Plasma Column with a Free End, *J. Geophys. Res.*, 112, A05S90 [DOI 10.1029/2006JA011995]

Kemp, J. C. 1970, Circular Polarization of Thermal Radiation in a Magnetic Field, *Astrophys. J.*, 162, 169

Kennel, C. F. & Coroniti, F. V. 1984, Confinement of the Crab Pulsar's Wind by Its Supernova Remnant, *Astrophys. J.*, 283, 694

Klimov, S., *et al.* 1986, Extremely Low Frequency Plasma Waves in the Environment of Comet Halley, *Nature*, 321, May 15, 292

Krasnosel'skikh, V. V., Kruchina, E. N., Volokitin, A. S., & Thejappa, G. 1985, Fast Electron Generation in Quasiperpendicular Shocks and Type II Solar Radiobursts, *Astron. Astrophys.*, 149, 323

Krause, O., Birkmann, S. M., Usida, T., Hattori, T., Goto, M., Rieke, G. H., & Misselt, K. A. 2008, The Cassiopeia A Supernova Was of Type IIb, *Science*, 320, 1195

Kronberg, P. P., Lesch, H., Ortiz, P. F., & Bietenholz, M. F. 1993, The Crab Nebula's Cosmic-Ray Acceleration Zone Revealed, *Astrophys. J.*, 416, 251

Lada, C. J. 1985, Cold Outflows, Energetic Winds, and Enigmatic Jets Around Young Stellar Objects, *Ann. Rev. Astron. Astrophys.*, 23, 267

Laming, J. M. 2001, Accelerated Electrons in Cassiopeia A: An Explanation for the Hard X-Ray Tail, *Astrophys. J.* 546, 1149

Landstreet, J. D. 1967, *Effects of a Large Magnetic Field on Energy Transfer in White Dwarf Stars*, Ph.D. Thesis, Columbia University (New York: Columbia University)

Melatos A. & Melrose, D. B. 1996, Energy Transport in a Rotation-Modulated Pulsar Wind, *MNRAS*, 279, 1168

Mestel, L. & Landstreet, J. D., 2005, Stellar Magnetic Fields, in *Cosmic Magnetic Fields*, ed. R. Wielebinski & R. Beck, Lecture Notes in Physics (Berlin: Springer-Verlag), 664,183

Morris, M., Uchida, K., & Do, T., 2006, A Magnetic Torsional Wave Near the Galactic Centre Traced by a 'Double Helix' Nebula, *Nature*, 440, 308

Mozer, F. S., Bale, S. D., McFadden, J. P. & Torbert, R. B. 2005, New Features of Electron Diffusion Regions Observed at Subsolar Magnetic Field Reconnection Sites, *Geophys. Res. Lett.* 32, L24102

Rees, M. J. & Gunn, J. E. 1974, The Origin of the Magnetic Field and Relativistic Particles in the Crab Nebula, *MNRAS*, 167, 1

Reynolds, S. P. 1998 Models of Synchrotron X-Rays from Shell Supernova Remnants, *Astrophys. J.*, 493, 375

Scargle, J. D. 1969, Activity in the Crab Nebula, *Astrophys. J.*, 156, 401

Schwartz, R. D. 1975, T Tauri Nebulae and Herbig-Haro Nebulae – Evidence for Excitation by a Strong Stellar Wind, *Astrophys. J.*, 195, 631

Seward, F. D., Gorenstein, P., & Smith, R. K. 2006, Chandra Observations of the X-Ray Halo around the Crab Nebula, *Astrophys. J.*, 636, 873

Stage, M. D. NASA photo

Stage, M. D, Allen, G. E., Houck, J. C., & Davis, J. E., 2006, Cosmic-Ray Diffusion Near the Bohm Limit in the Cassiopeia A Supernova Remnant, *Nat. Phys.*, 2, 614

Tobias, S. M., Diamond, P. H., & Hughes, D. W., 2007, β-Plane Magnetohydrodynamic Turbulence in the Solar Tachocline, *Astrophys. J.*, 667, L113

Tsinganos, K. 2007, Theory of MHD Jets and Outflows, in *Jets from Young Stars*, ed. J. Ferreira, C. Dougados, & E. Whelan, Lecture Notes in Physics (Berlin: Springer-Verlag), 723, 117

Uchiyama, Y. & Aharonian, F. A. 2008, Fast Variability of Nonthermal X-Ray Emission in Cassiopeia A: Probing Electron Acceleration in Reverse-Shocked Ejecta, *Astrophys. J.*, 677, L105

Uzdensky, D. & Goodman, J. 2008, Statistical Description of a Magnetized Corona above a Turbulent Accretion Disk, *Astrophys. J.*, 682, 608

Vink, J. & Laming, J. M. 2003, On the Magnetic Fields and Particle Acceleration in Cassiopeia A, *Astrophys. J.*, 584, 758

Wallace, B. J., Landecker, T. L., Kalberla, P. M. W., & Taylor, A. R. 1999, The Interstellar Environment of Filled-Center Supernova Remnants. III. The Crab Nebula, *Astrophys. J. Suppl.*, 124, 181

Weisskopf, M. C., *et al.* 2000, Discovery of Spatial and Spectral Structure in the X-Ray Emission from the Crab Nebula, *Astrophys. J.*, 536, L81

Whelan, E. T., Ray, T. P., Bacciotti, F., Randich, S., Jayawardhana, R., Natta, A., Testi, L., & Mohanty, S. 2007, Outflow Activity in Brown Dwarfs, in *Star-disk Interaction in Young Stars: Proceedings of the 243rd Symposium of the International Astronomical Union held in Grenoble, France, May 21–25*, ed. J. Bouvier & I. Appenzeller (Cambridge: Cambridge University Press), 357

5

Magnetic field configurations in large galaxies

5.1 Introduction

Theoretical understanding and observational clarification of galactic magnetic fields have made great progress since 1980. This chapter describes some diagnostic methods, important observational results, some relevant theory, and some magnetic field patterns observed in galaxies. The theory includes the "standard" dynamo theory of field amplification in galaxy discs. However, theory and observations do not yet interact at a level where the latter can clarify all key questions on the connections between interstellar magnetic field configurations and galactic dynamics.

5.1.1 *Some specific questions and puzzles*

(i) How did the large scale organisation of galactic magnetic fields evolve, and can we understand the presence or absence of large scale magnetic reversals?

(ii) How do patterns of magnetic fields correlate with other aspects of galaxies, such as star formation rate, overall dynamics, molecular gas distribution, and strength of outflow?

(iii) To what extent, and by what physical mechanisms, do disc galaxies amplify their large scale magnetic fields, and what determines the limits of field strengths over a galaxy's lifetime?

(iv) What are the original pre-galactic "seed" field strengths, and did these seed fields come from the post-Recombination Universe, or from early processes during Inflation – or both?

(v) Did galaxies like the Milky Way form in the presence of pervasive and significant magnetic fields?

(vi) Did the degree of ordering and the strength of the interstellar magnetic field have a strong influence on the locations and dynamics of star-forming regions?

(vii) What role did the early formation of galactic black holes play in providing a seed field for galaxies?

Some of these questions are interdependent, as will be obvious to the reader. Answers to some are beginning to emerge, and others, for example (iv), require deeper physical understanding and new generations of instruments.

The Milky Way and other late-type galaxies are permeated with (i) cosmic ray (CR) gas, (ii) magnetic fields, which are associated with polarised synchrotron radiation, (iii) ionised interstellar gas, which causes Faraday rotation in the radio, (iv) neutral and molecular gas, and (v) interstellar grains, which align with the interstellar field and induce linear polarisation in optical starlight through selective absorption or scattering. At least four of these ingredients make it possible to trace the large (and small) scale magnetic field structure in galaxies observationally. Component (i) (represented by Equation 1.1) has an intrinsic polarisation of up to $\approx 75\%$ in a completely aligned magnetic field. We note that the observed *degree* of linear polarisation in a 2-D projection (what is observed) does not necessarily reveal the true 3-D degree of alignment of the magnetic field, nor the *sign* of the local magnetic field vector. More specifically, observations of only polarised synchrotron emission do not distinguish between a zone of true unidirectional, aligned field and one where the magnetic field lines are stretched (or compressed) in *one* dimension, but systematically reverse their sign – a point elucidated by Laing (1981).

5.1.2 *Instrumental capabilities*

The improved sensitivity and polarimetric capability of both radio and optical telescopes has, over the past five to twenty years, revealed a great deal about the morphology and extent of magnetic fields in galaxies. In this section we explain observational methods and illustrate how they can reveal magnetic field morphologies and strengths in relatively nearby galaxies (see also M. Krause 1990). The latter have typical projected angular sizes of a few to a few tens of arcminutes. Thus, images with typically 5–15 linear resolution elements can be obtained with large, single-dish radio telescopes at 2–20 cm. Well-calibrated beam characteristics are required for polarisation mapping of extended objects, where the telescope's beam is embedded in an extended polarised source. The 100 m Effelsberg telescope of the Max-Planck-Institut für Radioastronomie, with a resolution of $4.0'(\lambda/10$ cm), has served as a prime instrument for such measurements at the shorter wavelengths. Longer wavelengths and more distant galaxies require the higher resolution of *interferometer arrays*, such as the NRAO VLA, the Westerbork Synthesis Radio Telescope, the newer LOFAR array (NL), the ASKAP (Australia) array, and the GMRT (India).

In the imaging of *diffuse* polarised emission in nearby galaxies some complicating difficulties can arise with interferometers. These result from their inherent tendency to filter out the largest scale components of diffuse emission. Also, in cases where the largest images – e.g. the Andromeda Galaxy, M31 – exceed a single dish beam size, it can be difficult to deconvolve the source's Stokes parameter distributions from the polarised side-lobe structure of a single interferometer element.

Some of these challenges can be overcome by combining images made with a single dish and interferometer, not necessarily in real time. The former can register all the large scale image Fourier components, and the latter can reproduce the smaller scale structures. Such techniques involve appropriate application of antenna aperture theory; see, for example, the books by Christiansen & Högbom (1969) and Thompson, Moran, & Swenson (1987). The combinations can be very effective in imaging all Stokes parameters, and hence magnetic structures on both large and small angular scales. An illustrative example, though applied only to Stokes I, involved a combination in Fourier transform space of the Arecibo 305 m single dish with the 1-km equivalent DRAO Synthesis Interferometer (image shown in Chapter 10, Fig. 10.1).

5.2 Our Milky Way – a spiral galaxy from within

5.2.1 *Basic features of the large scale magnetic structure*

In our Sun-centred view of the Galactic Faraday *RM* sky (Fig. 5.1), the Milky Way's magnetic field structure has a regular or prevailing component directed toward

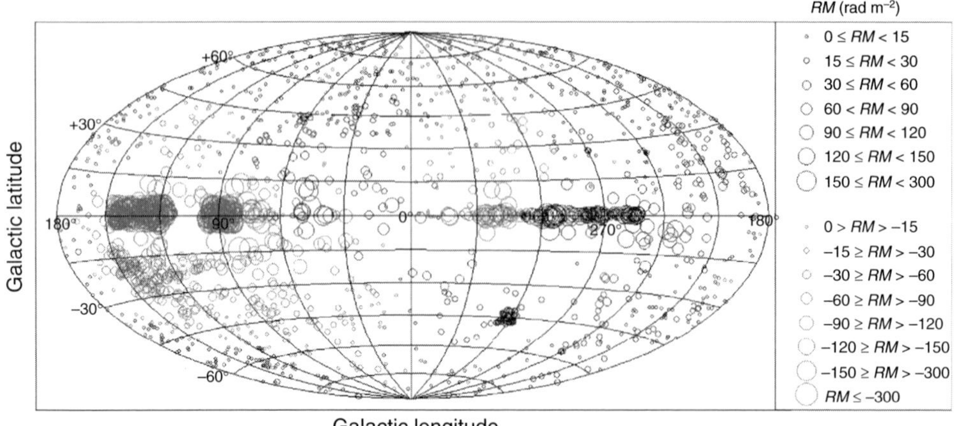

Figure 5.1 An Aitoff equal area projection of a smoothed representation of
2257 extragalactic source Faraday rotation measures in Galactic (l, b) coordinates from
Kronberg & Newton-McGee (2011). The smoothing method is described in Simard-
Normandin & Kronberg (1980) and Simard-Normandin *et al.* (1981). Filled and open circles
denote positive and negative *RM*s, respectively.

$l \approx 85°$. Patterns in the all-sky *RM* distribution of extragalactic sources appear generally cleaner over the southern Galactic hemisphere than that in the northern Galactic sky. Large scale magnetic field reversals were indicated in the earliest all-sky *RM* maps (Davies, 1966), based on pioneering discrete source polarisation measurements in Australia from the Parkes radio telescope. With more *RM* data, an abrupt reversal of *RM* sign at longitude $l \approx 60°$ near the plane suggested a large scale change of field direction in l in and near the Milky Way plane (Simard-Normandin & Kronberg 1980). This and other field direction reversals have been discussed in several papers since.

The interstellar magnetic field scale height of the Galactic disc is also of interest because it may indicate stellar-driven outflow away from the disc. Figure 5.2 shows the longitude-averaged, latitude-dependent *RM* variance for two simple large scale Galactic field models: (i) a uniform field with superimposed random fluctuations (upper line) and (ii) a similar field structure with no fluctuations (lower line). Two results emerge from Fig. 5.2: First, the estimate of the magnetic field scale height is relatively insensitive to details of the degree of superimposed disorder. Second, it is possible to convert the points in Fig. 5.2 to a physical scale height by combining them with independent data on the ionised hydrogen distribution (Cordes & Lazio 2003, Gaensler *et al.* 2008). These gave a magneto-ionic scale height, z_0, of 1–1.5 kpc. The number z_0 depends, of course, on the detailed form of the fall-off with scale height (z), which is less easy to specify, as well as the variation of the warm gas filling factor with z-height, discussed by Gaensler *et al.* (2008). It is generally concordant with scale heights of nearby edge-on galaxies similar to the Milky Way, such as NGC891 (Fig. 5.9). More generally, perhaps above some threshold, the magnetic field scale height is sensitive to the cosmologically recent level of star formation activity over the disc; galaxies with widespread, higher star formation activity, like NGC4631 (Fig. 6.1), have a thicker halo.

More extreme examples are starburst galaxies like M82, discussed in Chapter 6 (Fig. 6.2). These show a global "blowout" of magnetised gas well beyond the disc, along with CR and

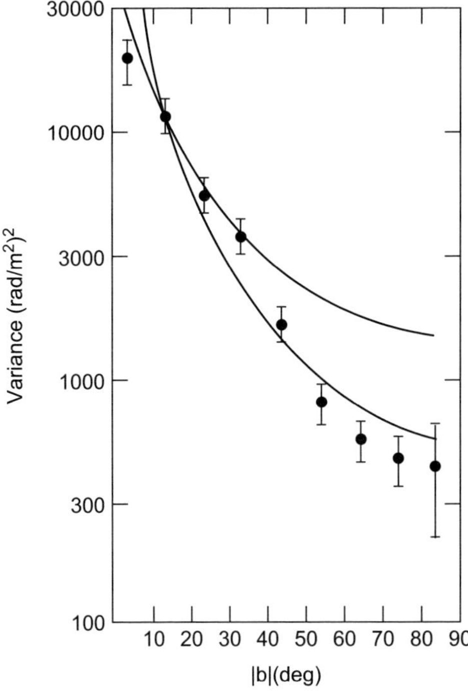

Figure 5.2 The variance of *RM* as a function of Galactic latitude gives an indication of the magneto-ionic *scale height* of the Milky Way. A full width of ~1.8 kpc was obtained using a 555-*RM* all-sky *RM* set. It is not strongly sensitive to the *degree* of magneto-ionic turbulence above the Galactic plane (source: Simard-Normandin & Kronberg 1980).

thermal ionised and neutral gas. In analysing z_0 for the Milky Way it is important to distinguish the scale height of the galaxy disc at a substantial galactocentric radius from the central halo height where there is enhanced outflow from the nuclear region. Because we sit at 8kpc from the Galactic centre it is not easy, for example, to distinguish the magnetic scale height in our vicinity from that closer to the Milky Way centre.

5.2.2 Magnetic field structure within the Galactic disc

The magnetic structure of the Galactic *disc* has been the subject of many observational studies using Faraday *RM*s of Galactic pulsars, extragalactic radio sources, and γ-ray emission. The first successful attempt using extragalactic source *RM*s by Gardner & Davies (1966) was followed by pulsar *RM* analyses (Manchester 1972, 1974), and all these studies revealed a prevailing magnetic field in the local spiral arm that is clock-wise as viewed from above. The interested reader could also consult papers by Simard-Normandin & Kronberg (1980), Rand & Lyne (1994), Frick *et al.* (2001), Han *et al.* (2006), Men *et al.* (2008), Vallée (2008), Noutsos *et al.* (2008, with references to earlier pulsar *RM* work), and Kronberg & Newton-McGee (2011). The Canadian Galactic Plane *RM* survey (CGPS; Brown *et al.* 2003) and the complementary Southern Galactic Plane Survey (SGPS; Brown *et al.* 2007) focus on very low galactic latitudes, $|b| < 4°$ and $|b| < 1.5°$, respectively.

Not all conclusions among the above *RM* analyses are in agreement, and this illustrates a principal difficulty in working with currently available *RM* data. One problem has been the sometimes inadequate angular density of *RM*s in important parts of the Galactic sky. This limitation has led to varying conclusions on the frequency of magnetic field reversals within and above the Galactic disc. Also, agreement between *RM*s from pulsars and extragalactic radio sources has often been poor, which reflects the even more limited angular coverage of pulsar *RM*s up to the time of writing. However, convergence between extragalactic radio source and pulsar *RM*s has improved, and a more recent work by Noutsos *et al.* (2008) provides a good sense of progress being made with pulsar *RM* probes of the Galactic disc magnetic field.

A longstanding question has been whether the global magnetic structure of our Milky Way resembles the clear underlying large scale patterns seen in other nearby spirals such as M51 (see Fig. 5.6). A 2247-extragalactic source *RM* compilation of accurate *RM*'s was used by Kronberg & Newton-McGee to produce the smoothed all-sky *RM* distribution in Fig. 5.1, and can be used to investigate this question. Their technique was to search for any underlying *RM* trend near the disc by "filtering" at a smoothing resolution comparable with the width and height of nearby spiral arms (~1–2 kpc). The result of this study (Kronberg & Newton-McGee 2011) was the (1) detection of a clear, underlying spiral pattern on our side of the Milky Way's disc (Fig. 5.3), (2) a transition to a different pattern at above ~1.5 kpc from the mid-plane, and (3) a continuity of *RM* sign through the mid-plane at lower Galactic latitudes for all longitudes.

Their "smooth, – fold, – reverse sign, – shift" analysis method is illustrated in the three panels of Fig. 5.3. For the inward semicircle of Galactic longitude, it gives a precisely determined average magnetic pitch angle of $-5.5° \pm 1°$ for $|b| \lesssim 12°$. At longitudes that point toward the Galactic anticentre, the average magnetic pitch angle is close to zero. Generally we can expect that, in a global plan view of the Milky Way plane, the spiral pitch angle will vary with (x, y) location over the plane.

5.2.3   *Magnetic field strength variation with Galactocentric radius*

The variation of total magnetic field, B_t, with R in Fig. 5.4 is worthy of comment, and our Milky Way may be one of the best suited galaxies to probe well beyond $R_\odot$ in this way. Nonetheless, to derive B_t vs. R curves for our own galaxy from the all-sky radio continuum brightness data (Fig. 5.4) does require some assumptions and modelling. This was done by Breuermann *et al.* (1985) based on the 408 MHz All-Sky Survey by Haslam *et al.* (1981, 1982) and subsequent observations of the Galactic plane emission from Reich & Reich (1988). B_t is the derived "total" *equipartition* field strength, with the usual assumptions such as the CR proton/electron ratio $k = 100$ for the Milky Way disc and adopted values for the disc thickness (though the latter is not very critical). The equipartition B_t values are claimed to be globally accurate to within ~25% in spite of the uncertainties and assumptions. There is good overall agreement with other B_t vs. R curves, determined by Broadbent *et al.* (1990), and also by Strong *et al.* (2000) from γ-ray emission-based models. The latter are independent of the usual CR/magnetic field equipartition assumptions.

The form of the B_t vs. R variation in Fig. 5.4 approximately fits an exponential decay,

$$B_t(R) = B_0 e^{-\frac{R}{R_{0B}}}, \qquad\qquad (5.1)$$

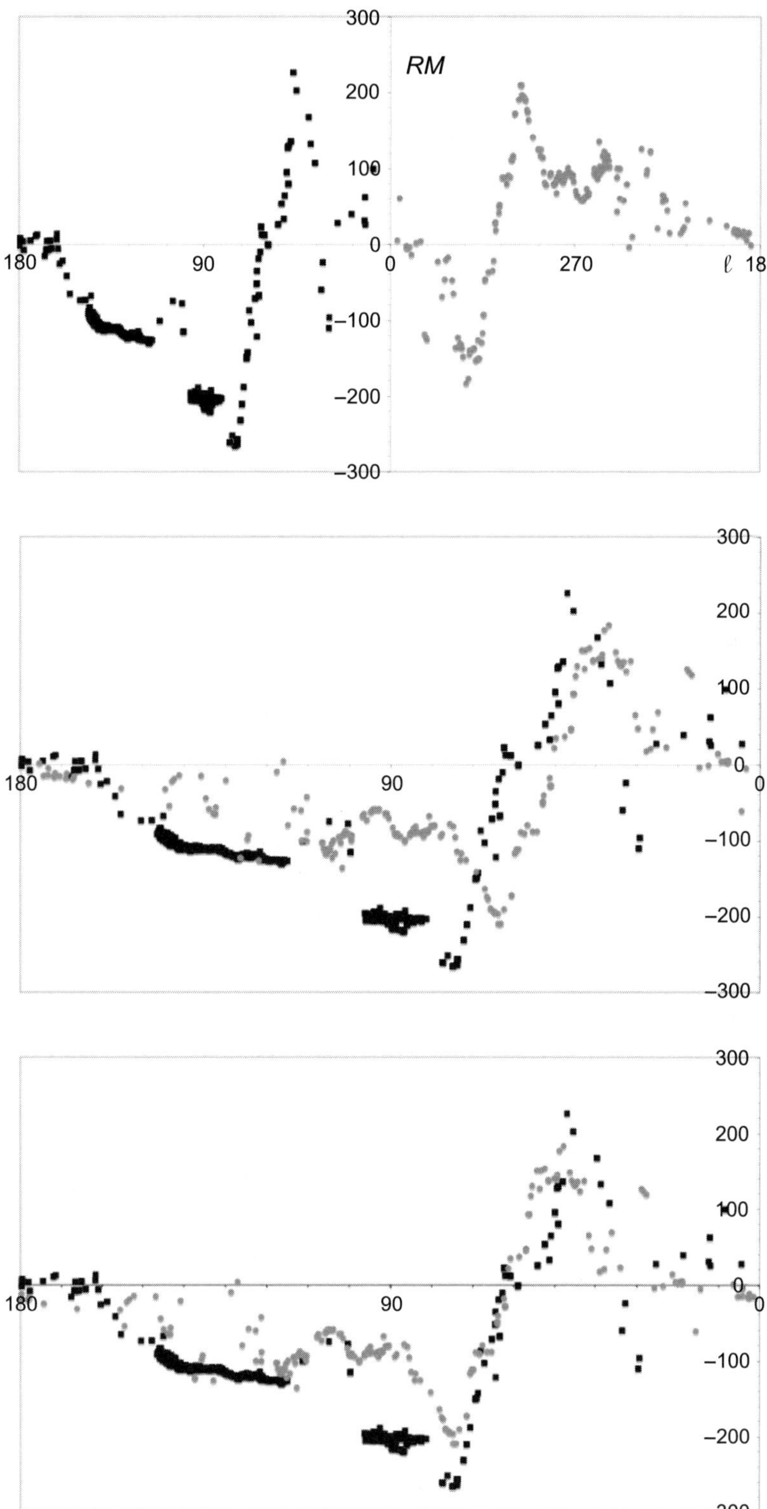

Figure 5.3 Variation of *RM*s within 15° of the Milky Way plane, when smoothed with a resolution comparable to the width of a spiral arm. The image shows a coherent underlying magnetic structure at, and inward of our galactocentric radius. This tilt is also reflected in the consistent *RM* sign change in Fig. 5.1 at $l < 180°$ at higher latitudes (Kronberg & Newton-McGee 2011). It demonstrates that an extra-Galactic observer would see our Milky Way's magnetic structure as another "grand-design" spiral such as M51 (Fig. 5.5).

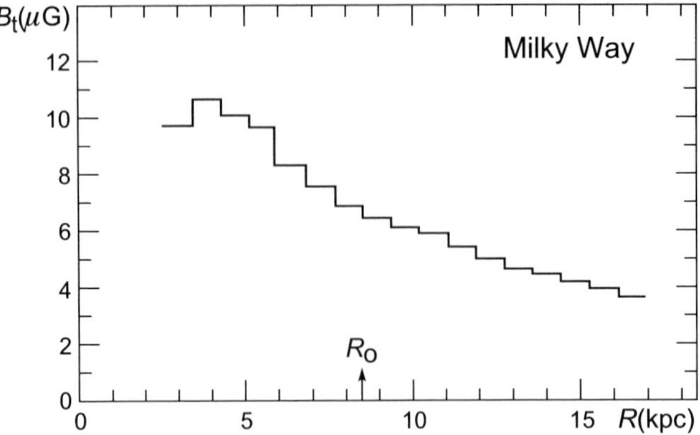

Figure 5.4 Variation of total interstellar magnetic field strength $<|B_t|>$ with galactocentric radius out to $R \approx 17$ kpc. This is scaled to an adopted $R_\odot = 8.5$ kpc. (Credit: E.M. Berkhuijsen, 2009.)

where R_{0B} is the magnetic exponential scale radius. For the range $3.4 < R < 13.6$ kpc in Fig. 5(c), Berkhuijsen (2009) obtains $B_0 = 15.1 \pm 0.5\ \mu G$ and $R_{0B} = 11.0 \pm 0.4$ kpc. For the range $6.90 < R < 13.6$ kpc, she obtains $B_0 = 13.7 \pm 0.4\ \mu G$ and $R_{0B} = 12.2 \pm 0.45$ kpc. Extrapolation of relation (5.1) to a 50 kpc galactocentric radius would give $B_t\ (50) = 0.16 \times 10^{-7}$ G. Although this cannot be considered a reliable extrapolation, the number illustrates the possibility that $|B|_{ISM}$ in an extended disc could approach, and/or match with, values typical of the IGM in galaxy filament zones of the Universe (see Chapter 10).

An independent investigation at $2 \lesssim R \lesssim 9$ kpc was undertaken by Han *et al.* (2006) using *RM*s of pulsars. Their estimate, over sightlines containing unknown reversals, is $\approx 2\ \mu G$ at $R \approx 9$ kpc, about a factor of 3 lower than $|B_t|$ above. To remove this discrepancy would require an upwards correction for pulsar line-of-sight small field reversals of $\approx 3\times$. More pulsar *RM*s and *DM*s near the disc and covering a larger range of R will be important. Additionally, uncertainties in the model-dependent B_t-scaling in Fig. 5.4 and a further elaboration of relation (5.1) may combine to alleviate this factor of 3 discrepancy.

The disc magnetic field strength at large galactocentric distances is an important quantity to understand for many purposes, including galaxy evolution, galactic dynamo theory, and its connections to "ambient" metagalactic and intergalactic magnetic fields. Figure 5.4 would lead us eventually toward the galactic/intergalactic interface zone, where field tracers both in our Milky Way and in external galaxies are difficult to measure because of the comparative faintness of the tracer quantities.

Greater levels of observational detail have led to varying conclusions about the structure of the large scale Galactic field – in particular about the small scale B reversals, which naturally tend to make integrated *RM*-determined fields smaller than B from synchrotron emission, which tends to reflect the total field strength. More definitive conclusions must await even higher angular densities of *RM* measurements, both for EGRSs and pulsars, and also related estimates of the 3-D distribution of interstellar ionised gas. Faraday depth tomography methods (see Chapter 2) offer promise here.

As we discuss below, most nearby external galaxies have not shown evidence for large scale field reversals across, and close to the mid-plane of their discs. Finally, as the instrumental angular resolution and sensitivity improve in future, smaller scale magnetic field reversals and other structures will be revealed in external galaxies.

Meanwhile latitude-dependent field reversals in the Milky Way can be searched for, given that we are within it. *RM* data at this time of writing indicate that the prevailing magnetic field direction does not change sign across the mid-plane within about 300 pc. However as we proceed to higher $|b|$, say above $20°$, there are large angular scale field reversals at some, but all longitude ranges. Further clarification must await higher angular densities of accurate EGRS *RM*s, and more pulsar *RM*s. Such close-in, cross-plane tests for field reversals are generally more difficult for external galaxies.

Galactic pulsars, through their combined measurements of Faraday rotation and dispersion measure, can give information on the 3-D structure of the ISM, especially when combined with independent data on Galactic *H*II regions, optical *H*α, radio recombination line observations, and other ISM species.

The local complexity of our "neighbourhood" ISM in the Milky Way disc is illustrated in Fig. 5.5 from the Canadian Galactic Plane Survey. The ISM contains a mixture of thermal bremsstrahlung from star-forming regions, thermal emission from heated dust, synchrotron radiation from supernova remnants at various evolutionary stages, and other energetic stellar events. All of these objects locally modify the ISM magnetic field structure. Specifically, they produce large local *energy overdensities* (by 10^2–$10^3 \times$) within the galactic disc. We should keep pictures such as these in mind when studying *external* galaxies for which we, necessarily, see only spatial averages over all this detail. Figure 5.5, which depicts remnants of supernovae at various stages, illustrates the additional factor of *temporal* averaging: Lifetimes of SNRs ($\lesssim 10^5$ year) are very short compared with the dynamical age of the Galactic disc.

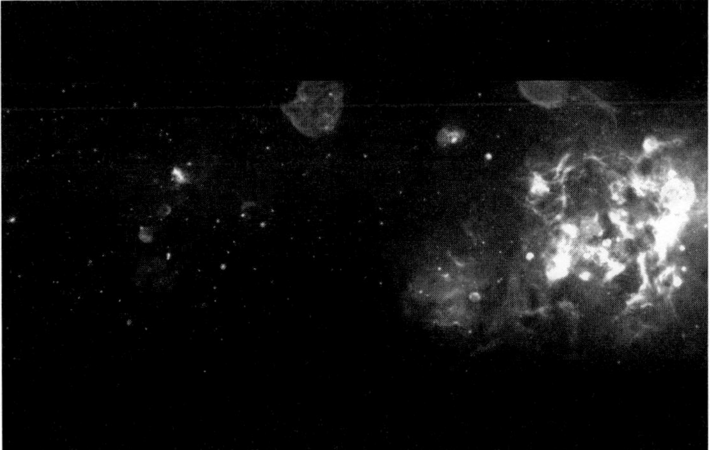

Figure 5.5 A combined radio/infrared image of a $20° \times 9°$ section ($78° \lesssim l \lesssim 98°$, and $-3.5° \lesssim b \lesssim +5.5°$) of the Canadian Galactic Plane Survey (Taylor *et al.* 2003). The Cygnus X region is at the right. The radiation is a combination of thermal bremsstrahlung, radio synchrotron radiation (from 408 MHz and from 1420 MHz), and the 60 μm dust emission from the IRAS survey. (Courtesy of Roland Kothes, Dominion Radio Astrophysical Observatory.)

Supernova remnants contain magnetic fields ranging from tens of microgauss to milligauss levels in the younger SNRs such as Cassiopeia A. If Fig. 5.5 could be made into a film with time steps of, say, $\sim 10^4$ years it would show diffusion of these fields into the wider disc ISM. Dramatic local overpressures are produced by SNRs and HII complexes relative to the average ISM ($\approx 10^{-12}$ erg cm^{-3}). Examples can be seen in part of the Cygnus X region on the right side of Fig. 5.5.

Such systems also produce magnetised outflow in galactic winds. The relative vigour of the outflow will depend on the star formation rate in a galaxy's cosmologically recent past. The strength of the outflow, in turn, affects the magnetic and gas scale heights discussed above. Where local rates of star formation are extreme, as in a galactic "starburst", major blowouts into the IGM will occur. An example is M82 in Figs. 6.2 and 6.3.

These starburst galaxies, discussed in Chapter 6, inject magnetic energy and associated metal-enriched gas into the local surrounding IGM. The area shown in Fig. 5.5 can be considered a benign, relatively quiescent microcosm of the interstellar environment that can produce a starburst outflow. Also, in such "milder" galaxy disc environments, the magnetic fields will inevitably connect to larger scale galaxy magnetic fields, which we discuss next. The details of how this happens have been widely discussed, but not yet completely clarified.

5.3   Magnetic structures of spiral galaxies

We define a "normal" galaxy as one that has some ordered disc or disc + bar structure, and no prominent outflow. Recall here that all late-type galaxies, whether "normal" or not, have *some* outflow. The star/supernova-driven outflow does not normally escape from the galaxy, but it may do so at sufficiently high redshifts, where the dimensions of the outflow material can be comparable with a galaxy–galaxy separation (see Chapter 6). The magnetic field structure, like the characteristic spiral patterns of such galaxies, can be traced in polarised radio synchrotron radiation and in Faraday rotation. The optical polarisation associated with interstellar dust grains is also coupled to the local interstellar magnetic field orientation. Both radio and optical polarisation images reveal large scale magnetic field patterns when we observe the galaxies at a small inclination angle (i.e. close to "face-on").

5.3.1   *Magnetic structure in "grand design" spiral galaxies*

A classic illustration of a galaxy's global disc magnetic field structure is given by the face-on galaxy M51 in Fig. 5.6. The projected magnetic field orientation, traced by the normals to the intrinsic (Faraday de-rotated) polarisation angle of interstellar synchrotron radiation, approximately follows the spiral structure. A closer comparison with the Hubble Space Telescope (HST) optical image superimposed in Fig. 5.6 shows particularly good magnetic alignment with the inter-arm dust lanes. This effect is also clearly seen in NGC6946 (Beck 2007). In general, the high degree of magnetic alignment, most prominent in the inter-arm zones in such "grand design" galaxies, exists out to the largest continuum-observable galacto-centric radius.

Typically, the magnetic field alignment appears stronger in the interarm regions and weaker in the spiral arms. The tight alignment in the interarm regions has been variously attributed to a large scale density wave (DW) compression, or to a decoupling of the synchrotron CR gas from the denser molecular gas. A lower average degree of linear polarisation is often observed within in the spiral arms, where star formation is prominent.

Figure 5.6 Contours of total 5 GHz radio emission (Stokes I) and polarisation vectors ($\chi = \frac{1}{2}$ arctan (U/Q)) for the galaxy M51. (Source: Beck 2006. Reproduced with permission of Rainer Beck).

This is probably caused by magnetic field tangling due to energy injection into the ISM by various stellar processes (see Fig. 5.5). Generally, attempts to reproduce grand design magnetic field patterns need to be able to approximate the large scale features, such as those in Fig. 5.6. However, success in simulating all the observed magnetic patterns has been limited so far. A recommended, informative overview has been given by Fletcher (2010).

Only in a few galaxies has it so far been possible to measure the *sense* of the magnetic field. This requires a Faraday rotation measurement over two dimensions, in addition to good surface brightness sensitivity at two, or preferably more wavelengths. It often necessitates the combination of a large single dish radio telescope (for surface brightness sensitivity and polarisation direction specification), and an aperture synthesis interferometer (for angular resolution).

An attempt to fit different 3-D models along with an instructive discussion of their limitations is presented in Elstner *et al.* (2000). Models based on the induction equation and extended with basic dynamical equations are important steps in understanding the complexities of the global magnetic field structure of galaxies. We can generally say that models based purely on a dynamo, or on other basic magneto-ionic concepts, are not easy to match to the complexities of a galaxy. This is partly because of the inherent non-linearity of these processes and the widely ranging timescales involved over a galaxy's lifetime. Therefore, detailed observations of various types are required to obtain appropriate constraints on theoretical models. In addition, the limits of instrumental and observational capability, indicated in this chapter and in Chapter 2, present challenges.

In spite of all these limitations and caveats, it is useful to create galactic magnetic field models. As the interplay between ever-improving observational detail and theoretical understanding develops, the models will be refined. The evolution of galactic magnetic fields may also have connections to metagalactic and intergalactic fields. A different motivation is to understand, for our Galaxy, the propagation of Galactic and intergalactic CRs from $\sim 10^{14}$–10^{20} eV that arrive in the Solar neighbourhood (see Chapter 11). For discussions of Galactic field models for this purpose, the reader can consult such references as Tinyakov & Thachev (2002) and Prouza & Smída (2003).

The total field strength, $|B_t|$, in the spiral arms needs to be estimated with higher resolution and sensitivity than that shown in Fig. 5.6 and similar images. An obvious physical reason is that smaller magnetic turbulence scales are generally associated with the "microcosmical" supernovae and star forming complexes, as illustrated in Fig. 5.5. According to basic energy equipartition arguments, $<|B_t|>$ in the more turbulent spiral arm ISM should be stronger than in the interarm regions. Finally, as typified in M51, large scale magnetic field ordering appears coherent out to the largest visible radius, although larger radii can sometimes be traced by some neutral hydrogen (HI) images. Detailed descriptions of observed polarisation and magnetic structures for a variety of spiral galaxies can be found in Braun *et al.* (2010); these are summarised in Section 5.3.2.

For the bisymmetric field case in Fig. 5.7, we can model the global magnetic field as a spiral as justified, for example, in Fig. 5.6, and plot the observed *RM* as a function of azimuth angle within the de-projected plane of the galaxy's disc. For an inclination angle $i > 0°$ (face-on = $0°$), a plot of *RM* vs. θ has four zero-crossings for a bisymmetric field galaxy, as opposed to two in the case of an axisymmetric (non-reversing) global field morphology. Additional geometric details of the disc field can be specified, such as the pitch angle, p, of the spiral field direction, and position angle, Φ. Incorporating these basic parameters, the model *RM*-θ curve for a galaxy's disc can be expressed as

$$RM(\theta) = 0.5 RM_0 \tan i \{ \cos(2\theta - p - \Phi) + \cos(p - \Phi) \} \tag{5.2}$$

(Sofue *et al.* 1986, Tosa & Fujimoto 1978).

As more detail emerges, the situation becomes more complex. A case in point is M31 (the Andromeda galaxy), where we have good linear resolution due to its proximity to us. A more complex pattern has also emerged in detailed studies of M83 (Neininger *et al.* 1993), in which, for example, the pitch angle, p, is not constant over the entire galaxy's disc; consequently, the simple model assumptions of Equation (5.2) and Fig. 5.7 do not always suffice. In general the underlying magnetic field structure is better aligned in the *outer*most regions of spiral galaxies. Patterns of Faraday rotation and field ordering on *smaller* scales

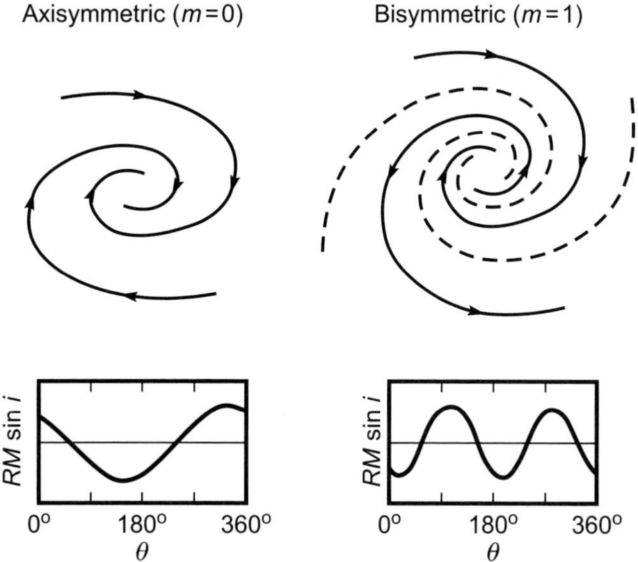

Figure 5.7 Illustration of the difference between unidirectional, axisymmetric ($m = 0$) (ASS) and bisymmetric (BSS) ($m = 1$) magnetic field structures. The latter likely characterises our Milky Way. As this indicates, large scale field reversals can be "diagnosed" by *RM* measurements over the tilted plane of the galaxy in question. The rotation axis is inclined at $i°$ from the vertical to the page (see also Sofue *et al.* 1986). The plots for the ASS and BSS cases show the azimuthal variation of *RM* at a constant galactocentric radius. This straightforward diagnostic of a "grand design" spiral galaxy was first introduced by Tosa & Fujimoto (1978).

often do not conform to either a simple ASS or BSS model (M. Krause *et al.* 1989, Beck *et al.* 1989). For physically understandable reasons, the magnetic pitch angle appears to vary somewhat with galactocentric azimuth and radius over a spiral galaxy disc. This also appears to be true for the Milky Way.

Although M81, like the Milky Way, shows evidence for a bisymmetric magnetic spiral structure (BSS, $m = 1$), most galaxies are thought to have the simpler, axisymmetric (ASS, $m = 0$) field configuration. This distinction is of interest for understanding the cosmic evolution of large scale fields in spiral galaxies. Ideally, the *RM* measurements illustrated in Fig. 5.10 should be made from measurements over a wide range of radio λ^2 – including radio wavelengths sufficiently short so that the far side of the galaxy's disc has not been depolarised by the medium in front of it.

Also, smaller scale systematic field reversals may occur. These may not, as above, conform to a large scale bisymmetric field; rather, they may be influenced by local effects, which may vary over a galactic evolution timescale. The models discussed above do not incorporate the halo or out-of-plane fields. We next discuss observational evidence and reasons that could explain magnetic fields directed *out of* a galaxy's plane and located above the plane.

5.3.2   *Off-plane and 3-D galaxy halo field configurations*

Two standard halo magnetic field configurations arising from galactic dynamo models are dipole and quadrupole configurations above the inner galactic zones. These are illustrated in Fig. 5.8.

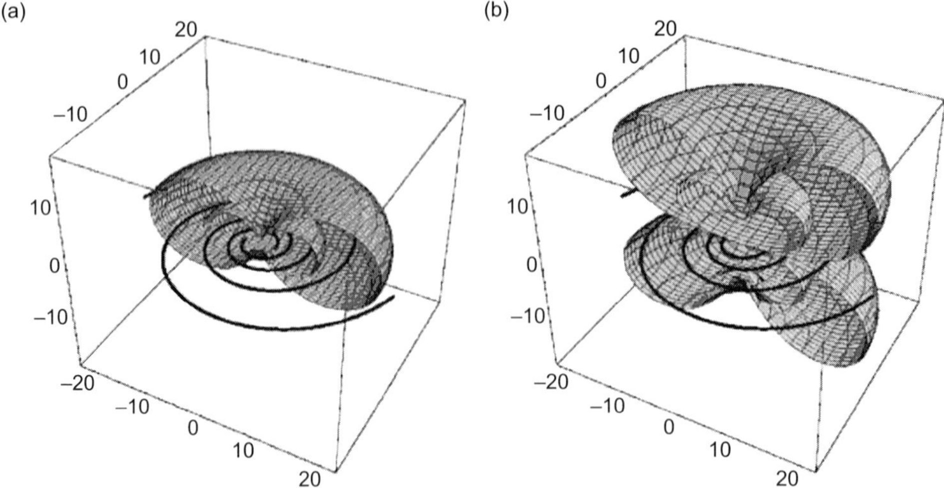

Figure 5.8. Idealised dipole (a) and quadrupole (b) halo field configurations superimposed on a logarithmic disc spiral arm layout. (Reproduced from Braun, Heald, & Beck 2010, courtesy of Robert Braun.)

In galaxy discs the prevailing magnetic field is usually aligned parallel to the plane. However, off the plane the galaxy magnetic fields have organised components that can be aligned away from the plane. The off-plane alignments tend to have common patterns, at least in a 2-D projection: Edge-on radio images of Stokes parameters I, Q, and U typically show a vertically aligned field component immediately above the centre of the plane. Proceeding further outward, and upward, an X-shaped pattern is typically visible edge-on in projection. An illustrative example is the nearby galaxy NGC891 in Fig. 5.9. There are some observational indications that magnetic fields seem "pushed up" into the halo, e.g. of NGC6946. These have been presented in Heald (2012). See also Section 5.5.5 below for a related discussion on CR-driven outflows out of a galaxy disc.

A projected "X"-halo pattern is approximately produced by some of the global galactic dynamo models produced by e.g. Elstner *et al.* (1992). Another step in understanding this puzzling projected X-pattern was made in an observational/geometrical analysis by Heesen *et al.* (2009) in the study of a similar, nearby "X"-halo galaxy, NGC253. By separating components in an ASS spiral disc magnetic field model with $25°$ pitch angle from the observed halo structure, Heesen *et al.* concluded the existence of a distinguishable, galaxy-scale *poloidal* magnetic field. A poloidal magnetic field is independently consistent with some galactic dynamo models, and can also be associated with an observed edge-on projected X-pattern. However, to establish a unique association may require similarly detailed studies of other galaxies, seen at different inclination angles (i).

The X-shaped projected magnetic field patterns now commonly seen in edge-on galaxy halos may be compatible with the vertical outflow-driven fields observed close to the galactic rotation axes. This has also been seen in *optical* polarisation above the central disc of NGC891 (Scarrott & Draper 1996). In this picture, the "ballooning", overpressured poloidal halos widen with increasing z-height. One idea is that, due to their relative over-pressure they might force an originally vertical halo field "down" toward the disc at higher

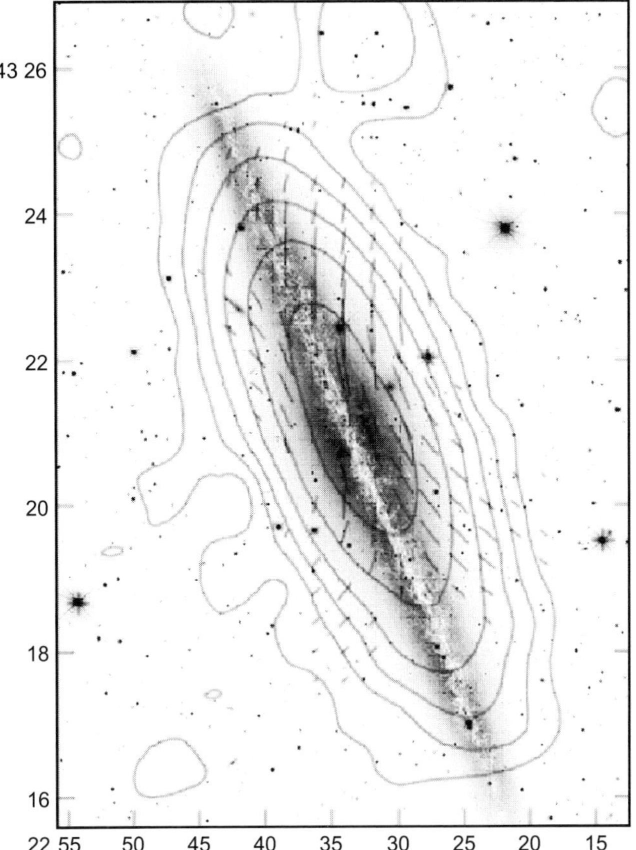

Figure 5.9 NCC891 imaged at λ 3.6cm with the Effelsberg radio telescope, showing the galaxy's Faraday *RM*-corrected, X-shaped projected magnetic field pattern. (Source: Marita Krause (2009).)

galactocentric radii. One configuration of the structure at $|z| > 0$ could be thought of as a "standing" halo structure, illustrated by the examples in Fig. 5.8, which is coupled to the overall disc dynamics. The disc's magnetic configuration might change over a galaxy's lifetime, as some models have predicted (e.g. Brandenburg *et al.* 1992). These two possibilities could, furthermore, be superimposed on a stellar-driven or AGN-driven vertical outflow field structure, which would likewise be epoch-variable, and coupled to cosmologically short-lived starburst events at lower galactocentric radii.

The above discussion illustrates the potential complexity in unravelling large scale galaxy field configurations. It demonstrates the value in attempting 2-D magnetic field imaging, extended by Faraday rotation synthesis and combined with 3-D magnetic field modelling. At this point it will be clear to the reader that the observational diagnostics and model parameter space may be much more complex than was illustrated in Fig. 5.7. Multi-frequency 2-D polarisation images can be used with the basic mathematical tools discussed in Chapter 2 (Faraday synthesis methods) to model a 3-D magnetic structure for the galaxy in question.

Illustrations of such complexities can be seen in Fig. 5.10 below. They show the result of 2-D, multi-frequency radio polarimetry of two galaxies. Ten further examples can be found in the paper by Braun, Heald, & Beck (2010). The two sample simulations are of: (1) a NGC6946-like galaxy having a similar inclination, and (2) the nearby edge-on galaxy NGC4631. An actual observed image of the latter appears in Fig. 6.1.

The simulated images in Fig. 5.10 illustrate how the projected polarisation plots are sensitive to many aspects of a galaxy's magnetic field structure. Depending on the inclination, the galaxy's disk and halo geometries, disc thickness, spiral pitch angle, star formation intensity, and other characteristics can all be probed by these methods. Larger samples of galaxies of the same type, but seen at differing inclinations can be studied in such ways to better study 3-D magnetic geometry, and to separate the disc magnetic structures from those in the halos.

The variety of galactic "magnetic" parameters that occur in nature, combined with the 2-D polarimetric images exemplified in Fig. 5.10, can serve as "templates" for the many-parameter complexity in galaxy magnetic field structures. An obvious procedure is to compute the observational signatures for model galaxies, and iterate with the observations to optimise the magnetic field and ionised gas configurations of the galaxies.

Interstellar magnetic field strengths in the discs of galaxies are in the range of 2–5 μG. However, $|B|$ has large excursions to the upside in the denser SNR shells and filaments that exist in a galaxy's mid plane. The latter are nicely illustrated in images of the turbulent mid-plane environment of the Milky Way (Fig. 5.5). A frequent feature of the quasi-3-D Faraday rotation models exemplified in Fig. 5.10 is that the organised polarisation structure immediately above, but on the far side of, a galaxy's mid-plane can be substantially depolarised at λ21 cm by the mid-plane magneto-ionic medium. That is, *at the resolution and frequency* illustrated in Fig. 5.10, much of the disc polarisation originates on the observer's (our) side of the galaxy mid-plane. Depolarisation of the far side has been model-interpreted as due to mid-plane ISM microturbulence and/or filamentation on physical scales as small as ~1pc (Braun *et al.* 2010).

For radio polarimetric diagnostics and 3-D modelling of galaxy field structures, the observational "parameters" of resolution and sensitivity need to be well-matched to the objects of study. This includes use of a suitably chosen *radio frequency range* and *channel bandwidth*. These are based on the Faraday rotation law (Equation 2.7) showing the λ^2 dependence of $\Delta\chi$, and the phenomenon of bandwidth depolarisation (Section 2.9).

Because the above Faraday rotation diagnostics can only be done if $\Delta\chi$ is easily detectable, there must be at least a few degrees of rotation, i.e. the product of scale l, n_e, and $|B|$ must be sufficient to generate a detectable $\Delta\chi$. This illustrates the importance of choosing the appropriate radio wavelength range for polarimetric imaging when designing observations. In outer galactic halos, n_e might be two orders of magnitude lower than in the disc planes. For 3-D probes of galaxy halos it highlights the importance of moving to lower frequencies, say 0.2 or 0.1 GHz, in addition to frequencies near 1.4 GHz used in Fig. 5.10. Low frequency radio telescopes with sufficiently high resolution will be important for galaxy halo field specification using the "*RM* synthesis" simulations of the kind used in Fig. 5.10 and described in Chapter 2. In another context, the same criteria for the product $l.n_e. |B|$ make observations with low frequency telescopes of great value for exploring intergalactic magnetic fields.

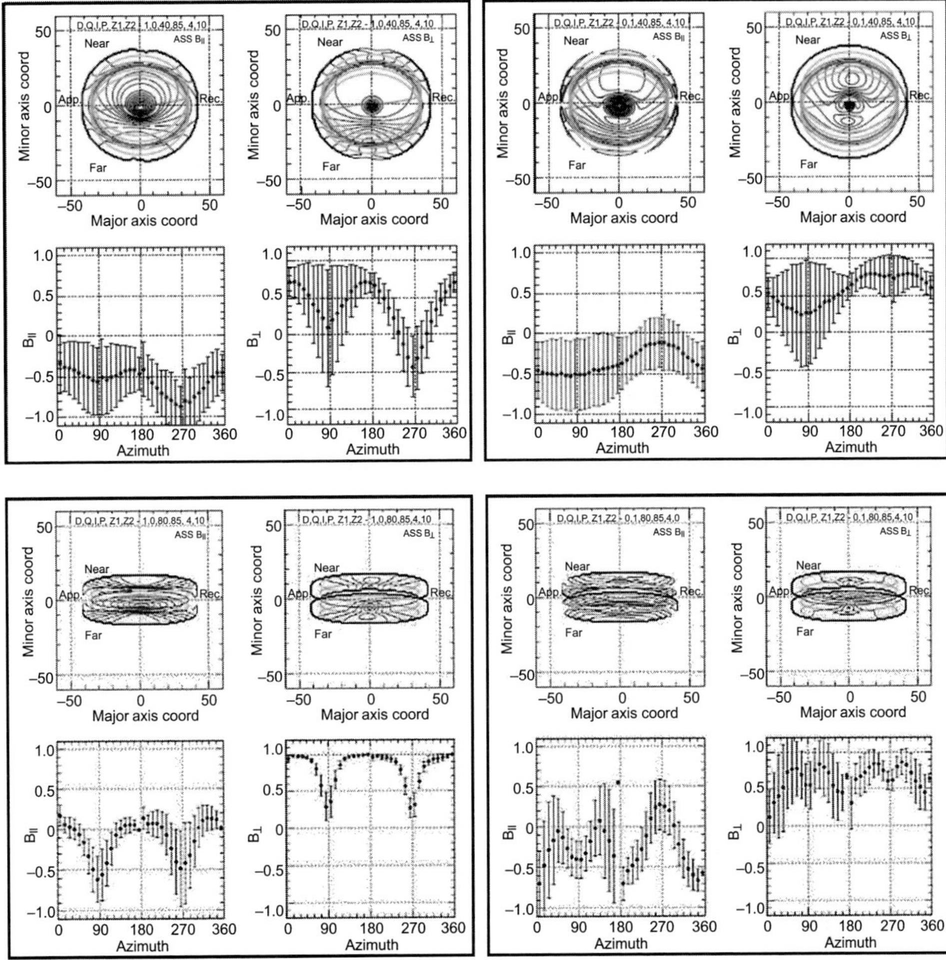

Figure 5.10 Sample simulations of the observed polarisation (smoothed) for an ASS spiral galaxy at two different inclinations to the line of sight, are shown separately for a dipole (left) and quadrupole (right) magnetic field configuration (see also Fig. 5.8). The galaxy parameters listed at the top of each panel are defined as D,Q, the dipole (left large panels) and quadrupole components, (right large panels) I, the inclination, P the spiral pitch angle, and I Z1I, IZ2I the lower and upper z-height limits of the magneto-ionic layer. In all cases shown here, the magneto-ionic layer was between z-heights (Z1 and Z2) of +4 to +10kpc on the near side, and −4 to −10kpc on the far side of the galaxy disc. *Upper* – a modestly inclined galaxy ($i = 40°$, close to the 33° inclination of NGC6946). *Lower* – a galaxy seen more edge-on ($i = 80°$, and close to the 85° inclination of NGC4631). The simulations were guided by images made near 1.4 GHz with the Westerbork Synthesis Radio Telescope (WSRT). The interested reader could consult Braun, Heald, & Beck (2010), which shows many more parameter combinations. These two images are drawn from figure 12 of that paper. Reproduced with the kind permission of Robert Braun. The characteristic edge-on X-shaped image of the halo configuration is discernable in an independent Effelsberg λ3.6 cm image of NGC4631 – see figure 6.1.

5.3.3   *Optical polarisation as a magnetic field tracer in galaxies*

The magnetic disc field structure can be independently traced through the interstellar polarisation of starlight in our vicinity of the Milky Way (Mathewson & Ford 1970, Heiles 1996). A large scale, organised field is revealed by polarimetry of the optical surface brightness, as shown in the image of NGC 1068 in Fig. 5.11. Here the field alignment is striking, as also in the radio polarisation maps discussed above. Although no information on the *sign* of B can be inferred from optical polarisation, the angular resolution of structure is limited only by the sensitivity of the detector/telescope combination, and the optical limit of resolution – of order 1 arcsec for a ground-based telescope. Because of the multiple causes of optical polarisation, and problems of optical extinction in inclined galaxy discs, the most yielding of information on the magnetic field morphology comes from nearly face-on, galaxies like the one shown. The lack of any Faraday *RM* information in Fig. 5.11 means that there is no optical signal of whether the local field has a component directed into or out of the plane of the sky.

5.4   Some dynamical and energetic aspects of galaxies

At smaller galactocentric radii, the rotation of spiral disc galaxies approximates "solid body" rotation ($v_\varphi \propto r$) which, due to the galaxy's distributed mass, merges into differential rotation at larger radii. The latter is normally flatter than Keplerian and can be explained by a quasi-spherical distribution of dark matter. Beyond the inner solid "body" zone, large scale velocity shear ($\partial v/\partial r$) occurs over most of a spiral galaxy's radius.

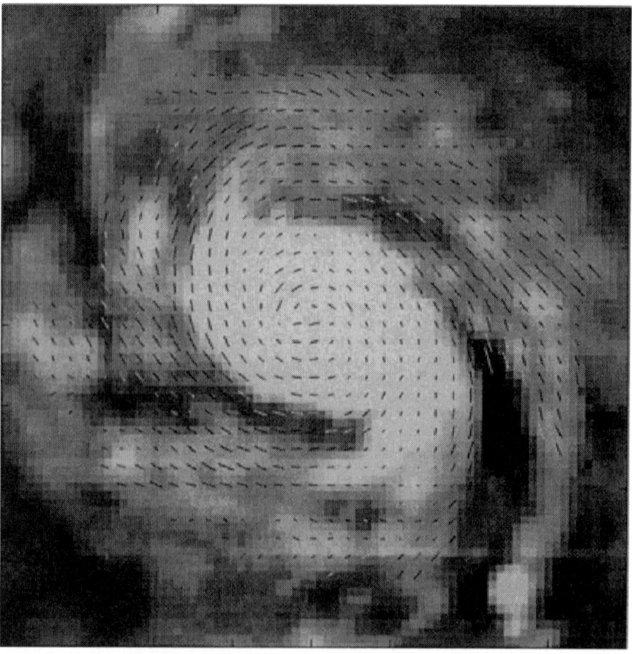

Figure 5.11 Polarised optical surface brightness of the spiral galaxy NGC1068. Typical degrees of optical polarisation are in the range 0–5%. (Scarrott 1991.)

Superimposed on these global dynamics are streaming motions, possibly induced by DWs or by winds from OB associations, supernova remnant shells, etc. For some galaxies, we can identify non-axisymmetric motions due to gravitational interactions with satellite galaxies or other nearby galaxies. It is of interest to know whether magnetic fields could also influence the global dynamics in some places. In all galaxies, there are interstellar turbulent motions, typically a few kms^{-1}. Typical (kinetic) energy densities due to global rotation, ε_{grav}, are of order 3.3×10^{-10} erg cm^{-3} g^{-1}, but these decrease significantly at large galactocentric radii. By contrast, energy densities of the interstellar synchrotron-emitting CR gas ε_{cr} are of order 10^{-12} erg cm^{-3} and the interstellar magnetic energy density is $\varepsilon_{m} = ((B/5\ \mu G)^2 \times 10^{-12}$ erg cm^{-3}. The energy density of the smaller-scale turbulent motions, $\varepsilon_{t} = \frac{1}{2}\rho_{is}v^2$, is $\approx 0.6 \times 10^{-12}$ erg cm^{-3}. It is worth noting that ε_{grav} dominates the others, except at large galactocentric radii, and that

$$\varepsilon_{cr} \approx \varepsilon_{m} \approx \varepsilon_{t}. \tag{5.3}$$

Large scale rotation in spiral galaxy discs and bars is influenced primarily by gravity. This means that gravitational energy is available to be tapped into, given some coupling mechanism by which it can be transferred into magnetic fields. CRs (e.g. due to gravitationally driven interstellar shocks, or thermal heating of the interstellar gas, etc.) are all ultimately driven by gravity, where ε_{grav} overwhelms other energy forms. Larger scale velocity perturbations, Δv, due to such factors as DWs can be of order 5–10 kms^{-1}, so that $\Delta \varepsilon_{grav} \approx 3 \times 10^{-13}$ erg cm^{-3}. This is not much different from typical values for ε_{cr}, ε_{m}, and ε_{t}.

It can be noted here that the total interstellar gas pressure, $P_{T} = 1/3\ \varepsilon_{T}$, includes both the thermal gas and CR pressures *and* the (sometimes ignored) relativistic protons and heavier nuclei. The latter particles may possess up to $\approx 10^2$ times the energy of the "visible" synchrotron-emitting relativistic electrons.

Another important energy component to consider is that from supernovae (e.g. multiple supernovae within a small galactic region), and dense clusters of young stars. These can, temporarily and locally, deposit significant energy densities (in the form of ε_{cr}, ε_{m}, and ε_{t}) into an otherwise "quiescent" interstellar disc. The main energy source in these sub-galaxy scale systems is thermonuclear, and so it is useful to denote this component as ε_*. These underlying energetic facts set an overall context for understanding how magnetic fields can be produced, and evolve in galaxy discs.

5.5 Basic principles of the galactic α-ω dynamo

5.5.1 A very brief history

Based on studies of stellar magnetic fields (Steenbeck & Krause 1969), a theory of galaxy scale field amplification called the α-ω dynamo was worked out in the 1970s for galactic magnetic fields by Parker (1971), Stix (1975), and White (1978). A *mean field approximation* was developed to accommodate the fact that we are dealing with averages over a range of correlation scales $<l_i>$, and timescales $<t_i>$ of the velocity field, and fluctuations in the magnetic field $<B_i>$. In an ideal world in which we could measure these quantities and had access to massive computing power, we could in principle calculate the non-linear time evolution of Galactic magnetic fields on all scales. Because this has not previously been possible, a successful mean field approximation has been developed as the basis for the equations governing models of large scale galactic magnetic field evolution.

The mean field approximation, although not trivial, has been worked out and/or discussed by Knobloch (1978), Moffatt (1978), Krause & Rädler (1980), Zel'dovich *et al.* (1983), Ruzmaikin *et al.* (1988), and others since then. It might be considered, very roughly speaking, as an MHD analogue of averaging methods used in the statistical mechanical study of gases.

Related to the mean field theory is the subject of the magnetic *fluctuation spectrum* in the interstellar medium. In fluid mechanics the energy density-wavenumber (k) distribution of turbulence typically follows a Kolmogorov spectrum, whose logarithmic slope is –5/3 up to some maximum k. At this point, viscosity effects take over and cause a small-scale (high k) cutoff. The question of the turbulence scale, both for weak- and strong-field limits is more complex. In magnetohydrodynamic theory, the specification of a definitive MHD fluctuation spectrum is more complex, but it can be connected to dynamo theory. The interested reader might wish to consult two papers on the fundamentals of interstellar MHD turbulence by Sridhar & Goldreich (1994) and Goldreich & Sridhar (1995).

A significant question affecting the mean field dynamo is, whether the buildup of fluctuation energy on the smallest scales is sufficiently fast that it is shorter than the mean field dynamo timescale. In that case, as Kulsrud & Anderson (1992) argued, the kinematic assumption of the mean field dynamo might be called into question. Fortunately, the observational details of interstellar turbulence and improved computational power promise to give better understanding of galaxy scale dynamos. Since the early 1990s evidence gathered from galaxy systems seen earlier along the cosmic time scale, i.e. at larger redshifts, indicates that microgauss-level galactic magnetic fields already existed at Proper Times (T) $\lesssim 10^8$ year. This is less than a galaxy rotation time, and much less than a galactic field amplification time scale due to a large scale slow α-ω dynamo. Therefore the assumption of very weak initial fields would appear to necessitate magnetic amplification to μ gauss levels in a very short, $\lesssim 10^8$ year. This seems unlikely in this model. In view of the fundamental importance of astrophysical dynamo theory and the complex behaviour of magneto-plasmas, we next briefly summarise of some of the physical principles and historical development of the subject of galactic dynamos.

5.5.2   *Some basics of the mean field galactic dynamo theory*

In a stratified, rotating, and turbulent (or convective) section of the galaxy disc, the first induction action is due to the velocity shear, dv/dr. Given an initially radial field, B_r, an azimuthal component, B_φ, is produced. By approximating the galaxy disc as a flattened, oblate spheroid of height b along the rotation (z–) axis, with a radial scale of a, we can, following Krause (1987), define an ω parameter

$$C_\omega = \frac{a^2}{\eta_{\mathrm{T}}} \left\{ r \frac{d\omega}{dr} \right\} \qquad (5.4)$$

where r is the radial coordinate and η_{T} is the turbulent diffusivity. When *turbulence* is injected into this stratified, magnetised layer, the turbulent motions will be subject to the Coriolis force in this "rotating platform". As a bubble of gas expands, it will create an expanding "loop" that is consequently twisted by the Coriolis force as it moves out of the plane (Fig. 5.12). The twist is in the right-handed sense above the plane (if ω is directed upwards) and in the opposite sense below the plane, thereby introducing a natural helicity to

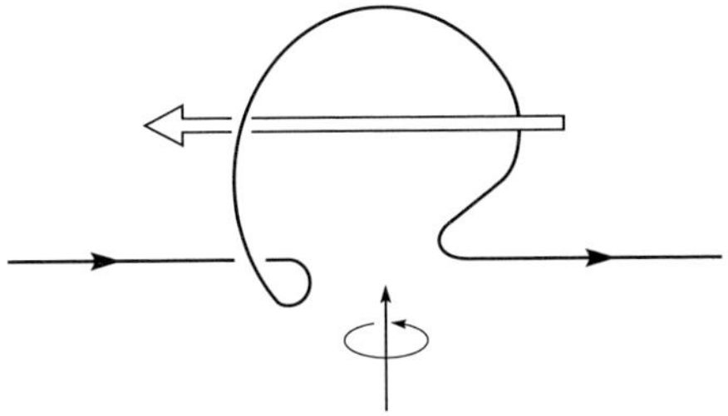

Figure 5.12 Illustration of the α-effect due to the Coriolis force in a rotating galaxy disc.

the outflowing magneto-plasma. This gives rise to the so-called α-effect. The B_z component will induce a current antiparallel to B_φ (see also below).

A natural consequence of the α-effect is the production of a poloidal magnetic field component. Because the energy density of the rotational motion is normally higher than ε_t in a typical spiral galaxy, the repeated twisting and stretching effects can amplify the field in the disc. In mean-field electrodynamics, α is related to the helicity $<v{\cdot}(\nabla \times v)>$, and is defined as

$$\alpha = -\frac{1}{3}\tau <v_T{\cdot}(\nabla \times v_T)> \tag{5.5}$$

(Steenbeck et al. 1966, Steenbeck, & Krause 1969), where τ is the correlation time of the turbulent velocity field represented by v_T. The helicity, expressed in terms of ω, τ, and L, is

$$<v_T \left(\nabla \times v^2\right)> = <v^2> (\omega\tau)/L. \tag{5.6}$$

Here L $(\gtrsim\lambda)$ is the scale height, a value for which can be taken from the measured magneto-ionic thickness of the disc of the Milky Way (~1.5 kpc) and other nearby galaxies.

We are now in a position to calculate the evolution of a galaxy's global magnetic field after inserting the parameter values, as measured for the disc of the Milky Way. The rotational shear, $r\,d\omega/dr$ (Oort's constant), is 30 kpc-kms^{-1} near the Sun, v_T is $\approx$10 kms^{-} ^{1}, and $a \approx 15$ kpc. The turbulent elements have a size, λ, $\approx$100 pc. The only non-measurable is the lifetime of the turbulent elements, τ. However it can be reasonably estimated from λ, giving $\approx$3 $\times$ 10^{14} s (=10^7 year) for the Milky Way. A key point is that, specifically due to the Coriolis force, the average helicity, $<v_T.(\nabla \times v_T)>$, is non-zero. When we incorporate α into the magnetic induction Equation (3.7), we see that the α term affects the *growth* rate of the magnetic field.

$$\frac{\partial B}{\partial t} = \alpha\nabla \times B + \eta\nabla^2 B. \tag{5.7}$$

Here, both α and the diffusivity, η, are taken as constants (up to this point). A typical galactic value for the magnetic diffusivity is $\eta \approx 0.1$ λv, $\approx$10^{25} cm^2 s^{-1}.

This illustrates how a differentially rotating galaxy disc can amplify a pre-existing magnetic field on some timescale. For a large spiral galaxy this timescale, based on the above arguments and parameter values, is typically 10^{17}s, a few Gyr. It corresponds to a few galaxy disc rotations, which is of the order of a galaxy's dynamical lifetime.

5.5.3   *Some simple solutions to the Mean Field Dynamo Equation*

Solutions to Equation (5.7) can be written in the form

$$B = (\pm \sin kz, \cos kz, 0)\, e^{\gamma t} \qquad (5.8)$$

where $\gamma = -\eta k^2 \pm \alpha k$, k being the wavenumber. $|B|$ grows exponentially with time if the helicity is non-zero, whether positive or negative, and if the scales are sufficiently large ($k < |\alpha|\eta$).

Our discussion at this point is limited to showing how the solutions can provide a first-order confirmation of the growth of magnetic fields in disc galaxies.

Figure 5.13 depicts the general qualitative result for the α-ω dynamo. The right hand side shows the azimuthal component and the left hand side the poloidal component. This and other configurations can be produced by solutions to Equation (5.7). The B-solution shown in Fig. 5.13 has symmetry with respect to the galaxy's rotation axis, and reflection symmetry about the equatorial plane. Other solutions can produce field geometries that are antisymmetric about the galactic plane (e.g. Krause 1987).

A general requirement of Equations (5.7) and (5.8), however, is that characteristic field growth times must be longer than τ. Other restrictive statements made above can be modified under different assumptions; for example, our statement above connecting the mean helicity with the α parameter is subject to the assumption that turbulence associated with α is *isotropic*. In the most general case, because of stratification, gravity, and rotational and small-scale motions in galaxy discs, α is actually a tensor, rather than a constant as assumed above. Nonetheless, the mean field dynamo is relatively robust, in the sense that *some* solutions of the form $|B| \propto e^{\gamma t}$ ($Re\,\{\gamma\} > 0$) are also possible when the turbulence is *an*isotropic.

The azimuthal dependence of B is modulated by the term $e^{im\theta}$, where $m = 0$ corresponds to the axisymmetric and $m = 1$ to the bisymmetric azimuthal configurations illustrated in Fig. 5.7. The B solution shown in Fig. 5.13 is axisymmetric with respect to the galaxy's rotation axis, and it has reflection symmetry about the equatorial plane. Other eigensolutions

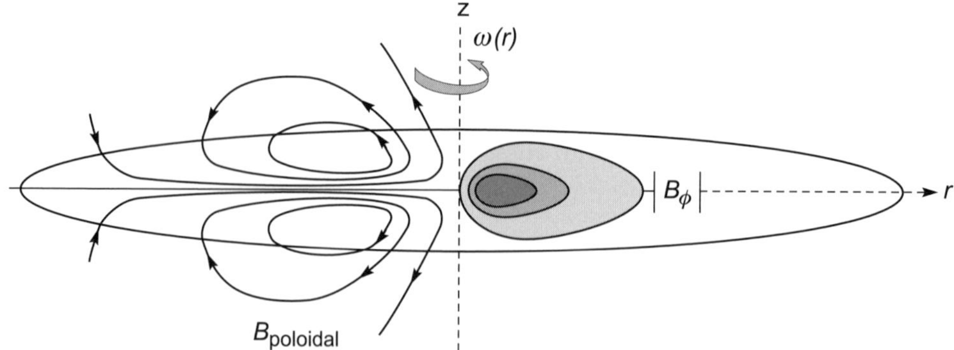

Figure 5.13 The 3-D magnetic field configuration produced by the α-ω dynamo in the flat ellipsoid model of Stix (1975), adapted from White (1978), and showing the morphology of the mean-field solutions for the azimuthal (right-hand side) component, and a 2-D slice of the quadrupolar poloidal component (left-hand side).

of the regeneration equation can also produce field geometries that are *antisymmetric* about the equator, as well as other azimuthal modes. In particular, the azimuth dependence of B can be tested against observations – those of the Faraday RM distribution, which is sensitive to the sign of B.

For the purpose of creating "model" galactic field distributions it is instructive to parameterise the α and ω effects using the convenient "C" parameters following Krause (1987); C_ω in Equation (5.4), and C_α , defined by

$$C_\alpha = \frac{a\alpha}{\eta}. \tag{5.9}$$

For typical disc galaxy conditions, it can then be established which modes dominate in what part of the galaxy. The C-parameters are thus convenient for exploring the solutions for model galaxy discs, which were first obtained by Stix (1975). Some numerical values are: $C_\omega \approx -7000$ (the minus sign from the negative $d\omega/dr$) and $C_\alpha \approx 50 \; \lambda/L$. It is immediately evident that $|C_\omega|/C_\alpha \gg 1$, meaning that differential rotation is the dominant induction action for amplification of the azimuthal (toroidal) field in the disc. Another useful parameter is $C_1 = C_\alpha C_\omega$. For the Milky Way, $C_1 \approx -10^{-5}$ (Krause 1987). For more detailed treatises on mean field solutions of the induction equation, and their application to kinematic turbulent dynamos and large scale galactic magnetic fields, the reader is commended to Krause & Rädler (1980), Elstner *et al.* (1992), and Zel'dovich *et al.* (1983), some of whose terminology has been followed above.

Solutions exist which give a quasi-rigid rotation which is close to that of a spiral DW pattern, ω_{DW}. This makes it possible for a *resonant coupling* to be set up between the two phenomena, as suggested by Fujimoto & Sawa (1990), Chiba & Tosa (1990), and others. We recall (Section 5.4) that the energy associated with the density wave-driven velocity perturbations can be comparable to that in the magnetic field perturbations. Such an effect has at least two interesting implications for galaxy evolution: (i) Large scale field amplification, or re-distribution could, in the right circumstances, be enhanced by DW patterns or even *vice versa*; (ii) the diffusion times of galactic disc DWs, $\approx 10^9$ years, might be substantially modified, in particular lengthened, by magnetic reinforcement, especially at larger galactocentric radii where the relative energy densities of DWs and ordered magnetic fields alter in favour of the latter.

An important point emphasized by Krause & Meinel (1988) is that, as long as C_α exceeds the critical value for field growth, saturation effects probably limit the importance of the relative linear growth rates in deciding what the final field configurations will be. Also, non-linear effects due to back-reactions can erase a galaxy's memory of the relative growth rates of different field amplification modes. This is an important point to keep in mind given, as we see later, that galactic magnetic fields, as far as we have able to probe them into the past, possibly saturated at the present-epoch levels (a few μG) at cosmological lookback times well past 4 Gyr ($z \sim 1$).

Moss & Tuominen (1990) investigated solutions for "thick disc" models in the *non-linear* régime (caused by an "α-quenching" mechanism). The conclusion was that long term oscillations are possible between solution modes, where bisymmetric and axisymmetric field geometries can alternate over a galaxy's lifetime. The possibility of such non-linear, "transient" phenomena has also been discussed by Brandenburg *et al.* (1992) and by Poezd, Shukurov, & Sokoloff (1993). They suggest that non-axisymmetric field patterns, including large scale field reversals in galaxy discs, are possibly *transient* phenomena, rather than stable solutions. This possibility, in addition to the saturation effects of non-linear B-growth mentioned above can further complicate the interpretation of the early magnetic history of galaxies. We are not permitted the privilege of direct time evolution observations of large scale galactic magnetic features as for the Solar corona and the Geosphere!

5.5.4   *Some limitations of galactic mean field dynamo theory*

A general point to be aware of is that Equation (5.7) does not contain, for example, pressure terms arising from energetic stellar events such as supernovae and HII associations, etc. whose plasma velocity fields can overwhelm the assumed quiescent galactic disc field. Also, as Parker (1992) emphasised, there is no known way, given $<B>$ as strong as several μG and $|\Delta B|$ ($\approx <B>$) fluctuating on a scale of ≈ 100 pc, that the tension in such a strong $<B>$ would permit the free swirling and mixing of magnetic flux that is needed to explain an eddy diffusion corresponding to $\eta_t \approx 10^{25}$ cm^2 s^{-1}. Such swirling and mixing are what is required to make large scale fields dissipate in 10^8 years. Also, turbulent mixing down to these small scales would imply the generation of much stronger small scale fields, for which there is not yet firm observational evidence. Similar doubts about the effectiveness of the diffusivity term in the standard dynamo theory have been raised by Rosner & DeLuca (1989).

The original galactic α-ω dynamo does not examine the detailed spatial and temporal nature of the (relativistic) *cosmic ray* gas, provided by supernovae and pulsars, and the non-relativistic hot gas ($10^5 - 10^7$ K) from active sites of star formation, i.e. large HII complexes and associations of O and B stars. Also not included is the existence of *organised outflows* from such sites – especially in the nuclear regions. Such complexes can inject an additional energy density $\varepsilon_{CR} + \varepsilon_*$ (the latter due to ionising radiation and non-relativistic hot gas from young star complexes) which is $\approx 10^{-11}$ erg cm^{-3} over ≈ 100 pc or more. This can be an order of magnitude or more *larger* than the global energy densities mentioned in Section 5.4. Symmetric outflow could destroy the symmetry of field *sense* above and below the plane. Also, over a few $\times 10^8$ years, several of these "transient" ($\tau_* \approx 10^7$ years) events will occur at "random" over a galaxy's disc, and their effect could be to influence any long-term monotonic mean-field dynamo buildup of a large scale field. A related point is that the combination of the supernovae and O & B stars can produce "galactic fountains" (Shapiro & Field 1976, Bregman 1980, Kahn 1981), and more generally, *galactic winds* (cf., Mathews & Baker 1971, Bardeen & Berger 1978, Habe & Ikeuchi 1980, Völk *et al.* 1990, Breitsch-werdt *et al.* 1991). (See Chapter 6). All of these phenomena will naturally tend to drive significant outflow from galactic discs, and are not directly included in the standard disc dynamo picture described above.

The effectiveness of the standard α-ω dynamo has been questioned from another stand-point by Kulsrud (1986), who pointed out that the galactic disc field is tied largely to the densest clouds, which are *ipso facto* the most neutral clouds. It has been argued that, consequently the cloud-field coupling may not be strong enough to prevent the field from simply passing through the clouds with the largest kinetic energy density. Following this argument, the competing effects just mentioned would, prevent a tight wind-up, or significant dynamo field generation (cf. also Kulsrud 1990 on the possible connection of a galaxy disc to a primordial magnetic field). Another questioning of the effectiveness of cloud-field coupling comes in the context of explaining the apparently strong and uniform inter-arm strengths in some nearby magnetic "grand-design" galaxies such as in Fig. 5.6. An interstellar MHD analysis of Fletcher *et al.* (2009) suggests that the dense interstellar clouds can detach from the regular magnetic field.

The above considerations, and other evidence discussed in later chapters, led to the conclusion (e.g. Kronberg 1994) that μG-level galactic magnetic fields existed at much earlier lookback times, and the required growth times are significantly shorter than a galaxy disc rotation time. Indeed, there is increasing evidence to suggest that μG-level fields may have existed at the protogalaxy stage.

5.5.5   *Modifications of the galactic dynamo when incorporating disc outflow*

The ongoing generation of CRs in the gaseous disc will produce ballooning field loops that are inflated outward at >30 kms^{-1} The waviness of these loops, caused by discrete energy sources in a galaxy's disc (e.g. Fig. 3.2), can cause upwardly transported heavier gas to fall back down the loops. In contrast, the CR gas tends to inflate the outward-going bulges, which are produced by the ongoing generation of CRs of supernovae and other energetic stellar events. These ideas and related discussion can be found in Mouschovias (1975), Parker (1979), and Shibata *et al.* (1989, 1990).

Some of the above ideas were advanced further by Parker (1992), who incorporated the effects of outflow in a modified α-ω dynamo. The outflow, driven by supernovae, produces outwardly inflating loops. These can be sufficiently close-packed that opposing, vertically aligned field lines can reconnect on a relatively rapid timescale. This, plus the rapid outward diffusion of the close-packed magnetic loops provide the turbulent diffusivity (η_t), which is otherwise difficult to physically justify on the "standard" dynamo model in Section 5.5.1. Another "natural" feature of Parker's model is that the outwardly inflated magnetic loops can sever themselves from the disc by reconnection at their footprints. At that point they can behave as "freely rotating" loops under the influence of the Coriolis force. Further reconnection fuses the large number of loops into a large scale poloidal field. A key and attractive aspect of Parker's modified α-ω dynamo is that it is a *fast dynamo*, whose "speed" increases with disc activity (viz. outflow force). Whereas the conventional mean-field dynamo amplifies the azimuthal disc field slowly over $\approx 10^9$ years or more, Parker's fast acting dynamo thrives on the outflow: Indeed, it is the source of the turbulent diffusivity in that the more vigorous the outflow, the faster and more effective the dynamo action. In consequence, the galactic dynamo growth time can be dramatically reduced to ~10^8 year. This, being of the order of approximately one galaxy rotation period, removes the need for shearing over several galaxy rotations and may, at least qualitatively, remove the contradiction with observational evidence for microgauss-level galactic magnetic fields at early Proper Times $\lesssim 10^8$ year (see Section 5.5.1, 5.5.2, and Chapter 12).

A more recent understanding of gaseous, in particular CR outflows in shaping galactic magnetic fields has been developed by Hanasz & Lesch (2003), and in subsequent papers on the "cosmic ray-driven dynamo". Incorporation of the CR dynamo into 3-D galactic field evolution models of large disc galaxies can produce $\eta_B \sim 3 \times 10^5$ cm^2 s^{-1} (Hanasz *et al.* 2009). It appears that B – amplification saturates when relation (5.3) is approximately satisfied, i.e. $\varepsilon_{cr} \approx \varepsilon_m \approx \varepsilon_t$. Thus B at some level is determined by this approximate equality so that, beyond some level of ε_m we would not expect the local supernova rate to correlate precisely with the magnetic field in the outflow.

Simulations of the above kind rely on an anisotropic diffusion of the CRs. A ratio of perpendicular-to-parallel diffusion ratio of ~5% as proposed by Giacalone & Jokipii (1999) appears consistent with the B amplification range obtained in the 3-D simulations of Hanasz *et al.* 2008.

5.5.6   *The régime of very strong stellar/SN-driven galactic outflows*

The effects of star- and supernova-driven outflow on the magnetic history of large disc galaxies discussed in this chapter have many, wider implications. In particular, if the final state reaches the equilibration $\varepsilon_{cr} \approx \varepsilon_m \approx \varepsilon_t$, the basic magneto-plasma processes might apply to equipartition situations in other, larger astrophysical systems that are discussed in later chapters.

Finally, an important extension of CR dynamo calculations on large galaxies is the application of similar principles to galaxies with extreme outflow, including dwarf galaxies and the so-called "starburst" galaxies. This leads us to the next chapter.

References

Bardeen, J. M. & Berger, B. K. 1978, A Model for Winds from Galactic Disks, *Astrophys. J.*, 221, 105

Beck, R. 2006, Das Square Kilometre Array, *Sterne und Weltraum*, 45, 22

Beck, R. 2007, Magnetism in the Spiral Galaxy NGC6946: Magnetic Arms, Depolarization, Dynamo Modes and Magnetic Fields, *Astron. Astrophys.*, 470, 539

Beck, R., Loiseau, N., Hummel, E., Berkhuijsen, E.M., Gräve, R., & Wielebinski, R. 1989, High resolution Polarization Observations of M31. I. Structure of the Magnetic Field in the SW Arm, *Astron. Astrophys.*, 222, 58

Berkhuijsen, E. M. 2009 (*private communication*)

Brandenburg, A., Donner, K. J., Moss D., Shukurov, A., Sokoloff, D. D., & Tuominen, I. 1992, Dyanamos in Discs and Halos of Galaxies, *Astron. Astrophys.*, 259, 453

Braun, R., Heald, G., & Beck, R. 2010, The Westerbork SINGS Survey III. Global Magnetic Field Topology, *Astron. Astrophys.*, 514, 42

Bregman, J. N. 1980, A Wind in the Galaxy, *Astrophys. J.*, 237, 280

Breitschwerdt, D., McKenzie, J. F., & Völk, H.-J. 1991, Galactic Winds. I – Cosmic Ray and Wave-Driven Winds from the Galaxy, *Astron. Astrophys.*, 245, 79

Breuermann, K., Kanbach, G., & Berkhuijsen, E. M. 1985, Radio Structure of the Galaxy – Thick Disc and Thin Disc at 408 MHz, *Astron. Astrophys.*, 153, 17

Broadbent, A., Haslam, C. G. T., & Osborne, L. J. 1990, A Detailed Model of the Synchrotron Radiation in the Galactic Disc. *Proc. 21st Int'l. Cosmic Ray Conf.*, 3, 229

Brown, J. C., Haverkorn, M., Gaensler, B. M., Taylor, A. R., Bisunok, N. S., McClure-Griffiths, N. M., Dickey, J. M., & Green, A. J. 2007, Rotation Measures of Extragalactic Sources Behind the Southern Galactic Plane: New Insights into the Large-Scale Magnetic Field of the Inner Milky Way, *Astrophys. J.*, 663, 258

Brown, J. C., Taylor, A. R., & Jackel, B. J. 2003, Rotation Measures of Compact Sources in the Canadian Galactic Plane Survey, *Astrophys. J. Suppl.*, 145, 213

Chiba, M. & Tosa, M. 1990, Swing Excitation of Galactic Magnetic Fields Induced by Spiral Density Waves, *MNRAS*, 244, 714

Christiansen, W. N. & Högbom, J. A. 1969, *Radiotelescopes* (Cambridge: Cambridge University Press)

Cordes, J. M. & Lazio, T. J. W. 2003, NE2001. II. Using Radio Propagation Data to Construct a Model for the Galactic Distribution of Free Electrons

Davies, R. D. 1966, Magnetic Fields in Galaxies, *Science*, v2 No. 10, p62

Elstner, D., Meinl, R., & Beck, R. 1992, Galactic Dynamos and Their Radio Signatures, *Astron. Astrophys.*, 94, 587

Elstner, D., Otmianowska-Mazur, K., von Linden, S., & Urbanik, M. 2000, Galactic Magnetic Fields and Spiral Arms. 3D Dynamo Simulations, *Astron. Astrophys.*, 357, 129

Fletcher, A. 2010, The Dynamic Interstellar Medium: A Celebration of the Canadian Galactic Plane Survey, *ASP Conf. Ser.* 438, 197

Fletcher, A., Korpi, M., & Shukurov, A. 2009, Dynamically Dominant Magnetic Fields in the Diffuse Interstellar Medium, *Proc. IAU Symp.*, 259, 119

Frick, P., Stepanov, R., Shukurov, A., & Sokoloff, D. 2001, Structures in the Rotation Measure Sky, *MNRAS*, 325, 649

Fujimoto, M. & Sawa, T. 1990, Generation and Maintenance of Bisymmetric Spiral Magnetic Fields and Interaction with Spiral Density Waves, *Geophys. Astrophys. Fluid Dyn.*, 50, 159

Gaensler, B. M., Madsen, G. J., Chatterjee, S., & Mao, S. A. 2008, The Vertical Structure of Warm Ionised Gas in the Milky Way, *Publ. Astr. Soc. Aust.*, 25, 184

Gardner, F. F. & Davies, R. D. 1966, Faraday Rotation of the Emission from Linearly Polarized Radio Sources, *Aust. J. Phys.*, 19, 129

Giacalone, J., & Jokipii, R. J. 1999, The Transport of Cosmic Rays across a Turbulent Magnetic Field, *Astrophys. J.*, 520, 204

Goldreich, P. & Sridhar, S. 1995, Toward a Theory of Interstellar Turbulence. 2: Strong Alfvenic Turbulence, *Astrophys. J.*, 418, 763

Habe, A. & Ikeuchi, S. 1980, Dynamical Behavior of Gaseous Halo in a Disk Galaxy, *Prog. Theor. Phys.*, 64, 1995

Han, J. L., Manchester, R. N., Lyne, A.G., Qiao, G. J., & van Straten, W. 2006, Pulsar Rotation Measures and the Large-Scale Structure of the Galactic Magnetic Field, *Astrophys. J.*, 642, 868

Hanasz, M. & Lesch, H. 2003, Incorporation of Cosmic Ray Transport into the ZEUS MHD Code: Application for Studies of Parker Instability in the ISM, *Astron. Astrophys.*, 412, 331

Hanasz, M., Otmianowska-Mazur, K., Kowal, G., & Lesch, H. 2009, Cosmic-Ray-Driven Dynamo in Galactic Discs. A Parameter Study, *Astron. Astrophys.*, 498, 335

Haslam, C. G. T., Klein, U., Salter, C. J., Stoffel, H., Wilson, W. E., Cleary, M. N., Cooke, D. J., & Thomasson, P. 1981, A 408 Mhz All-Sky Continuum Survey. I – Observations at Southern Declinations and for the North Polar Region, *Astron. Astrophys.*, 100, 209

Haslam, C. G. T., Salter, C. J., Stoffel, H., & Wilson, W. E. 1982, A 408 Mhz All-Sky Continuum Survey. II – The Atlas of Contour Maps, *Astron. Astrophys. Suppl. Ser.*, 47, 1

Heald, G.H. 2012, Magnetic Field Transport from Disc to Halo via the Galactic Chimney Process in NGC6946, *Astrophys. J. Lett.*, 754, 35

Heesen, V., Krause, M., Beck, R., & Dettmar, R.-J. 2009, Cosmic Rays and the Magnetic Field of the Nearby Starburst Galaxy NGC 253, *Astron. Astrophys*, 506, 1123

Heiles, C. 1996, The Local Direction and Curvature of the Galactic Magnetic Field Derived from Starlight Polarization, *Astrophys. J.*, 462, 316

Kahn, F. D. 1981, Dynamics of the Galactic Fountain, in *Investigating the Universe: Papers Presented to Zdenek Kopal on the Occasion of his Retirement, September 1981* (Dordrecht: D. Reidel), 1

Knobloch, E. 1978, Turbulent Diffusion of Magnetic Fields, *Astrophys. J.*, 225, 1050

Krause, F. 1987, Dynamo Excitation in Very Large Scales, in *Interstellar Magnetic Fields*, ed. R. Beck & R. Gräve (Berlin: Springer), 8

Krause, F. & Meinel, R. 1988, Stability of Simple Nonlinear 2-Dynamos, *Geophys. Astrophys. Fluid Dyn.*, 43, 95

Krause, F. & Rädler K.-H. 1980, *Mean Field Electrodynamics* (Berlin: Academic Press)

Krause, M. 1990, Multi-Frequency Radio Observations of Spiral Galaxies and their Interpretation, in *Galactic and Intergalactic Magnetic Fields; Proceedings of the 140th Symposium of IAU, Heidelberg, Federal Republic of Germany, June 19–23, 1989*, ed. R. Beck, P. P. Kronberg, & R. Wielebinski (Dordrecht: Kluwer), 187

Krause, M. 2007, Large-Scale Magnetic Field in Spiral Galaxies, *Mem. Soc. Astron. Ital.*, 78, 314

Krause, M. 2009, Magnetic Fields and Star Formation in Spiral Galaxies, in *Magnetic Fields in the Universe II: From Laboratory and Stars to the Primordial Universe* ed. A. Esquivel *et al.*, *Rev. Mex. Ast. Astrof., Ser. de Conf.*, 36, 25

Krause, M., Beck, R., & Hummel, E. 1989, The Magnetic Field Structure in Two Nearby Galaxies. II, The Bisymmetric Spiral Magnetic Field in M81, *Astron. Astrophys.*, 217, 17

Krause, M., Wielebinski, R., & Dumke, M. 2006, Radio Polarization and Sub-millimeter Observations of the Sombrero Galaxy (NGC 4594). Large-Scale Magnetic Field Configuration and Dust Emission, *Astron. Astrophys.*, 448, 133

Kronberg, P. P. 1994, Extragalactic Magnetic Fields, *Rep. Prog. Phys.*, 57, 325

Kronberg, P. P. & Newton-McGee, K. J. 2011, Remarkable Symmetries in the Milky Way Disc's Magnetic Field, *Publ. Astr. Soc. Aust.*, 28, 171

Kulsrud, R. 1986, The Wrap Up Problem of the Galactic Magnetic Field, in *Plasma Astrophysics: Proceedings of the Joint Varenna-Abastumani International School and Workshop, held in Sukhami, USSR, 19–28 May 1986*, ed. I. M. Zeleny (Paris, France: European Space Agency), 531

Kulsrud, R. M. 1990, The Present State of a Primordial Galactic Field, in *Galactic and Intergalactic Magnetic Fields; Proceedings of the 140th Symposium of IAU, Heidelberg, Federal Republic of Germany, June 19–23, 1989*, ed. R. Beck, P. P. Kronberg, & R. Wielebinski (Dordrecht: Kluwer), 527

Kulsrud, R. M. & Anderson, S. W. 1992, The Spectrum of Random Magnetic Fields in the Mean Field Dynamo Theory of the Galactic Magnetic Field, *Astrophys. J.*, 396, 606

Laing, R. D. 1981, Magnetic Fields in Extragalactic Radio Sources, *Astrophys. J.*, 248, 87

Manchester, R. N. 1972, Pulsar Rotation and Dispersion Measures and the Galactic Magnetic Field, *Astrophys. J.*, 172, 43

Manchester, R. N. 1974, Structure of the Local Galactic Magnetic Field, *Astrophys. J.*, 188, 637

Mathews, W. G. & Baker, J. C. 1971, Galactic Winds, *Astrophys. J.*, 170, 241

Mathewson, D. S. & Ford, V. L. 1970 Polarization Observations of 1800 Stars, *MNRAS*, 74, 139

Men, H., Ferrière, K., & Han, J. L. 2008, Observational Constraints on Models for the Interstellar Magnetic Field in the Galactic Disc, *Astron. Astrophys.*, 486, 819

Moffatt, H. K. 1978, *Magnetic Field Generation in Electrically Conducting Fields* (Cambridge: Cambridge University Press)

Mouschovias, T. Ch. 1975, On Cosmic Rays and Final Equilibrium States for the Parker Instability, *Astron. Astrophys.*, 40, 191

Neininger, N., Klein, U., Beck, R., & Wielebinski R. 1993, The Magnetic Field of M83, in *The Cosmic Dynamo IAU Symposium No. 157*, ed. F. Krause, K.-H. Rädler, & G. Rüdiger, 315

Noutsos, A., Johnston, S., Kramer, M., & Karastergiou, A. 2008, New Pulsar Rotation Measures and the Galactic Magnetic Field, *MNRAS*, 386, 1881

Parker, E. N. 1971, The Generation of Magnetic Fields in Astrophysical Bodies. II. The Galactic Field, *Astrophys. J.*, 163, 255

Parker, E. N. 1979, *Cosmical Magnetic Fields* (Oxford: Clarendon)

Parker, E. N. 1992, Fast Dynamos, Cosmic Rays, and the Galactic Magnetic Field, *Astrophys. J.*, 401, 137

Poezd, A., Shukurov, A., & Sokoloff, D. D. 1993, Nonlinear Dynamo in a Disk Galaxy, *Proceedings of IAU Symposium 157, Potsdam, 1992*, eds. F. Krause, K.-H. Rädler, & G. Rüdiger (Kluwer: Dordrecht), 349.

Prouza, M. & Šmída, R. 2003, The Galactic Magnetic Field and Propagation of Ultra-High Energy Cosmic Rays, *Astron. Astrophys.*, 410, 1

Rand, R. & Lyne, A.G. 1994, New Rotation Measures of Distant Pulsars in the Inner Galaxy and Magnetic Field Reversals, *MNRAS*, 268, 497

Reich, P. & Reich, W. 1988, Spectral Index Variations of the Galactic Radio Continuum Emission – Evidence for a Galactic Wind, *Astron. Astrophys.*, 196, 211

Rosner, R. & Deluca, E. 1989, On the Galactic Dynamo, in *The Center of the Galaxy: Proceedings of the 136th Symposium of the International Astronomical Union, Held in Los Angeles, U.S.A., July 25–29, 1988*, ed. M. Morris (Dordrecht: Kluwer), 319

Ruzmaikin, A. A., Sokoloff, D. D., & Shukurov, A. 1988, Magnetism of Spiral Galaxies, *Nature*, 336, 341

Scarrott, S. M. 1991, Optical Polarization Studies of Astronomical Objects, *Vistas Astron.*, 34, 163

Scarrott, S. M. & Draper, P. W. 1996, Further Evidence for Vertical Magnetic Fields in the Galaxy NGC 891, *MNRAS*, 278, 519

Shapiro, P. R. & Field, G. B. 1976, Consequences of a New Hot Component of the Interstellar Medium, *Astrophys. J.*, 205, 762

Simard-Normandin, M. & Kronberg, P. P. 1980, Rotation Measures and the Galactic Magnetic Field, *Astrophys. J.*, 242, 74

Simard-Normandin, M., Kronberg, P. P., & Button, S. 1981, The Faraday Rotation Measures of Extragalactic Radio Sources, *Astrophys. J. Suppl.*, 45, 97

Sofue, Y., Fujimoto, M., & Wielebinski, R. 1986, Global Structure of Magnetic Fields in Spiral Galaxies, *Ann. Rev. Astr. Astrophys.*, 24, 459

Sridhar, S. & Goldreich, P. 1994, Toward a Theory of Interstellar Turbulence. 1: Weak Alfvenic Turbulence, *Astrophys. J.*, 432, 612

Steenbeck, M. & Krause, F. 1969, On the Dynamo Theory of Stellar and Planetary Magnetic Fields I: AC Dynamos of Solar Type, *Astron. Nachr.*, 291, 49

Steenbeck, M., Krause, F., & Rädler, K.-H. 1966, Berechnung der mittleren LORENTZ-Feldstärke $v \times B$ für ein elektrisch leitendes Medium in turbulenter, durch CORIOLIS-Kräfte beeinflußter Bewegung, *Z. Naturforsch.*, 21, 369

Stix, M. 1975, The Galactic Dynamo, *Astron. Astrophys.*, 42, 85

Strong, A. W., Moskalenko, I. V., & Reimer, O. 2000, Diffuse Continuum Gamma Rays from the Galaxy, *Astrophys. J.*, 537, 763

Taylor, A. R., Gibson, S. J., Peracaula, M., Martin, P. G., Landecker, T. L., Brunt, C. M., Dewdney, P. E., Dougherty, S. M., Gray, A. D., Higgs, L. A., *et al.* 2003, The Canadian Galactic Plane Survey, *Astron. J.*, 125, 3145

Thompson, A. R., Moran, J. M., & Swenson, G. W. 2007, *Interferometry and Synthesis in Radio Astronomy* (New York: John Wiley & Sons)

Tinyakov, P. G. & Tkatchev, I. I. 2002, Tracing Protons through the Galactic Magnetic Field: A Clue for Charge Composition of Ultra-High-Energy Cosmic Rays, *Astroparticle Phys.*, 18, 165

Tosa, M., & Fujimoto, M. 1978, The Configuration of Magnetic Fields in the Spiral Galaxy M 51, *Publ. Astron. Soc. Japan*, 30, 315

Vallée, J. P. 2008, An Improved Magnetic Map of the Milky Way, with the Circularly Orbiting Gas and Magnetic Field Lines Crossing the Dusty Stellar Spiral Arms, *Astrophys. J.*, 681, 303

Völk, H.-J., Breitschwerdt, D., & McKenzie, J. F. 1990, Galactic Winds and Magnetic Fields from Spiral Galaxies, *IAU Symp 140, Heidelberg*, eds. R. Beck, P. P. Kronberg, & R. Wielebinski (Kluwer: Dordrecht), 177

White, M. P. 1978, Numerical Models of the Galactic Dynamo, Astron. *Nachrichten*, 299, 209

Zel'dovich, Ya., Ruzmaikin, A. A., & Sokoloff, D. D. 1983, *Magnetic Fields in Astrophysics* (New York: Gordon and Breach Science Publishers)

6

Magnetic field outflow from dwarf and starburst galaxies

6.1 Introduction

Where galaxies have elevated star formation activity, and consequent strong outflow, the z-heights of halos are more pronounced and larger at cm and metre radio wavelengths and in optical emission lines. Polarised emission in the halos can trace the magnetic field and its structure, sometimes out to z heights of ≈ 20 kpc and more – i.e. dimensions approximating those of the parent galaxy discs. These phenomena are particularly pronounced in small galaxies, where outflow velocities are similar to those in larger galaxies, but the galaxy's gravitational potential can be a few orders of magnitude smaller. Apart from this, dwarf galaxies are the most numerous per intergalactic co-moving volume. Some classes of dwarfs can be considered fossils of the early post-Recombination primeval universe, in the sense that some of them have avoided galaxy mergers and "cannibalism" over most of a Hubble time. From the standpoint of magnetic history of the Universe, they can be useful tools for probing both galactic and intergalactic magnetic field strengths back to $z \gtrsim 10$. By virtue of this, they can provide information on the cosmic evolution of extragalactic magnetic fields, as we illustrate later in the chapter.

Different degrees of outflow, corresponding to the star formation and supernova rates in the underlying galaxy plane, are illustrated in the comparison between NGC 891 (Fig. 5.9), a galaxy slightly more "active" than the Milky Way, and galaxies with progressively more star/SN-driven outflow. See, for example, Fig. 6.1 (NGC 4631), and Figs. 6.2 and 6.3, which show a nearby classic starburst galaxy, M82.

The surface brightness sensitivity of the radio telescope and the halo's spectral index strongly influence the maximum z-height that is detected when comparing relative heights of outflow halos. This is because the synchrotron radio spectrum often steepens sharply with distance from the galaxy due to relativistic electron "aging" – thereby making the halo less visible at greater z-heights and higher observing frequencies. To compensate for this, instruments used to probe the extent of magnetised stellar/SN-driven outflows must be capable of surface brightness-sensitive radio imaging. The imaging needs to include Stokes parameters I, Q, and U and extend to the lowest possible radio frequencies where magnetic field strengths in starburst outflow halos can be sensitively measured to their highest z-heights. A review of some of these issues can be found in Beck (2010).

The outflow material is also traceable in optical thermal emission such as free–free bremsstrahlung and emission lines of hydrogen. This is nicely illustrated in some Hubble Telescope images, which show outflow gas in detailed filamentary structures to heights similar to those seen in radio synchrotron emission. Although the optical images do not

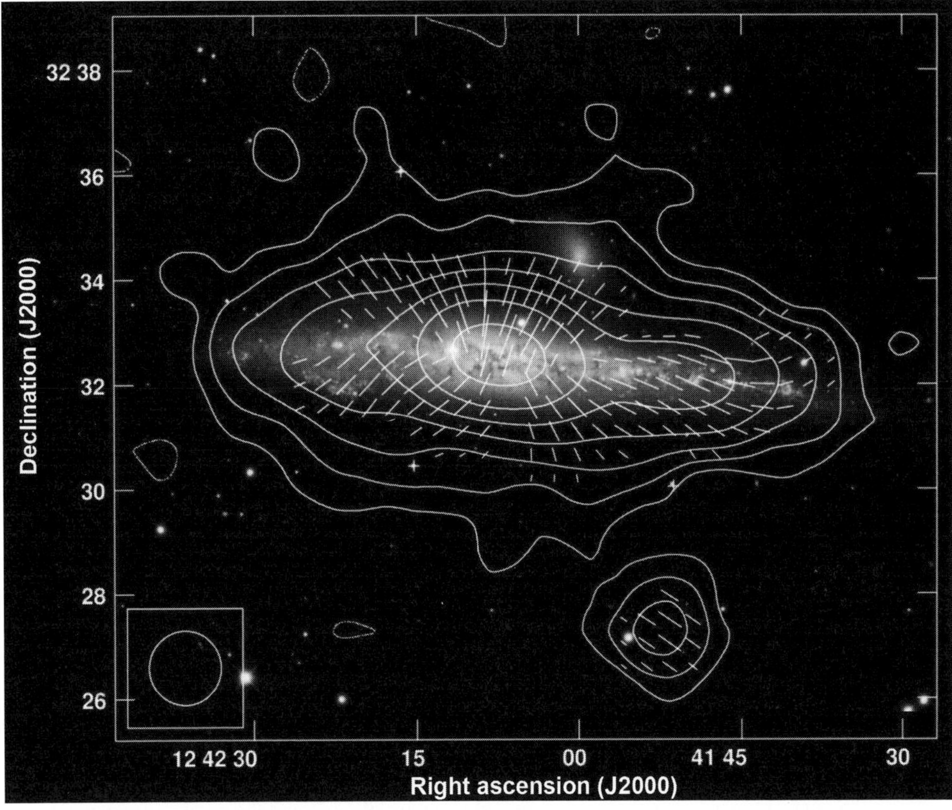

Figure 6.1. Radio and optical images of the edge-on galaxy NGC4631, showing the outflow halo. Contours show the radio surface brightness at $\lambda 3.6$ cm and $1.4'$ resolution and lines show the intrinsic (Faraday-*RM* corrected), projected magnetic field directions (M. Krause, 2009). The radio image was made with the Effelsberg 100 m telescope, and the optical image at the Misti Mountain Observatory. (Courtesy of Dr. Marita Krause).

directly probe magnetic field strength, they provide rich complementary information, especially in the velocity field, via the emission line Doppler shifts, which probably have some coupling to the magnetic field. Similarly, the higher energy photons, such as in *Fermi LAT* γ-ray images, are revealing hot gas bremsstrahlung and the inverse Compton component of the outflow halo emission. Optical/spectroscopic UV, X-ray, and γ-ray data on outflows are becoming increasingly sensitive and detailed, and will provide a wealth of information for defining the complete physical state of magnetised outflow halos in galaxies.

6.2   Star formation in galaxies and associated magnetised outflows

6.2.1   *Outflows measured in edge-on galaxies*

Halos of spiral galaxies are usually best seen if the galaxy happens to be edge-on – for example NGC891 in Fig. 5.9. Progressively more "extreme" examples include NGC 4631, the edge-on spiral galaxy depicted in Fig. 6.1, which has widespread, enhanced star

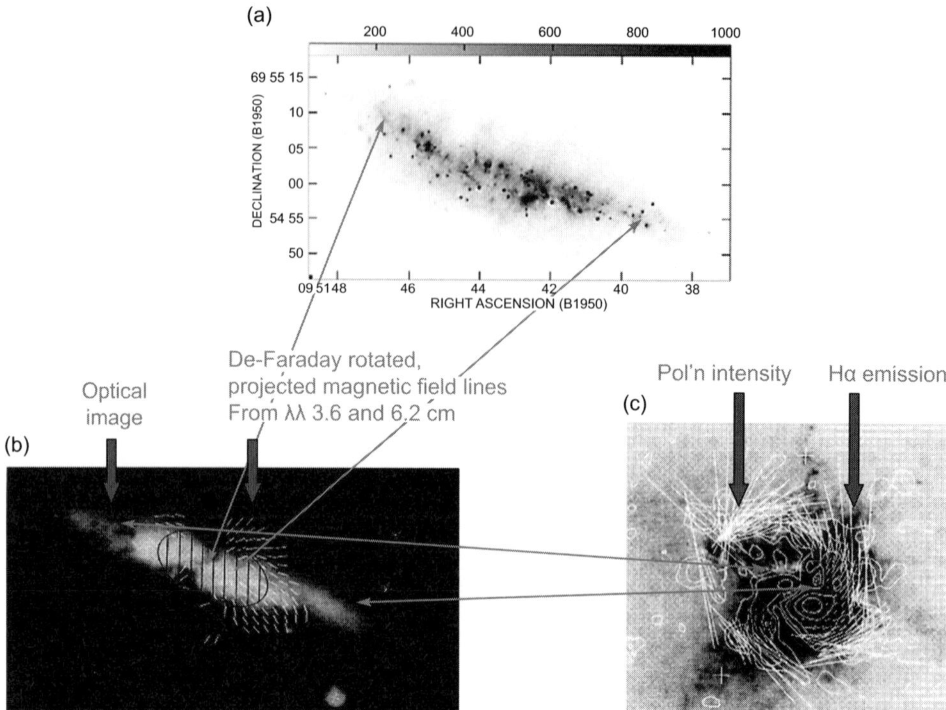

Figure 6.2. (a) M82's inner 600pc zone (above) contains over 100 radio supernova remnants (previously unpublished 8 GHz VLA image by M. L. Allen and the author). The lower left image, (b) shows the amorphous optical structure of M82 and the projected magnetic field structure of the inner radio halo. The starburst outflow halo (c) seen at λ3.6 cm (right) and 15″ resolution extends into the surrounding IGM (Reuter *et al.* 1994). The frame width in (c) is 4.2′= 4.8 kpc. The halo zones are highly polarised at 5 and 8 GHz, but the inner part is highly depolarised at these wavelengths due to a combination of stronger magnetic field and high gas density. The superimposed *H*α emission from observations by McCarthy *et al.* (1987) shows some of the same gas that causes the Faraday rotation and depolarisation.

formation within its innermost 2 kpc. NGC 4631 shows a magnetic field radially aligned away from the central zone of the disc. Here, in contrast to large scale field-ordering *parallel* to galaxy discs, the above-plane magnetic fields are typically directed *out* of the plane above the central zones of more "active" spiral galaxies. Using standard assumptions of equipartition, estimates of $\approx$8 μG have been made for the magnetic field *strength* in the disc of NGC 4631 and $\approx$5 μG in its halo. Proceeding further from the moderately active nuclear region we can see the characteristic X-pattern (Fig. 6.1), reminiscent of NGC 891 in Fig. 5.9. Recent studies of the nearby, more quiescent large spiral M31 (the Andromeda galaxy) reveal zones where the field appears to turn up out of the galaxy's plane (Beck *et al.* 1989).

Even violent outflow systems such as M82 (next section) are moderately small galaxies, hence relatively numerous per co-moving space volume. They, and the even more spatially numerous dwarf galaxies have the potential to contribute to *extragalactic* magnetic fields, as we discuss below. This can be anticipated on *a priori* energetic grounds given that halo magnetic fields are associated with outflowing winds, the consequence of the calculable overpressure in regions of enhanced supernova and star formation activity in galaxy discs.

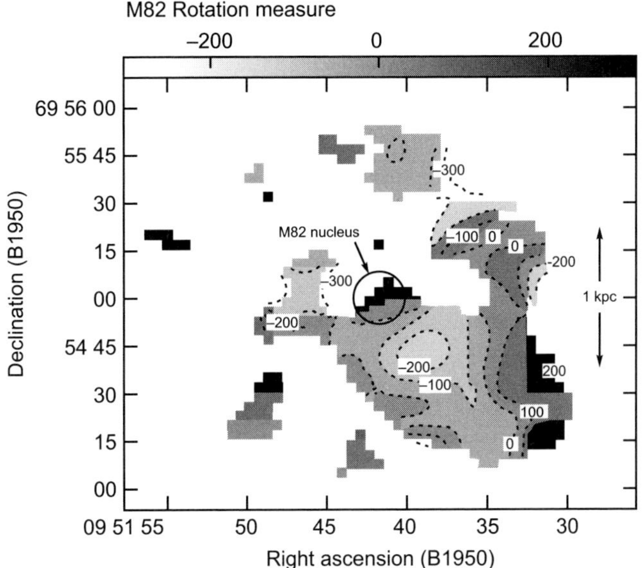

Figure 6.3. The Faraday *RM* distribution within the inner outflow halo of M82, from two-frequency polarisation images having sensitivity of ≈ 200 μJy/beam and angular resolution of 15″. The *RM* in the inner halo varies from –200 to +300 rad m^{-2}. The *RM* range indicates a halo magnetic field strength B, $\mathcal{O}$ 15 μG (adapted from Reuter *et al.* 1994). The outer boundaries of detectable *RM* (and elevated B) are limited by the signal/noise, so that the true extent of large B zones in the outflow halo is likely $\gtrsim 5$ kpc, indicated by the optical emission line zones in Fig. 6.2(c).

6.2.2　*Magnetic fields in strong outflows and "starbursts"*

An extreme case of halo outflow occurs when a "cauldron" of rapid star formation occurs within a limited volume, for example within ≈ 1 kpc of a galaxy's nucleus. These "starburst" regions also indirectly produce abnormally intense infrared emission, where most of the luminosity is between ≈ 20 and 100 μm, and represents re-radiation by dust of the energetic photons from massive hot stars and supernovae. The same regions emit non-thermal radio emission from the aggregate synchrotron electrons and magnetic fields that originate from supernovae and pulsars. The nucleus may or may not have a detectable massive black hole, and there are usually no prominent jets or outer "double" radio lobes in these systems. The total *radio* luminosity of these starburst galaxies is typically smaller than that of jet-lobe radio galaxies.

The first indication that starburst galaxies were directly powered by conventional stellar sources such as SN came from subarcsecond 5 GHz radio imaging of the nearby (3.8 Mpc) galaxy M82 (Kronberg & Wilkinson 1975, Kronberg, Biermann, & Schwab 1985). A more recent, all-configuration VLA image at 8GHz is shown in Fig. 6.2(a). It illustrates ~100 radio supernova remnants at 0.3″ resolution, all within the inner 600 pc of M82's nucleus. It also reveals vertically directed "fibrils" of radio emission close in to the starburst zone, where the outflow halo is being launched. This system is presumed to be prototypical of similar, more distant and observationally less accessible systems.

This inner zone of M82's nucleus is undergoing intense star formation and is generating a total luminosity of about $4 \times 10^{10} L_\odot$, mostly in the infrared, and due to reprocessed stellar radiation. The high luminosity density makes it inevitable that outflow will occur. The active region also typically emits lines of H_α, [OII], etc., and molecular lines, as well as X-rays, along with the aforementioned *infrared* continuum radiation, and synchrotron radiation.

The vigorous stellar activity of the inner zone of M82 and the associated outflow of cosmic rays (CRs) (Seaquist & Odegard 1991, Reuter *et al.* 1992, 1994), along with the X-ray emission (Schaaf *et al.* 1989) and strong halo magnetic field, combine to make M82 a key test bed for investigating magnetic field generation in this and other galaxies with starburst-like nuclear activity. Figure 6.2(c) shows near normals to the projected magnetic field pattern seen in the outflow halo around M82 – obtained from VLA observations at $\lambda\lambda$ 6.2 and 3.6 cm, and $15''$ resolution (Reuter *et al.* 1994). An extensive magnetised halo is also seen around the active spiral galaxies NGC 253 (Carilli *et al.* 1992), and NGC 3079.

The α^2-dynamo (e.g. F. Krause, 1987), proposed for the inner co-rotating zones of "quiescent" galaxies may not apply here, though this point is not yet clear. Observations show that magnetic fields are expelled along with some plasma at more than M82's escape velocity. If Parker's (1992) idea of reconnecting detached magnetic loops from a spiral disc is correct, we may be seeing, in M82, a more powerful version of a similar physical process. The starburst phenomenon produces high velocity outflows, which are sustained for $\sim 10^8$ years. This means that the magnetic fields around M82 are produced by, and probably *in*, the outflow phenomenon. All of this happens on a much shorter timescale than the dynamical age of the galaxy.

In the study by Reuter *et al.* (1994 – Figs. 6.2(b) and (c) and 6.3), images were made with the VLA at $\lambda\lambda$ 3.2 and 6 cm in Stokes parameters I, Q, and U. This permitted Faraday rotation and depolarisation images over the outflow halo. Where a strong outflow proceeds outward from the nuclear region in a vertical cone, the projected magnetic field is closely aligned with the vertical outflow. Below a certain threshold of starburst activity the projected magnetic field is parallel to galaxy's disc above the nuclear region, and then appears to flare upwards at larger $|x,y|$ distances from the galaxy's rotation axis. This same pattern is nicely illustrated in NGC891 (Fig. 5.9), and in several other edge-on galaxies. In general vertically aligned halo fields above the nucleus are a common characteristic of starburst galaxies.

The M82 outflow halo in Fig. 6.2 has an overall extent of ~ 6 kpc. Correlations among the observed $H\alpha$ emissivity, synchrotron emissivity, radio depolarisation rate, and RM structure indicate the following physical parameters for the halo: (1) The outflow velocity is up to ~ 1000 kms^{-1} and higher, (2) the magnetic field orientation is poloidal near the outflow axis, (3) the magnetic field and RM coherence scales are ~ 1 kpc, (4) the halo $|RM|$ is up to ~ 200 rad m^{-2} (Fig. 6.3), and (5) the magnetic field strength is ~ 10 μG.

The RM in M82's halo zone shown in Fig. 6.3 varies between -200 and $+300$ rad m^{-2}. This is greater than or comparable to the RM in a typical sightline through a large spiral galaxy disc. The implied halo magnetic field strengths are up to 10's of μgauss – higher than in the denser disc of the Milky Way! These interesting facts emphasise the strong constraints on processes of magnetic field regeneration. They demonstrate that mechanisms other than the slow galactic-scale dynamo *are* able to amplify and, as discussed elsewhere, perhaps also create magnetic fields.

An example of a halo with similar magnetic field strength – not a strong starburst galaxy, but rather a hydrogen-poor one – is the Virgo cluster galaxy NGC 4569, analysed by

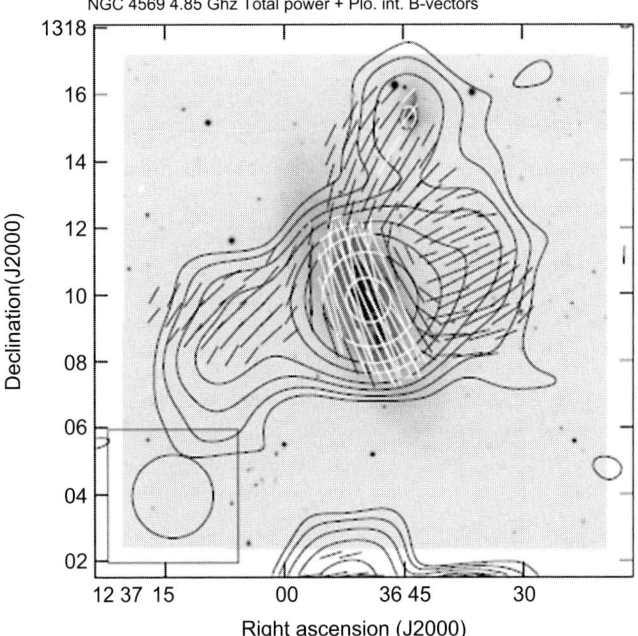

Figure 6.4. Magnetised outflow halo of the Virgo Cluster galaxy NGC 4569 (Chyży *et al.* 2006). Its large overall dimension of 47 kpc seems unusual, given its membership in a galaxy cluster.

Chyży *et al.* (2006). As shown in Fig. 6.4, it has one of the largest outflow halos so far seen: ~48 kpc, roughly comparable to the diameter of a large disc galaxy.

6.2.3   *Magnetic structures in dwarf galaxies*

Dwarf galaxies are important test objects for understanding the origin, evolution, and geometry of magnetic fields within late-type galaxies. This is because, given their total mass down to $\lesssim 10^{-3}$ of the Milky Way, very little galactic rotation has occurred over a Hubble time, i.e. far less than would be required to drive a large, galaxy-scale dynamo. This is well illustrated in dwarf galaxies such as NGC 4449 in Fig. 6.5 (Chyży *et al.* 2000). The halo magnetic field strength of this galaxy is ~14 μG, despite the fact that little global rotation seems to have occurred. This seems to rule out a large scale galactic dynamo as the field amplifier for such relatively low mass galaxies. Another example not shown is NGC 1656 (Kepley *et al.*, 2010), imaged at four radio frequencies. It reveals regular field strengths in the halo of 10–15 μG, similar to the M82 halo in Fig. 6.3.

In general, field strengths up to tens of μgauss are observed in some dwarf galaxies, and up to several kpc from the host galaxy nucleus. These dwarf galaxy properties alone are sufficient to rule out a scenario in which a primordial field strength of $\ll 10^{-10}$ G was amplified by the internal galaxy dynamics of dwarf spirals (Kronberg 1994). So how did they generate such strong interstellar and halo magnetic fields? Relative to their more massive spiral cousins, dwarf galaxies permit stellar-driven outflows to proceed more easily to a given halo radius. Thus, the halos of dwarf galaxies could more easily contribute to the

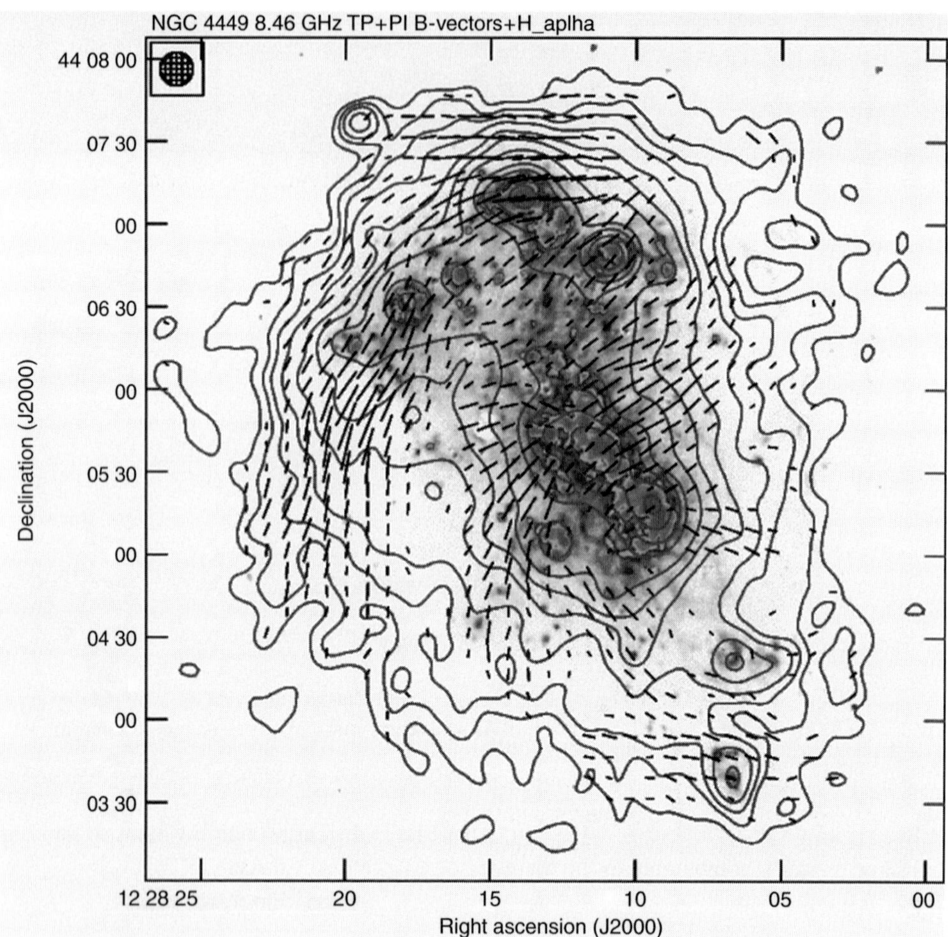

Figure 6.5. An 8 GHz image of the 7×10^{10} M$_\odot$ dwarf galaxy NGC 4449 from Chyży *et al.* (2000). Contours are in units of total radio brightness (Stokes I), and the lines show polarised intensity where the orientation gives the projected magnetic field orientation. (Reproduced by permission of Ulrich Klein).

global magnetisation of the IGM in an earlier, more compact Universe, perhaps aided by stronger inter-galaxy tidal effects at the lower galaxy–galaxy separations expected at large redshifts – see Section 6.3.

The outflow velocities, $v_{\mathrm{outf}} \simeq v_z$, are measured at ~150 to $\gtrsim 1000$ km/s, and these same numbers also apply roughly to starburst outflows in larger galaxies. The outflow has a heterogeneous composition, ranging from the heavy ions produced by nucleosynthesis in the SN and massive stars that drive the outflow, to the light magnetised CR fluid that is detected by synchrotron radiation. On average, v_{outf} for the latter will be greater than for the ions, and the slower and faster fluids are magnetically coupled. The resulting ∇v are typically very high, ∂v being up to hundreds of kms^{-1}. The implied strong velocity shear provides ideal conditions for coupling the faster moving, magnetised CR gas with the hot thermal gas. Energy is transferred from the powerful supernova-driven CR wind to the bulk flow of the heavier thermal gas. In this scenario, the $\sim 10\,\mu$G halo fields can be generated over a few $\times 10^7$ years, and mainly from the aggregate energy of SN and other stellar sources.

The above discussion outlines one scenario for generating magnetic fields in dwarf galaxy systems. But alternative possibilities have also been proposed: We could hypothesise that μgauss-level magnetic fields in dwarf galaxies were entrained and/or compressed as fossils of a pre-existing primeval magnetic field. This pre-existing field originated in the earliest Pop III stars, a class of stellar system that could have generated, or seeded, substantial fields in a short space of cosmic time. Another variant of the latter idea is that the very first stars might have seeded their own magnetic fields via a Biermann battery-like process, and these were subsequently amplified in the ejecta of supernova-like explosions and spread into the compact primeval IGM at $z \gtrsim 10$ (see Section 6.3).

What can optical and IR observations tell us about dwarf galaxies as fossils of the primeval Universe? "Late"-type dwarfs formed early and have an undisturbed disc – an indicator of having avoided tidal disruption. Other dwarf types, such as the dwarf spheroidals, likely lost their interstellar gas in very early starburst outflows. This can be chronologically "calibrated" in well-known ways e.g. by verifying their aggregate stellar main-sequence turnoff times.

A further possibility is that the hypothesised strong "pre-existing" fields might have been seeded and/or pre-energised by early black holes. This hypothesis may be supported by discoveries of surprisingly massive black holes in some small galaxies. The general question of black hole seeding of the IGM magnetic fields is discussed later in the book.

All of the above underscores the value of dwarf galaxies as testbeds for understanding cosmic magnetic fields, and indeed the early evolution of galaxy systems. To explore these possibilities, we need detailed computational models that incorporate all the known physical processes and relevant observations.

6.2.4 *Summary of the star-driven magnetic outflow story*

The unanticipated high magnetic field strengths in diffuse outflows indicates the potential for starburst outflows to contribute to intergalactic magnetic fields. Particularly significant is that outflow halos with strong magnetic fields occur over a wide range of galaxy masses down to $\lesssim 10^8$ M$_\odot$. The uniform component of halo fields is typically 10–18 μG, an impressively high number, especially in terms of the corresponding magnetic energy density, ε_{m}. It is stronger than ordered components seen in the much denser discs of some large galaxies.

It is instructive to develop a global galaxy co-moving density model over all masses down to $M_{\mathrm{G}} \lesssim 10^8 M_\odot$. The high outflow velocities are not very sensitive to parent galaxy mass and are sometimes above v_{esc}. Some simple model simulations in Section 6.3 indicate that a significant volume of the IGM can be magnetised by star-driven outflows, even if we ignore the possible additional role of supermassive black holes in these galaxies. This IGM magnetic "seeding" process will be accompanied by baryons that are enriched with SN nucleosynthesis products. A vertical field alignment above the galactic nuclear region occurs in many (perhaps even most) galaxies, ranging from large spirals like our Galaxy to smaller dwarfs.

6.2.5 Could a galaxy self-seed its own large scale magnetic field?

An ingenious scenario for the *generation* and the *amplification* of large scale magnetic fields involving the molecular ring or torus of a starburst galaxy was posited by Lesch *et al.* (1989). It proposes that M82 creates a kpc-scale magnetic field by a *galaxy scale* version of the Biermann battery, caused by the interaction of the galactic plane component of a nuclear starburst wind with the rotating, dense molecular gas torus which surrounds the nucleus – e.g. in M82. The latter was revealed in both $\lambda21$ cm HI observations (Weliachew *et al.* 1984), and in $\lambda2.7$ mm CO (1-0) observations (Lo *et al.* 1985, Nakai *et al.* 1987, Loiseau *et al.* 1988). A hot, subsonic outflow excites sound waves that propagate at

$$c_s = \left(\frac{5}{3}\right)^{\frac{1}{2}} \left(\frac{kT_e}{m_p}\right)^{\frac{1}{2}} \tag{6.1}$$

(k is Boltzmann's constant, T_e the electron temperature, and m_p the proton mass). In this model, the sound waves become compressional waves which, on arrival at the rotating torus push the radially streaming plasma into the azimuthally rotating molecular gas. Because the electrons couple more easily to the gas than the protons, the *differential* stopping time,

$$t_{sp} = \left(\frac{m_p}{m_e}\right) t_{se}, \tag{6.2}$$

causes a current to build up in the ring, which could generate a poloidal magnetic field. The initial magnetic field is amplified by a combination of $v \times B$ compression and turbulence, a process described in more detail by Alfvén & Fälthammer (1963), Parker (1979), and others. In M82, the wavelength of sound waves, λ_s, is comparable with the system's dimensions, and the waves will reach the radius of the torus, $R_T \simeq 225$ pc, before they are damped; in this model they get reflected back from the inner edge of the torus on a time scale $t_R \approx 2\, R_T/c_s$, which is $\simeq 10^7$ years. This is the time scale over which the battery current breaks down, so that a current is maintained as long as t_R exceeds the diffusion time,

$$t_D = L^2/4\eta_t \, s \tag{6.3}$$

(Parker 1976). L is the characteristic system scale and η_t is the turbulent diffusivity. A consequence, in this model, is that the ring current will prevail over the tendency of the plasma to cancel it. Lesch *et al.* (1989) estimate $t_D \approx 10^6$ years for M82's inner molecular ring, in which case a current, and hence a poloidal field, will be built up, consistent with observation.

Given the creation of a nuclear poloidal field by this galactic version of a battery, the molecular ring, which marks the end of the co-rotation zone, might also "feed" the field outward into the differentially rotating disc. In this way, over longer times, an α-Ω dynamo might be able to propagate and amplify the galaxy's disc field, as discussed in the preceding sections. An implication of this model is that it might enable a galaxy to generate, and propagate its own, large scale field.

6.2.6 Magnetic amplification within outflow winds due to strongly shearing flows

One mechanism by which outflow winds could tap into the substantial kinetic energy of starburst outflows is by amplifying the field in Kelvin–Helmholtz instabilities in the shearing of the organised outflows. The twisting and stretching in the outflow boundary layers can produce a substantial $\nabla \times (v \times B)$ term in the induction equation (e.g. Clarke 1993, Frank *et al.* 1996, Birk *et al.* 2000). An important point is that starburst outflow winds will contain not just the hot gas and CRs produced by concentrations of hot stars and supernovae: They might also entrain significant quantities of ambient interstellar gas, including neutral gas. A realistic calculation of B-regeneration in shear instabilities of outflow winds must consequently account for the fact that the gas is only *partially* ionised.

The basic equations describing the dynamics of a partly ionised, dissipation-free plasma must be modified to incorporate conservation of both mass and momentum for the neutral *and* ionised components. Extending the discussion in Section 3.2, they become

$$\frac{\partial \rho}{\partial t} = -\nabla \cdot (v\rho) \tag{6.4}$$

$$\frac{\partial \rho_n}{\partial t} = -\nabla \cdot (v_n \rho_n) \tag{6.5}$$

$$\frac{\partial}{\partial t}(\rho v) = -\nabla \cdot (\rho v v) - \nabla p + \frac{1}{4\pi}(\nabla \times B) \times B - \rho v(v - v_n) \tag{6.6}$$

$$\frac{\partial}{\partial t}(\rho_n v_n) = -\nabla \cdot (\rho_n v_n v_n) - \nabla p_n + \rho v(v - v_n) \tag{6.7}$$

(Birk *et al.* 2000), where ρ, v, p, and B denote the density, velocity, thermal pressure and magnetic field. The subscript n refers to the neutral gas component, and v is the neutral-ionised gas collision rate. The terms $\rho v(v - v_n)$ specify the rate of momentum transfer between the neutral and ionised components.

Numerical simulations of starburst outflow winds that relax the restriction of compressibility show that, for realistic parameters, Kelvin–Helmholtz vortex features develop after about 100 Alfvénic transit times, and that they become non-linear after about 250 transit times. While some of the kinetic energy is converted into compressional heating, a large fraction is converted into magnetic energy, where computed B-field levels are comparable to those observed within time frames of a few $\times 10^7$ years. The field growth rates and k-scale distributions seem relatively insensitive to the neutral-ionised gas ratio in the outflow winds. These types of simulation are encouraging, in that they could be integral to understanding the rapid regeneration of strong magnetic fields that are measured in starburst-driven galaxy outflows (Figs. 6.2, 6.3). Since this phenomenon was probably universal, and took place in the early phases of galaxy evolution, it may have been a significant factor in early universe magnetic seeding of the intergalactic medium by starburst outflows.

6.3   IGM seeding due to "conventional" stellar processes in galaxies

The scenario in which the first stars and galaxies formed is still incomplete, though steadily improving. More clearly understood is the density and distribution of galaxies in the current-epoch Universe at lower redshifts. Proceeding from $z = 0$ to $z \approx 5$, that of distant quasars and imaged galaxies, evidence for the merging history of galaxies is becoming clearer (e.g. Lavery *et al.* 1996), as is the global history of star formation, illustrated in Fig. 12.7. For the latter, the specific density of the star formation rate peaks at $z \approx 5$ to 6, at a level of $\gtrsim 0.1$ M$_\odot$ year^{-1}Mpc^{-3} (Springel & Hernqvist 2003). The specific co-moving star formation density has been roughly estimated out to $z \approx 20$. Also, stellar population chronology of globular clusters has established the approximate epoch range at which the earliest stellar systems formed (*ca.* 12–13 $\times$ 10^9 years).

Complementing this information, the timescales of outflow halos of nearby starburst galaxies and their associated gas and magnetic fields have recently been measured for well-studied objects near $z = 0$. For starburst-driven galaxy outflows it is evident from magnetic field strengths in starburst outflow halos that they can fill a substantial volume close to the host galaxy with hot gas, CRs and magnetic fields up to 10 µG and more – all in a few times 10^7 years. Vigorous outflow halos are seen to emanate from star-forming dwarf galaxies at the current epoch. This is significant in the sense that dwarf galaxies, though more difficult to image at large distances, are the most numerous class of galaxy in the Universe. The results of several volume-limited dwarf galaxy surveys since 1992, summarised by e.g. Kronberg *et al.* (1999), lead to global estimates for a smoothed-out density at $z \simeq 0$ of $\rho_0^G \approx 0.5$ Mpc^{-3} for *all* galaxies brighter than $M_{\mathrm{B}} \approx -18$. This number density is probably conservative, given that the data are not complete at the faint end of the luminosity scale.

It is remarkable that observed outflow volumes can be infused out to ~10 kpc radius with magnetic fields up to 10 μG or more on a timescale of 10^8 years or less. This represents a magnetic energy of $\gtrsim 10^{56}$ ergs in a field structure that, as we saw in Chapter 5, has some degree of large scale coherence. Starburst wind-generated magnetic fields have more recently emerged as a common phenomenon (Chapter 5). This also provides input to simulations of IGM magnetisation that we discuss below.

A first order quantitative calculation of the IGM magnetic field-filling effect due to star formation can be carried out as follows: We parameterise the intergalactic volume at a given cosmological epoch as the sum of contiguous, touching spheres, each of which is "proprietary" to one galaxy. In a simple toy model, when two or more galaxies merge at some point, so do their respective "proprietary" spheres. As the sketch in Fig. 6.6 illustrates, each sphere, being an elemental volume of all intergalactic space, represents a volume available (V_{A}) to be filled with magnetised outflow gas. After a time, $\Delta\tau$ from some "beginning", a fraction $\Delta f \cdot V_{\mathrm{A}} = \Delta V_{\mathrm{F}}$ becomes filled with magnetised gas.

Just as ρ_0^G is directly based on observations, we parameterise V_{F} as closely as possible to quantities that could actually be observed in the nearby universe: Thus, one can write

$$\Delta V_{\mathrm{F}} = \sigma \cdot v_{\mathrm{outf}} \cdot \Delta\tau \tag{6.8}$$

where v_{outf} is the average outflow velocity, observable from optical spectral lines, σ is the effective galactic chimney cross section, and $\Delta\tau$ the duration of an episode of starburst-driven outflow. A combination of starburst models, infrared data and supernova remnant counts leads to $\Delta\tau$ values ranging from a few $\times 10^6$ year to a few $\times 10^7$ year (see an important related article by Heckman et al. 1990).

Here and elsewhere in the book, the term "galaxy zones" of LSS is used to denote the intergalactic volume occupied by galaxy clusters, groups and field galaxies, as distinct from the giant voids. For the purposes of asking what fraction of the IGM gets "filled" with a magnetised starburst driven outflow material, the relevant total "available" V_{A} is that of the galaxy zones that, in turn, are a fraction $\eta(z)$ of all space at that redshift. The z-dependence of

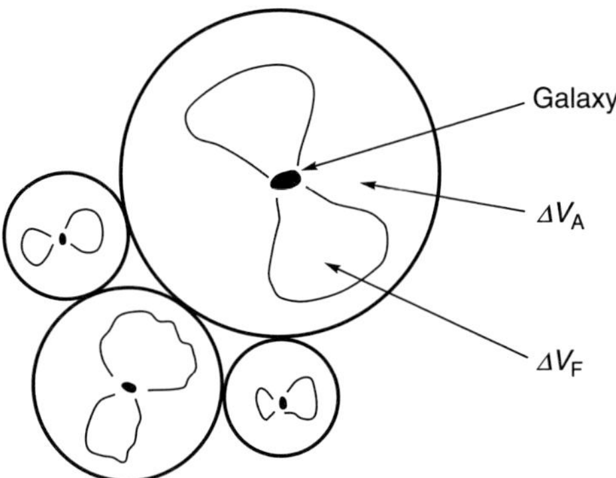

Figure 6.6. A sketch illustrating the contiguous spheres model of Kronberg, Lesch & Hopp (1999), where a galaxy exists at the centre of each proprietary sphere, characterised by V_F and V_A at any given time.

η allows for the gradual evolution of the galaxy zone/void ratio, as large scale structure develops over proper time. The galaxy merging rate can be parameterised by the exponent m up to some modest z, such that

$$\rho^c(z) = \rho^c(0) \cdot (1+z)^m,$$
(6.9)

where $\rho^c(z)$ is the co-moving galaxy density. Estimates of the global merging index m out to $z \approx 2$ range from about 1.5–3 (cf. Lavery *et al.* 1996 and references therein). Figure 6.7 shows simple model calculations of the cumulative increase of the fraction $f(z)$ of the galaxy zones (LSS filaments) of intergalactic space that has been filled, on the simple assumption that all galaxies began their evolution at $z = 10$.

$$f(z) = \frac{\sum \Delta V_F(z)}{\sum V_A(z)}.$$
(6.10)

This illustrative set of $f(z)$ curves in Fig. 6.7 assumes that all galaxies formed in a hierarchical merging scenario beginning with dwarf galaxies, which had formed just prior to $z = 10$. In one model, the aggregate burst time, τ, over a Hubble time is <250 Myr, and it is distributed with $\Delta\tau = 8$ Myr at intervals of $\Delta z = 0.25$, beginning at $z = 10$ and ceasing at $z = 7.5$. Another calculation assumes burst intervals at $\Delta\tau = 3$ Myr, but their frequency decreases to one episode per $\Delta z = 0.5$ after $z = 0.7$. The large effect of varying m (purely a convenient empirical parameter and poorly known for $z > 2$) is also illustrated. In all variations of this simple model, note that $f(z)$ saturates by $z \approx 7$, i.e. starburst outflow winds cease to be effective in filling more V_A after that time. But they can be very effective towards $z = 10$. In all cases shown, a significant fraction of the IGM gets filled with magnetised plasma after $z \approx 7$. The above calculations do not incorporate subsequent amplification, $|\partial B/\partial t|$, which depends upon the subsequent evolution of large scale structure and galaxies, and intergalactic magneto-plasma effects.

Even without some IGM "post-amplification" (Chapter 12, Section 12.7), the galaxy zones of the universe (within $\eta(z)$) will be volume-filled with a non-trivial magnetic field,

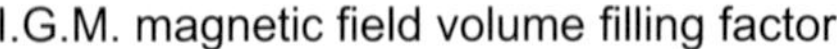

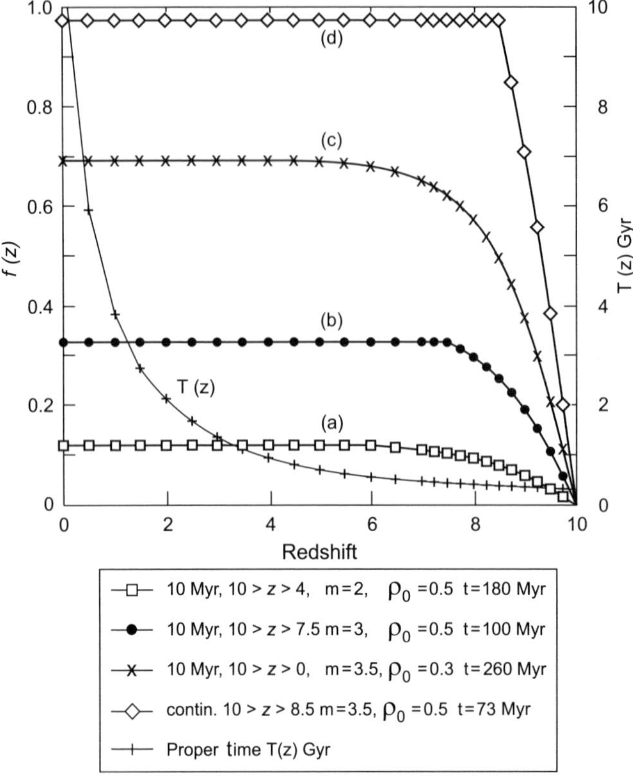

Figure 6.7. Growth of the magnetic field filling factor for four different scenarios, illustrating the effect of different outflow time distributions in redshift and merging indices, m. In all curves, η (assumed to be z-independent) = 0.2, ρ_0 = 0.5 galaxies Mpc^{-3}, σ = 1.5 kpc^2, and v_{outf} = 500 kms^{-1}. The increase of $T(z)$ (right scale) is also shown, adopting the cosmological parameters H_0 = 75, Ω_{M} = 0.3, and Ω_{Λ} = 0.7 in a flat universe model. (Source: Kronberg, Lesch, & Hopp (1999).)

~10^{-9} G. Subsequent IGM amplification might be expected to increase $<B_{\mathrm{IG}}>$ further. This possibility appears consistent with some simulations, and with initial estimates of $<B_{\mathrm{IG}}>$ within some nearby galaxy filament zones.

Since galaxy clusters are largely gravitationally bound, most of the magnetic flux seeded into the IGM since the clusters' formation probably remains in the cluster – a detached sub-Universe for our purposes. One consequence is that clusters, once formed, may not be very effective in spreading their intracluster medium (ICM) magnetic fields beyond their bounds into the wider intergalactic medium. In the same connection we note that that the aggregate volume of galaxy clusters is only a small fraction of the total intergalactic volume.

In summary, a significant fraction of all intergalactic space may have been seeded within the first ≈1 Gyr by starbursting dwarf galaxies, as the simple models in Fig. 6.7 suggest. These were the precursors of the large numbers of now gas-poor dwarf galaxies seen in the present universe and the "building blocks" of present day large galaxies, assuming hierarchical merging. As intergalactic matter was drawn into clusters, the ICM would likewise have been seeded from the same intergalactic magnetic field.

ICM and IGM magnetisation is also provided by jets and lobes and energised by the central black holes of more massive galaxies; this is discussed further in Chapters 7–9. The large black holes feed very large amounts of magnetic and particle energy ($\gtrsim 10^{61-62}$ ergs *per* supermassive BH) into the intergalactic medium (Fig. 8.1). Their magnetic energy input will *add* to the galaxies' star/SN stellar wind. In fact, IGM magnetisation by BH-energised jets and lobes could be the primary source of intergalactic fields in LSS filaments, and of at least a significant fraction of the B_{IGM}.

At sufficiently early times, the ambient photon energy density ($\propto (1 + z)^4$) will exert increasing counter-pressure to the ejection and outward diffusion of magnetised gas. These conditions become more "extreme" as we proceed toward epochs beyond $z \approx 10$. Partially compensating the higher ambient pressure at very high z is the fact that starburst-driven outflow winds may include an unresolved "forest" of very high velocity "galactic spicules" having υ of several thousand kms^{-1}, each with a smaller cross-section, σ, than the entire outflow wind. Hints of this can be seen in the outflow structure of M82 (Fig. 6.2(a). In this refinement, the winds would more effectively penetrate the ambient environment than would a homogeneous wind.

As z approaches ~10 where average inter-galaxy spacings $<|r_{ij}|>$ are small, galaxy–galaxy tidal interactions will also play a role. They are too complex to include in the simple model outlined here and are best determined through computer simulations. One would expect that tidal removal of interstellar matter, which requires detailed simulation, would be more prevalent at these early epochs, and would tend to further enhance the star/supernovae-driven outflows' ability to spread magnetised (and metal-enriched) gas into the IGM. One observational check on these effects might be the excitation state of the primeval gas, which is becoming more accessible to observation. Tidally dispersed material may contain more cool gas than the direct SN-driven outflow material, but to confirm this needs detailed observations and modelling.

Several factors could modify these calculations and there is considerable room for refinement through more detailed simulations and better observational data. These include better observations of individual galaxies at large z, as well as improvement of our present incomplete knowledge of the beginnings of gravitational collapse, star formation, tidal interactions among primeval galaxies, and the first stellar luminosity functions (when the metallicity was still low). Computational models along similar lines have since been developed by Bertone *et al.* (2006), resulting in similarly high volume filling factors, and present-epoch IGM fields $10^{-12} \lesssim B \lesssim 10^{-8}$G. Simulations of this sort have a promising future, as the physical details are better clarified in future, particularly at redshifts extending into the primeval galaxy era. Examples of such observational detail are spectroscopy of the IGM, including Lyα forest and related data, high z Faraday *RM* probes, sub-mm imaging and spectroscopy of high-z galaxies, and the merging histories in cosmological LSS evolution.

As discussed earlier the strength of magnetic fields in the halos of low-z dwarf galaxies is found to be typically ~10–15 μG. It is additionally possible, beginning with $f(z = 0)$ to make a "baseline" estimate of the IGM magnetic field strength in galaxy filaments that is solely due to stellar outflows. If we assume a simple adiabatic expansion, no retention of the outflow gas by the host galaxies, and no B-dissipation, we obtain $<B>_{IGM} \sim 10^{-9}$ G at $z = 0$.

This value does not include the IGM magnetic energy that is deposited by galactic black holes – a topic is discussed separately in Chapters 7 and 8. Nor does it include post-outflow amplification in the IGM. On the other hand, the assumption of no halo gas retention is overly optimistic.

Other processes will work to enhance the magnetic energy of the stellar/SN-driven outflow over $10 > z > 0$, during which matter continues to fall into large scale intergalactic filaments, galaxies, galaxy groups and galaxy clusters. The infall velocities and accompanying shearing motions allow for conversion of some of this large scale (gravitational) infall energy into magnetic energy, enhancing that which has already been seeded. This large scale intergalactic gravitational-to-magnetic energy conversion was first investigated by Ryu, Kang, & Biermann (1998). An instructive treatment of basic relevant MHD calculations for this general phenomenon is contained in Frank *et al.* (1996) and references therein. A more recent refinement of such simulations can be found in e.g. Ryu *et al.* (2008). Here, the seed intergalactic magnetic field is proposed to be amplified in turbulent, shearing eddies of infalling intergalactic gas streams (see Section 12.7). The calculated field strengths, which are scaled by many parameter assumptions, are claimed to approach 10^{-8}–10^{-6} G within and near filaments at $z \approx 0$.

Somewhat analogous processes appear to happen within large galaxy clusters (Chapter 9). The *intracluster* medium, can be considered a "sub-universe" of the wider IGM. Here, the ICM becomes infused over a cluster lifetime with starburst-driven outflows of the cluster member galaxies. Völk & Atoyan (2000) were the first to investigate this phenomenon, and calculated the contribution of galactic outflow winds to the magnetisation of the IGM *within* galaxy clusters, based on observed galactic wind outflow properties. They arrived at similarly high magnetic field filling factors for the ICM within clusters. Extensive modelling based on more recent detailed simulations and input data can be found in Donnert *et al.* (2009). As before, these ICM magnetisation scenarios are in addition to those caused by SMBH-driven jets and lobes discussed in Chapters 7 and 8. In the case of cluster-internal BH-powered *lobes*, the story is somewhat different, as discussed in Chapter 9.

It will be evident from the previous discussion that extrapolations and scalings between ICM and IGM magnetic fields have limited usefulness. Cluster cores have, over cosmic time, become gravitationally detached (partially or completely) from the wider IGM. When this happens, the ICM magnetic fields are largely determined by intra-cluster dynamics and outflows, and these are not trivially scalable to the IGM. Unlike dwarf field galaxies, the ICM supernova-driven magnetic fields from cluster members do not normally get ejected outside a gravitationally bound cluster. Galaxy clusters act, to first order, as "sinks" to magnetised infall material from the IGM. These considerations form a brief prelude to Chapter 9.

References

Alfvén, H. & Fälthammar, C.-G. 1963, *Cosmical Electrodynamics* (Oxford: Clarendon)
Beck, R., Loiseau, N., Hummel, E., Berkhuijsen, E. M., Grave, R., & Wielebinski, R.1989, High-Resolution Polarization Observations of M31. I – Structure of the Magnetic Field in the Southwestern Arm, *Astron. Astrophys.*, 222, 58
Beck, R. 2010, Galactic and Extragalactic Magnetic Fields, arXiv0810:2923v4 [astro-ph]
Bertone, S., Vogt, C., & Enßlin, T. E. 2006, Magnetic Field Seeding by Galactic Winds, *MNRAS*, 370, 319
Birk, G. T., Wiechen, H., Lesch, H., & Kronberg, P. P. 2000, The Role of Kelvin-Helmholtz Modes in Superwinds of Primeval Galaxies for the Magnetization of the Intergalactic Medium, *Astron. Astrophys.*, 353, 108

Carilli, C. L., Holdaway, M. A., Ho, P. T. P., & de Pree, C. G. 1992, Discovery of a Synchrotron-Emitting Halo Around NGC 253, *Astrophys. J.*, 399, 59

Chyży, K., Beck, R., Kohle, S., Klein, U., & Urbanik, M. 2000, Regular Magnetic Fields in the Dwarf Irregular Galaxy NGC 4449, *Astron. Astrophys.*, 355, 128

Chyży, K. T., Soida, M., Bomans, D. J., Vollmer, B., Balkowski, Ch., Beck, R., & Urbanik, M. 2006, Large-Scale Magnetized Outflows from the Virgo Cluster Spiral NGC 4569: A Galactic Wind in a Ram Pressure Wind, *Astron. Astrophys.*, 447, 465

Clarke, D. A. 1993, 3-D MHD Simulations of Extragalactic Jets, in *Jets in Extragalactic Radio Sources*, ed. H. J. Röser & K. Meisenheimer, *Lecture Notes in Physics* (Berlin: Springer-Verlag), 421, 243

Donnert, J., Dolag, K., Lesch, H., & Müller, E. 2009, Cluster Magnetic Fields from Galactic Outflows, *MNRAS*, 392, 1008

Frank, A., Jones, T. W., Ryu, D., & Gaalaas, J. B. 1996, The Magnetohydrodynamic Kelvin-Helmholtz Instability: A Two-Dimensional Numerical Study, *Astrophys. J.*, 460, 777

Heckman, T. M., Armut, L., & Miley, G. K. 1990, On the Nature and Implications of Starburst-Driven Superwinds *Astrophys. J. Suppl.*, 74, 833

Kepley, A. A., Mühle, S., Everett, J., Zweibel, E. G., Wilcots, E. M., & Klein, U. 2010, The Role of the Magnetic Field in the Interstellar Medium of the Post-Starburst Dwarf Irregular Galaxy NGC 1569, *Astrophys. J.*, 712, 536

Krause, F. 1987, Dynamo Excitation in Very Large Scales, in *Interstellar Magnetic Fields: Observation and Theory: Proceedings of a Workshop held at Schloss Ringberg, Tegernsee September 8–12, 1986*, ed. R. Beck & R. Gräve (Berlin: Springer-Verlag), 8

Krause, M. 2009, Magnetic Fields and Star Formation in Spiral Galaxies, in *Magnetic Fields in the Universe II: From Laboratory and Stars to the Primordial Universe, Cozumel, Mexico, January 28th to February 1st, 2008*, ed. A. Esquivel et al., Rev. Mex. Ast. Astrof., Ser. de Conf. (Mexico, D.F.: Instituto de Astronomia, Universidad Nacional Autonoma de Mexico), 36, 25

Kronberg, P. P. 1994, Extragalactic Magnetic Fields, *Rep. Prog. Phys.*, 57, 325

Kronberg, P. P., Biermann, P., & Schwab, F. R. 1985, The Nucleus of M82 at Radio and X-Ray Bands – Discovery of a New Radio Population of Supernova Candidates, *Astrophys. J.*, 291, 693

Kronberg, P. P., Lesch, H., & Hopp, U. 1999, Magnetisation of the Intergalactic Medium by Primeval Galaxies, *Astrophys. J.*, 511, 56

Kronberg, P. P. & Wilkinson, P. N. 1975, High-Resolution, Multifrequency Radio Observations of M82, *Astrophys. J.*, 200, 430

Lesch, H., Crusius, A., Schlickeiser, R., & Wielebinski, R. 1989, Ring Currents and Poloidal Magnetic Fields in Nuclear Regions of Galaxies, *Astron. Astrophys.*, 217, 99

Lo, K. Y., Cheung, K. W., Masson, C. R., Phillips, T. G., & Woody, D. P. 1985, Aperture synthesis observations of CO emission from M82, in *ESO-IRAM-Onsala Workshop on (Sub)Millimeter Astronomy, Aspenas, Sweden, June 17–20, 1985, Proceedings (A86-39226 18–90)*, ed. P.A. Shaver & K. Kjär, ESO Conf. Proc. (Garching: European Southern Observatory), 22, 197

Loiseau, N., Reuter, H.-P., Wielebinski, R., & Klein, U. 1988, A C-13O(2-1) Map of M82, *Astron. Astrophys.*, 200, 1

McCarthy, P. J., van Breugel, W., & Heckman, T. 1987, Evidence for Large-Scale Winds from Starburst Galaxies. I – The Nature of the Ionized Gas in M82 and NGC 253, *Astron. J.*, 93, 264

Nakai. N., Hayashi, M., Handa, T., Sofue, Y., Hasagawa, T., & Sasaki, M. 1987, A Nuclear Molecular Ring and Gas Outflow in the Galaxy M82, *Publ. Astr. Soc. Japan*, 39, 685

Parker, E.N. 1976, Ring Currents and Poloidal Magnetic Fields in Nuclear Regions of Galaxies, *Astron. Astrophys.*, 217, 99

Parker, E. N. 1979, *Cosmical Magnetic Fields* (Oxford: Clarendon)

Parker, E. N. 1992, Fast Dynamos, Cosmic Rays, and the Galactic Magnetic Field, *Astrophys. J.*, 401, 137

Reuter, H. P, Klein, U., Lesch, H., Wielebinski, R., & Kronberg, P. P. 1992, Gaps and Filaments in the Synchrotron Halo of M82 – Evidence for Poloidal Magnetic Fields, *Astron. Astrophys.*, 256, 10

Reuter, H. P., Klein, U., Lesch, H., Wielebinski, R., & Kronberg, P. P. 1994, The Magnetic Field in the Halo of M 82. Polarized Radio Emission at Lambda-Lambda(6.2) and 3.6 CM, *Astron. Astrophys.*, 282, 724

Ryu, D., Kang, H. S., & Biermann, P. L. 1998, Cosmic Magnetic Fields in Large Scale Filaments and Sheets, *Astrophys. J.*, 335, 19

Ryu, D., Kang, H., Cho, J., & Das, S. 2008, Turbulence and Magnetic Fields in the Large-Scale Structure of the Universe, *Science*, 320, 909

Schaaf, R., Pietsch, W., Biermann, P. L., Kronberg, P. P., & Schmutzler, T. 1989, X-Ray Observation of the Starburst Galaxy M82, *Astrophys. J.*, 336, 722

Seaquist, E. R. & Odegard, N. 1991, A Nonthermal Radio Halo Surrounding M82, *Astrophys. J.*, 369, 320

Völk, H. J. & Atoyan, A. M. 2000, Early Starbursts and Magnetic Field Generation in Galaxy Clusters, *Astrophys. J.*, 541, 88

Weliachew, L., Fomalont, E. B., & Greisen, E. W. 1984, Radio Observations of H I and OH in the Center of the Galaxy M 82, *Astron. Astrophys.*, 137, 335

7

Extragalactic jets and lobes – I

7.1 How much energy and from where?

7.1.1 Some background and earlier history

The supra-galaxy scale morphology of double radio sources was observationally established during the 1950s. G. R. Burbidge (1956) called attention to the large energy content of the radio emitting lobes outside of the parent galaxy. Because central black holes of 10^6–10^9 M$_\odot$ had not then been discovered, and realising that the thermonuclear energy of all host galaxies' stars was not sufficient to explain the 10^{60+} erg energy release, Burbidge originally considered $e^+ e^-$ annihilation as a possible energy source.

A remarkable fact of such systems is the conversion of energy into a form that is highly collimated, along with the transfer of energy to $\approx 10^{11}$ times larger scales – AU to megaparsecs! In this chain of processes, gravitational energy is converted into magnetic fields and high energy particles and photons – gravity is required as the primary energy source because no other known source can explain the enormous energy released (see below and Chapter 8). Relativistic cosmic ray (CR) particles are also accelerated in the process, and these "illuminate" the magnetic fields via synchrotron and inverse Compton (IC) emission in the jets and lobes. The acceleration of CR nuclei and leptons, which also involves magnetic fields, is a closely related topic.

In this and the following chapter we explore the connection between supermassive central black holes of galaxies, and galactic and intergalactic magnetic fields. Our ultimate understanding of all the physics involved in the sequence of processes will greatly advance our understanding of the high energy universe. Radio and X-ray images suggest that a large fraction of this converted gravitational energy is in magnetic form. Gravitational infall energy at the central black hole is converted by a remarkably efficient process, not yet fully understood, into magnetic fields and CR particles. Many of these are eventually dispersed into intergalactic space.

The global properties of radio galaxies suggest that, empirically, their power and energy content arise from the energy output at the galaxy nucleus. This has been demonstrated in several studies of radio galaxies; by Daly & Loeb (1990), Falke, Malkan, & Biermann (1995), Best *et al.* (1999), and Kronberg *et al.* (2001). Falke *et al.* noted a correlation between jet power and the accretion disc luminosity as revealed by the UV flux in low-luminosity AGNs. Best *et al.* found that the ratio of the radio "core" luminosity to that in the extended radio lobes is nearly constant, at least in a statistical sense, and that it only weakly depends on the overall radio power. It was found to apply over a large range of radio source sizes. Kronberg *et al.* analysed the energetics of very large radio sources to provide a calibration of the energy release by the galactic central black holes. All of these studies point to an energy flow path that closely couples the galaxy nucleus with the jets and luminous lobes.

7.1.2   *What creates the collimated high energy flows?*

The above analyses appear difficult to reconcile with a longstanding paradigm that envisaged radio source lobes being energised at a "working surface", or hotspot, at the lobe-IGM interface. In this paradigm the radio luminosity would be co-determined by the (independent) properties of the immediate external medium. The subtly different message from the analyses above and from studies of giant radio galaxies (e.g. Willis & Strom 1978, Strom & Willis 1980, Kronberg *et al.* 2001, 2004) is that, in a global sense, the main *energy flow* proceeds *directly* from the massive galaxy nucleus via a jet into the outer lobes, and not principally via a secondary acceleration site at an outer hotspot. This newer insight is connected with another question discussed below, namely how magnetic flux and energy get transported from the central black hole into intergalactic space. This process is an approximate analogue of the outflow processes for star-driven jets (e.g. Section 7.9). These phenomena of high energy jets and lobes lead to a different question, namely how are CRs accelerated to relativistic energies in this process? Particle acceleration seems to occur on many scales, from the BH accretion disk to galactic, and possibly even extragalactic scales.

In early important attempts to explain energy transport into jets and radio lobes (e.g. De Young 1971), attention was focused on a "working surface" at the lobe-IGM interface. The mechanisms involve hydrodynamic shearing, turbulence and supersonic shocks. Variations of these concepts were subsequently developed, partly guided by increasingly precise radio images, thanks in large measure to the Westerbork Synthesis Radio telescope (WSRT) and the NRAO Very Large Array (VLA). Models of particle acceleration involved Fermi acceleration (Fermi 1949) and cascades of magnetic turbulence often connected with Kelvin–Helmholtz instabilities at the IGM interface, and within the jets and lobes. An important constraint on many of these models is that the acceleration mechanism has to be very efficient, much more than $\mathcal{O}$ 1%. A prerequisite for efficient Fermi acceleration is that the strong shocks be supersonic. However, increasingly precise images show that only a few radio lobes show firm evidence for a supersonic shock at the lobe/IGM interface. Interestingly, the frequent absence of an outer peripheral shock applies not only to extragalactic radio lobes discussed here, but also to the interface between the expanding Crab Nebula and the interstellar medium discussed in Chapter 2. On the other hand, outer shock structures *are* evident in other well-imaged shell-type supernova remnants (SNR) such as Cassiopeia A, Tycho, 3C10, and other SNRs in the Milky Way. The Crab Nebula, is classed as a "plerion" type SNR, which is continuously powered by a central energy source, typically a pulsar. Plerions are not "edge-brightened" like the shell-type SNR. Both classes contain strong magnetic fields, up to as high as milligauss levels.

The observed characteristics of extragalactic jets have gradually become clearer with time (e.g. Bridle *et al.* 1994). They have been proposed as collimated beams of high energy particles (hadrons and/or an e^+e^- beam) originating from the nucleus, and which feed and inflate the lobes. An early advance in attempting to explain the remarkable collimation of jets used a "Laval nozzle" (Blandford & Rees 1974). Further developments of particle beam-driven jets have occurred, with improved observational description and computer simulation of both jets and lobes and their interconnection. 3-D MHD simulations have explored a variety of jet Mach numbers, in both uniform and stratified ambient media. The paper by O'Neill *et al.* (2005) is recommended to the interested reader, in part because it summarises important prior work on the foregoing and related categories of jet/lobe modelling. The reader could also consult the book by DeYoung, *The Physics of Extragalactic Radio Sources* (2002).

Most particle beams whose energy flow consists of particles are inherently susceptible to instabilities at some point, in particular Kelvin–Helmholtz instabilities at the interface to the ambient confining plasma. Added to this is the inevitable *heterogeneity* of the ambient pressure, especially in a dense intracluster medium. These factors usually make it difficult to realistically simulate the long and narrow, "undisturbed" beams that some observations increasingly show.

7.2 Jets as electromagnetically driven systems

Energy outflow models involving coherent, electromagnetic Poynting flux have been explored and discussed since the early 1970s by Lovelace (1976), Blandford & Znajek (1977), Benford (1978, 2006), Appl & Camenzind (1992), Lucek & Bell (1996), Lynden-Bell & Bolly (1994), Lynden-Bell (1996), Lesch *et al.* (1989), Lyubarskii (1999), Nakamura & Meier (2004), Li *et al.* (2006), and others.

Maxwell's equations proscribe that an electric current is inherently coupled to the magnetic field configuration of a Poynting jet. In this class of model, the energy flow, observationally derivable from the observed energetics and lifetimes of the jet/lobe systems, can be associated with a putative electric current, so far difficult to measure and which, based purely on a power flow calculation, would be in the range of 10^{17-19} ampères when scaled to the mass of a supermassive black hole and the energy deposited into the IGM.

Following the discovery of pulsars in 1967, but before the discovery of supermassive black holes, Rees (1971) explored the idea of extracting rotational energy from groups of collapsing stars or pulsars as an energy source for extragalactic radio sources. Rees proposed strong low frequency electromagnetic waves that could propagate away large amounts of energy. By virtue of their strength and low frequency they would also be capable of directly accelerating particles to relativistic energies before the wave's slowly oscillating E-field reversed direction. Although the idea was not fully developed, its intent was to naturally provide an efficient and non-stochastic mechanism for accelerating particles in jets to relativistic energies. Such ideas were important predecessors of the concepts of a Poynting flux produced by a rotating supermassive black hole as the electromagnetic energy source for powerful radio sources.

An important related phenomenon to understand is how relativistic particle *acceleration* can happen in the jets and lobes. Jets are an important candidate source of high energy CRs in the Universe, and CR acceleration can occur through a variety of magnetoplasma processes. These can include, for example via Fermi acceleration, via strong, long wavelength EM waves, in shocks generated in entrained matter along the periphery of the jets, or within the jet – or possibly some mixture of these. Inherent in the Poynting flux class of jet models is that direct electrodynamic acceleration can occur in strong coherent electric and magnetic fields. Thermal nuclei and leptons can also become entrained, and somehow tap into the electromagnetic energy flow "pipe".

Different lines of evidence appear to favour Poynting flux-dominated energy flow, as distinct from hadronic, or leptonic e^+e^- flows, etc., to explain energy transport in jets, at least on kiloparsec or intergalactic scales. Parenthetically, we add here that, for stellar jets with much lower power flows and in high density environments, this paradigm might be modified.

Since the early 1990s, knowledge of the physical conditions in some large scale jets has become better established through observations. In addition, computational simulations have advanced considerably. This is due to ever greater supercomputing power, and to better insights into the physics of MHD plasmas. As a result, the evidence is now even more persuasive that ordered magnetic fields and their associated currents are integral to the functioning of extragalactic jets. Their function is twofold: The interaction between magnetic and gas dynamical forces, described by solutions to the MHD equations (Equations 7.1–7.9), can be shown to make a current-dominated jet function as a highly collimated energy flow conduit over distances as large as intergalactic dimensions, *and* to be robust to disruption. Magnetically dominated jets carried by a current are the essential basis for the suite of jet and lobe models that are discussed below.

A return current, required to balance the outgoing axial "beam current", flows along or somewhere outside of the outer magnetic sheath. Rotation of the field lines gives rise to electric fields that are perpendicular to the magnetic surfaces. In a model simulation by Lesch, Appl, & Camenzind (1989), the beam terminates in a hot spot, or "working surface". In their model the outermost discontinuity of the resulting "magnetic tower" forms a bow shock if it moves supersonically relative to the ambient medium.

In supermassive BH-produced Poynting flux dominated jets, the instability problems of some hydrodynamic models largely disappear. Depending on model details of beam equilibria, relativistic electrons can be carried force-free over long distances, thereby avoiding synchrotron losses, in spite of the large jet fields. These can be as high as milligauss level (Benford 1978). Even if a dense plasma sheath surrounds the jet, the axial confinement due to B_ϕ associated with the current can be resilient to disruption by instabilities.

Current/magnetic field-dominated jets can also naturally accelerate CRs at the end of the beam, and possibly along it. For example, at the end of a current-driven beam boring its way into dense ambient plasma, beam-plasma instabilities in some cases might accelerate CR particles to high energies. This possibility also connects to themes discussed in Chapters 10 and 11. Most of the following sections are based on the premise that powerful extragalactic jets, and the lobes they feed, are electromagnetically dominated, or at least begin that way.

7.3   Representative model simulations of radio lobes fed by a Poynting flux jet

7.3.1   *Examples of computational frameworks*
The basic ideal MHD equations, in (x, y, z) coordinates, can be written as follows:

$$\frac{\partial \rho}{\partial t} + \nabla \cdot (\rho v) = \frac{\partial \rho_{\mathrm{inj}}}{\partial t} \tag{7.1}$$

$$\frac{\partial (\rho v)}{\partial t} + \nabla \cdot (\rho v v) + P_{\mathrm{g}} + P_B - BB) = \frac{\partial}{\partial t} \left(p_{\mathrm{inj}} \right). \tag{7.2}$$

The right side of Equation (7.2) represents an injection of additional momentum. This could be zero, depending on the chosen setup of physical conditions.

$$\frac{\partial E}{\partial t} + \nabla \cdot \left[(E + P_g + P_B) v - B(v \cdot B) \right] = \frac{\partial E_{\text{inj}}}{\partial t} - \rho v \cdot \nabla \Psi \qquad (7.3)$$

$$\frac{\partial B}{\partial t} - \nabla \times (v \times \boldsymbol{B}) = \frac{\partial B_{\text{inj}}}{\partial t} \qquad (7.4)$$

(Nakamura *et al.* 2006, whose terminology we have largely followed). The gas pressure, P_g, and the internal energy density, ε, are related by

$$P_g = \varepsilon(\gamma - 1) \qquad (7.5)$$

for an ideal gas having $\gamma = 5/3$. The magnetic energy density, P_B, is written as $B^2/2$, and the total energy is

$$E = \rho v^2 + P_g + P_B. \qquad (7.6)$$

The MHD equations above are, in general, not conservative in that they allow for the external injection of gas, energy, and magnetic flux. The latter needs to be subject to the constraint $\nabla \cdot (\partial B_{\text{inj}}/\partial t) = 0$. In a realistic computational simulation, these injection terms must be specified in a physically plausible and, preferably, computationally convenient form. Examples of a set of injection parameters for B, ρ, and E are

$$\partial B_{\text{inj}}/\partial t = \gamma_b B_{\text{inj}} \qquad (7.7)$$

and

$$\partial E_{\text{inj}}/\partial t = \gamma_b B_{\text{inj}}.B \qquad (7.8)$$

where B and B_{inj} are functions of (x, y, z, t)
and

$$\frac{\partial \rho_{\text{inj}}}{\partial t} = \gamma_\rho \rho_0 \exp \left[-\frac{(r^2 + z^2)}{r_\rho^2} \right] \qquad (7.9)$$

(Li *et al.* 2006).

Here, the coefficient γ_ρ specifies the rate of mass injection, r_ρ is the characteristic radius within which the mass injection occurs, and z is the distance along the jet axis.

The injection of mass deserves extra mention at this point. It corresponds to the insertion, e.g. entrainment, of matter downstream from the central BH/accretion disc. This loading of additional mass onto the magnetic field lines violates ideal MHD conditions. However, it can make the simulation more realistic, since gas entrainment will inevitably occur, beginning with the BH's host galaxy environment, and it may also be associated with an ambient external pressure. Without it, the large energy injection might, given certain model parameters, cause the system to expand prematurely by a large volume factor. The simulations would consequently enter a régime of very low density magnetised regions. This can cause numerical – though not necessarily physical – problems. The precise mathematical form of Equation (7.9) does not have a detailed physical justification but it provides a numerical approximation to a solution to the problem. For simplicity, the injected gas (ρ_{inj}) might be assumed to have the same temperature and velocity as the pre-existing gas in the injection region.

In view of the above, and to maintain sufficient pressure where it otherwise might become negative, it is convenient, following Ryu *et al.* (1993) and Balsara & Spicer (1999a, b), to solve two auxiliary equations. These solve respectively for the internal energy, ε, and a modified entropy, S:

$$\frac{\partial \varepsilon}{\partial t} + \nabla \cdot (\varepsilon v) = -P_{\mathrm{g}} \nabla \cdot v \tag{7.10}$$

and

$$\frac{\partial S}{\partial t} + \nabla \cdot (Sv) = 0, \tag{7.11}$$

where

$$S = \frac{P_{\mathrm{g}}}{\rho^{\gamma - 1}}. \tag{7.12}$$

Continuing the simulation setup, B_{inj} is assumed axisymmetric. The adopted form of the poloidal component of magnetic flux, ψ, is

$$\Psi(r, z) = r^2 e^{-\left(r^2 + z^2\right)}. \tag{7.13}$$

The azimuthal component of the injected magnetic flux is

$$B_{\mathrm{inj},\phi} = \frac{\alpha \Psi}{r}, \text{ or } \alpha r e^{-\left(r^2 + z^2\right)}, \tag{7.14}$$

where α scales with the ratio of the toroidal to poloidal magnetic flux. These are roughly equal when $\alpha \approx 2.6$. If we assume that the magnetic field footprints are coupled to the rapidly rotating central BH accretion disc, B_ϕ will initially dominate, i.e. $\alpha \gg 1$. The associated toroidal and poloidal currents are given by

$$I_z = \frac{2\pi\alpha}{e} \approx 2.3\alpha \tag{7.15}$$

and

$$I_\phi = \iint J_\phi \, dr dz \sqrt{2} = 2\sqrt{\pi}, \tag{7.16}$$

where J_ϕ is the current density ($\nabla \times B$). From the last two equations we see that the ratio of poloidal to toroidal currents, I_z/I_ϕ, is 0.65α, so that I_z will dominate when $\alpha \gg 1$.

7.3.2   *Extensions to classical, non-relativistic MHD simulations*

Maxwell's equations are fully applicable in the relativistic limit. However, the classical MHD equations are often used only to apply Maxwell's equations within the non-relativistic limit, and combined with the dynamical equations of a plasma. When written in fully conservative form, the MHD equations provide for strict conservation of energy, mass and momentum, and they also contain the important constraint of $\nabla \cdot B = 0$. In practice, as a magnetised jet-lobe system expands by several orders of magnitude, the dilution of plasma density in a purely conservative MHD model often results in very low plasma β values at large t and z (axial distance along the jet). This can make it computationally difficult, or impossible, to maintain the $\nabla \cdot B = 0$ condition. The problem can occur due to truncation errors, and may require some *ad hoc* adjustments to the code. The aim of a successful simulation is to make such *ad hoc* adjustments physically realistic. An example was the ρ_{inj}

term (Equation 7.9) in the kpc-scale tower jet simulation of Nakamura *et al.* (2006). The MHD equations expressed in purely conservative form can be found in Camenzind (2005).

7.3.3 Non-relativistic MHD simulations of a "magnetic tower", Poynting flux-dominated jet

Listed below are some typical physical parameters for the simulation of a massive BH-driven, Poynting flux-dominated jet, based on the above-mentioned simulation model of Li *et al.* (2006):

- Ambient density $n_0 = 3 \times 10^{-3}$ cm^{-3}
- Ambient temperature = 7 keV
- Sound speed, $c_s \approx 10^8$ cm s^{-1} ($\approx 0.3\%$ of c)
- System scale = 15 kpc
- Initial magnetic field strength $B_0 \approx 20$ μG
- Simulation run time ~9.2×10^7 year
- Total energy released ~6.2×10^{59} ergs
- Poloidal flux ~ 9.4×10^{40} G cm^2
- $I_z \sim 1.7\alpha \times 10^{18}$ ampères, where α is in the range ~1–20, depending on the simulation run.

Figure 7.1 illustrates a computed magnetic field structure produced in a sample simulation using numbers similar to those listed above. The outgoing magnetic field lines proceed away

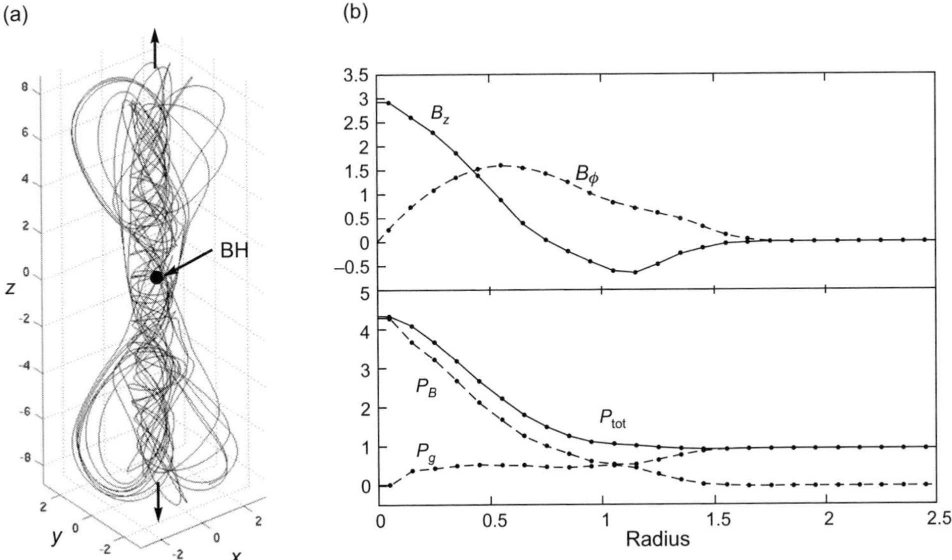

Figure 7.1 (a) Left – Simulated magnetic field structure of a magnetic tower jet after $\approx 10^8$ year from the initial injection, in which the magnetic energy dominates initially. The central BH is located at (0,0,0). (b) – right (upper). The B_z and B_ϕ field components as a function of radius $|r| = |x,y|$ from the jet axis at $z = 3$, ($\Rightarrow t = 9.2 \times 10^7$ year). Right (lower) The total pressure P_{tot}, magnetic pressure, P_B, and plasma pressure, P_g for the same time (t) and location (z) as above. Further details can be found in Li *et al.* (2006). The ordinates are in arbitrary units.

from the BH source (at $z = 0$, $r = 0$) in a collimated column ("tower") in a tightly wound central helix, which corresponds to the outgoing current. The return current corresponds to the more loosely wound inward and downward propagating helical magnetic field. Given the parameters of this model, the outgoing electrical current is concentrated to the axis, and the outer, return current is more spatially diffuse. In Chapter 8, we describe an attempt to observationally measure the axial current in a kpc-scale jet.

To produce a more complete and realistic simulation, it is important to be aware of what conditions, or components, of the Poynting flux simulation model need to be modified or added. One of these is variation in the external ambient pressure. This was assumed constant in the above sample simulation. It can sometimes be estimated from independent observations. An example is the gas pressure profile in the potential well of the host galaxy, or galaxy cluster. Especially in the latter case, X-ray emitting hot gas environments can be measured, and then used as a "test environment" to investigate different models of jet–lobe systems. In particular, the radial (z) dependence (Fig. 7.1) of the ambient pressure and temperature in the galaxy, or galaxy cluster can be used to test the internal physics, such as the magnetic field conditions in the lobes. A few of these refinements and complications are discussed below.

The MHD magnetic tower simulation shown in Fig. 7.1 does not incorporate relativistic particle acceleration around and within the magnetic helix, nor does it include energy transfer to particles via relativistic shocks around the jet periphery, or ahead of the jet. Jet-internal shocks have been observationally identified, for example, in the well-imaged kpc-scale jet of the nearby radio galaxy M87 (Marshall *et al.* 2002).

The above discussion serves to illustrate some of the additional physics that must be incorporated into a realistic, *complete* 3-D simulation of an astrophysical jet system. Comprehensive simulations need to include relativistic shocks, particle acceleration, and other dynamical interactions. These require even more computing power and more detailed magneto-plasma physics. At this time of writing there is considerable scope for incorporating some of these more complex "model components".

In the current-dominated (I_z) jet simulation shown below, the jet often expands at some identifiable point to form a lobe. This jet-lobe transition point occurs at approximately the core radius of the host galaxy or galaxy halo. The return current corresponds to the wider helical structure that has the opposite pitch (Fig. 7.1).

In the expanded lobe of radius x (Fig. 7.2), the azimuthal field component, B_ϕ, and current density, J (as distinct from the total current), are correspondingly smaller in the lobe than in the jet. In the inner lobe, where ram pressure effects are unimportant, there is approximate pressure balance between the internal magnetic field and the ambient pressure, P_a. In these current-dominated MHD simulations, the former is the dominant internal pressure component (Nakamura *et al.* 2006, 2008), that is,

$$B_\phi^2 \geq P_a, \tag{7.17}$$

and these quantities are related to the axial lobe ($\approx$jet) current, I_z by

$$B_\phi^2 \sim \left(\frac{I_z}{r_{\text{lobe}}}\right)^2. \tag{7.18}$$

Typical numbers for the simulation shown in Fig. 7.2 are: current density $J_z \approx 4.5 \times 10^{-24}$ $A\,\text{cm}^{-2}$ corresponding to, $I_z \approx 5 \times 10^{17} A$, $P_a \approx 2.5 \times 10^{-11}$ dynes cm^{-2}, and $r_{\text{lobe}} \approx 156$ kpc.

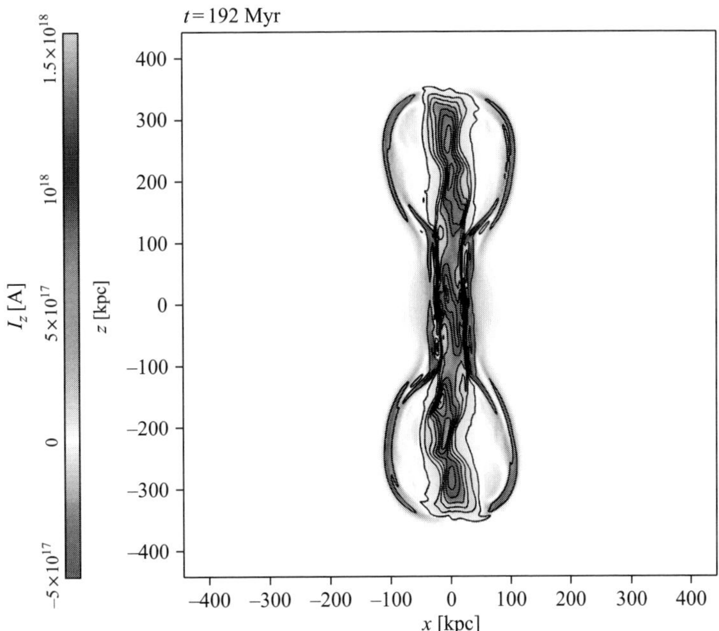

Figure 7.2 The distribution of current in a magnetic tower simulation in which the jet in Fig. 7.1 proceeds further and forms a radio lobe. Real physical dimensions and electric current are shown. The return current is shown in darker grey, and is associated with the outer helix in Fig. 7.1. This simulation differs slightly from Fig. 7.1 in that it contains kinks both within the jet and in the outer lobes – see text. (Courtesy of Masanori Nakamura; refer to Nakamura *et al.* (2007, 2008).)

The total lobe magnetic energy is ~1.4 × 10^{60} ergs. These numbers are roughly consistent with the parameters listed above, and they are typical for a BH-powered, extragalactic jet-lobe system. In Poynting flux-dominated models the jet *current* and its distribution are prime physical parameters that determine the jet and lobe characteristics.

Comparisons of 3-D simulations with real data can be done for different jet-lobe models, whether powered by a jet that is dominated by an e^+e^- beam, a p^+e^- beam, or a current-dominated pointing flux as we have used above. Exploration of the "parameter space" can be done for different models and can be tested against clear predictions – for example of the jet-to-lobe transition point that happens near the scale height of an ambient atmosphere.

7.3.4 *Instabilities and disruption in magnetic tower jets and lobes*

Other important observational and computational discriminators of the jet and lobe physics – whether hydrodynamic and/or electromagnetic – are the instabilities that can disrupt the jet-lobe system. In principle, several types of instabilities could, but may not necessarily, be important in testing whether the flows are hydrodynamic or electromagnetically dominated, and consequently for establishing the type of fluid involved (e.g. e^+e^- beam, a p^+e^- or partly neutral). Discontinuities or gradients in the power flow can also serve as helpful discriminators of the physical parameters of a jet/lobe "system".

Kelvin–Helmholtz (K–H) instabilities can develop where there is a gradient (∇v), or curl ($\nabla \times v$) in the 3-D velocity field. More specifically, K–H instabilities often occur where there are density discontinuities and velocity shear at the edge of a hydrodynamic flow. Another magnetic field-related instability is the Parker instability, in which magnetic loops can expand into an adjacent lower density region, as in the solar corona. They have also been proposed (Parker 1992) to be important above the Galactic disc as a key element of a Galactic dynamo. The Weibel instability in anisotropic velocity fields (Weibel 1959, Schlickeiser & Shukla 2003) can occur in high Mach number flows within an organised magnetic field perpendicular to an expanding relativistic shock front. Weibel instabilities could be important at, and beyond the outer edge of a magnetic tower jet (e.g. in Figs. 7.1 and 7.2), but they are not necessarily specific to a magnetic tower configuration. Another important type of instability is known as the Kruskal–Shafranov (K–S) instability. It applies especially to magnetic fingers, loops and towers. Since magnetically dominated jets and lobes can be relatively resistant to breakup, shredding, and dissipation, the K–S instability deserves a little more attention here.

The K–S instability can be prevented if the *azimuthal* magnetic twist, Φ, in a magnetic column is less, over a given length, L, than a critical value set by the Kruskal–Shafranov criterion (Shafranov 1958, Kruskal *et al.* 1958). For a magnetic flux tube of radius r, the K–S criterion for stability is

$$\Phi(r) < \Phi_{\text{crit}} = \frac{LB_\phi}{rB_z} \tag{7.19}$$

where B_ϕ and B_z are the azimuthal and poloidal field components (Fig. 7.1), and $\Phi_{\text{crit}} \approx 2\pi$, i.e. ≈ 1 turn. The K–S criterion is subject to modification depending on how the field lines are anchored, and other factors. For example, in solar coronal loops Φ_{crit} in the kink instability has been estimated in the range of $\sim 2\pi$–6π. For a force-free field having uniform twist the estimate is 3.3π (Hood & Priest 1979). Unlike solar coronal loops, tower jets have $L \gg r$. The K–S criterion was modified by Nakamura *et al.* (2006) who replaced L by a spectrum of wavelengths, λ_i along the z axis up to some λ_{max}, above which the K–S instability sets in.

In a Poynting flux jet, current-driven kink instabilities (CDI) can become important at some point, and examples of kinks can be seen in Fig. 7.2 in both the jets ("internal" kinks) and the lobes ("external" kinks). For the parameter setup of the simulation run in Fig. 7.2, both kink types have azimuthal mode $m = 1$ relative to the symmetry axis. Instabilities for the $m = 0$ mode would show no non-axial perturbations in the (x, y) plane. Here, the kinks are within the K–S criterion, that is, they have not "destroyed" the collimation of the jet or the basic structure of the outer lobe magneto-plasma.

7.4   Tests of kpc scale jet-lobe systems in different environments

Synchrotron radio-emitting lobes of extragalactic radio sources are made visible by relativistic electrons. For radio synchrotron emission these have $\gamma(=E_e/mc^2) \approx 10^3$–$10^4$ in the presence of some energetically equivalent magnetic field with some degree of orientational ordering. The presence of relativistic protons ($E_{p'}$), or an e^+e^- pair plasma, is less easy to verify, though this may not be impossible. Then, as mentioned earlier, a non-relativistic plasma component is revealed by Faraday rotation imaging – unless it was an equal mix of

e^+e^- pairs. X-ray data provide a useful complement to this information, as we discuss later. And of course, time evolution is not easily revealed in successive "snapshot" images on kpc-Mpc scales!

In summary, advanced computer simulations of these systems have recently become possible through better knowledge of the basic magneto-plasma processes, combined with improved computational methods and algorithms. Independently measurable gas pressure environments in galaxy clusters have also greatly helped the simulation models. Where observations can specify an external pressure environment, the simulations can come closer to fully describing magneto-plasma state of jet-lobe systems.

7.4.1 Radio lobes: The importance of magnetic pressure and stability

A study by Robinson *et al.* (2004) investigated, using computational models, the evolution of radio source bubbles in the hot gas environment of a galaxy cluster. The bubbles appear as "holes" or depressions in the X-ray images due to the relative absence of thermal gas within them. Using the FLASH adaptive mesh refinement code, it was found that the observed persistence of the radio bubbles (X-ray holes) in a cluster environment could be best explained if their internal pressure is largely supported by magnetic fields. Hydrodynamic instabilities, and gas vortex motions, in contrast, would normally shred, or substantially distort the bubbles before they can reach the simulated (and observed) dimensions of e.g. ~30 kpc radius within the intracluster medium. The conclusion is that the radio lobes must be essentially *magnetically* dominated structures (Benford 2006). We next describe some experimental tests of lobe-internal physics within galaxy clusters.

7.4.2 Galaxy cluster bubble tests for the role of magnetism in BH-powered radio/X-ray lobes

As mentioned, the radio lobes appear as X-ray "holes" in 2-D images of the thermal bremsstrahlung image of the cluster. At the time of writing, $\mathcal{O}$ 100 such bubbles have been verified. Six of them are associated with *one* BH-powered radio source: Hydra A in the Hydra cluster of galaxies (Wise *et al.* 2007). The bubbles will first expand due to their internal energy (gas, CR, and/or magnetic pressure). Hypothetically they could represent a hydrodynamically-driven adiabatic expansion, with, or without continuous energy supply from the source, or a current-powered expansion which is dominated by associated magnetic pressure, as discussed for Fig. 7.3. The sorting out of these possibilities provides opportunity for quantitative analyses in a galaxy cluster – using it as a kind of extragalactic "laboratory".

Because quantitative measurements of the lobe/hole evolution with varying cluster-centric radius are importantly connected to the lobe/hole physics, details of the detectability and "experimental" analysis of these (3-D) bubbles using the observed 2-D images are obviously important. Enßlin & Heinz (2002) provide an informative discussion of the details of bubble detectability, especially at larger radii from the cluster centre, and the effects of evolution of the radiating relativistic electron population. The X-ray sensitivity and contrast, particularly at larger radii from the cluster-centre, are important for producing a complete census of the bubble statistics. Of necessity, these are derived from the observed 2-D projected images.

A 3-D reconstruction for these purposes has used a standard King-type cluster gas model, which seems well-suited, although a two-component core model has also been used (Ettori 2000). In a straightforward cluster model, the hot gas pressure, P, and density, ρ, are expressed as

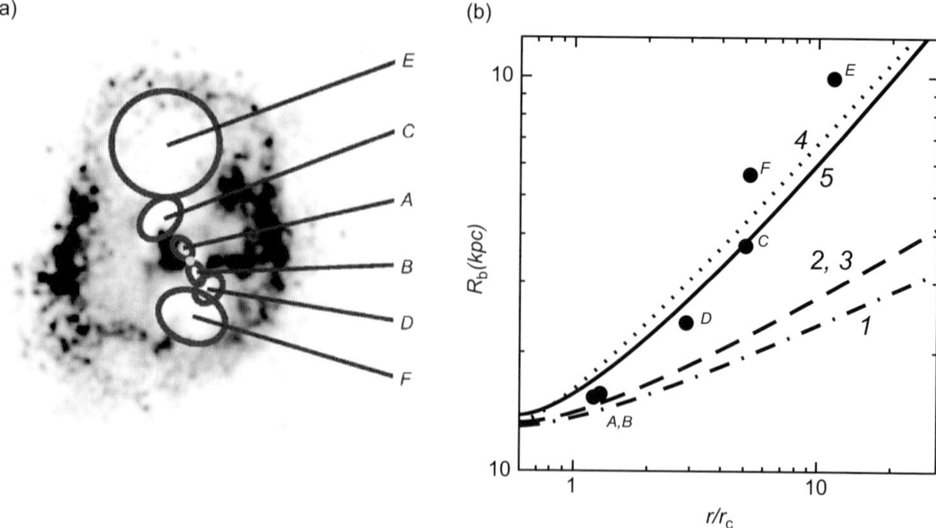

Figure 7.3. Comparison of five models of the radio source lobe (X-ray hole) size evolution with r/r_c for six measured X-ray holes/lobes (A–E) in the Hydra cluster of galaxies. The X-ray image in grey scale (a), and the thick dots in (b) show R_b (kpc) against r/r_c for each lobe/hole. These are reproduced from Wise *et al.* (2007). They are compared with predictions of five different physical models, from Diehl *et al.* (2008), and described in the text.

$$P(r) = P_0 \left[1 + \left(\frac{r}{r_c} \right)^2 \right]^{\frac{-3\beta}{2}}, \tag{7.20}$$

where

$$P_0 = \rho_0 \frac{kT}{\langle m \rangle} \tag{7.21}$$

and

$$\rho(r) = \rho_0 \left[1 + \left(\frac{r}{r_c} \right)^2 \right]^{\frac{-3\beta}{2}}. \tag{7.22}$$

P_0 and ρ_0 are the pressure and density of the ambient gas at the cluster core that has a mean molecular weight $\langle m \rangle$ and which are related by the ideal gas law with $\Gamma = 5/3$. r_c is the cluster core radius, and β, in this case, is the exponential scale parameter (not the plasma β).

Deviations from these ideal conditions will be mentioned later. Of interest here is how the internal physical state and constituents of the lobe/hole can be tested by observing the lobe radius evolution as it expands and rises in the cluster potential well. Variation of the bubble size, $R_b(r)$, with cluster-centered radius r can be written in the form

$$R_b = R_{b,0} \left[1 + \left(\frac{r}{r_c} \right) \right]^{\alpha}. \tag{7.23}$$

Different models are distinguished by the two parameters $[R_{b,0}, \alpha]$. They are based on the physical state and evolution of the lobes assuming the different, somewhat idealised models. These are (1) Adiabatic expansion of a pre-formed hydrodynamic bubble with $\Gamma = 5/3$, (2) The same, but using $\Gamma = 4.3$ for a relativistic gas, (3) A hydrodynamic bubble that is threaded with frozen-in magnetic loops. This follows a formalism developed by Thompson and Duncan (1993), which was based on a neutron star formation model. The growth in model (3) is essentially indistinguishable from the preceding model, (2), (4) A hydrodynamic bubble ($\Gamma = 4.3$) that is continuously inflated (energised) at a constant rate over the bubble lifetime, and (5) a current-dominated MHD jet fed bubble similar to that illustrated in Fig. 7.2.

The curves, from Diehl *et al.*'s (2008) study, in Fig. 7.3(b) show two versions of an adiabatically expanding hydrodynamic bubble model. They are curves 1 and 2, representing bubbles (lobes) that are inflated by a hot, magnetised CR gas. These invariably produce a slower than observed increase in the bubble (lobe) size with increasing cluster-centered radius. The same also occurs in a model in which a hydrodynamic bubble is threaded with a magnetic "web" – model 3, and which is difficult to distinguish from model 2 in Fig 7.3(b). Hydromagnetic models *can*, however, be made to agree with the observed R_b vs. r/r_c *if* they are continuously pumped with energy at just the right rate, e.g. by a jet from the AGN. This is case (4) in Fig. 7.3(b).

However, model (4) seems physically difficult to justify in practice, especially in a multi-lobe system. For this to work, the required continuous feeding of energy means that the energy needs to flow through the narrow lobe-to-lobe connections without disrupting the interconnected multi-lobe system. Furthermore, the timescales for various instabilities are generally much shorter than the dynamical age of the multi-lobe system. A similar problem occurs with models (1) and (2), where *deus ex machina* mechanisms need to be invoked to suppress instabilities that emerge in the computations. Apart from the above inter-lobe instability problem, model (4) also requires that the lobes are constantly energised over $\gtrsim 2 \times 10^8$ year at the improbable rate of $\sim 10^{44}$ erg/s in the case of the Hydra bubbles.

In summary, hydrodynamic bubble models seem to require "over-contriving" to agree with the data and the energetics. By contrast, the evolution of a current-dominated lobe with r/r_c (curve (5) does appear able to reproduce the observed lobe size vs. r/r_c relation. The conclusion of Diehl *et al.* (2008) is that all models attempted, *except* for the current-dominated one, fail the test in one way or another.

In a separate, detailed study of 63 measured bubbles in 32 host clusters, Diehl *et al.* (2008) similarly investigated the variation of parameters such as bubble luminosity, size, and energy content, as a function of r. This wider investigation reinforces the conclusion above from the six bubbles in the single Hydra A system in Fig. 7.3(b). It is described in some detail by Diehl *et al.* (2008) along with graphical displays of the results. The collective properties of the lobe radii vs. cluster-centred distance likewise seem to require magnetically-dominated lobes (bubbles), whose size and internal pressure are scaled to a jet current in the range $\sim 5 \times 10^{17}$ to $\sim 5 \times 10^{19}$ A. This range of current is independently consistent with calculations and measurements described elsewhere in the book.

This latter 63 bubble analysis required a careful determination of the lobe (hole) detectability, and the ability to deproject the 2-D images such as in Fig 7.3(a) to true 3-D sizes and pressures, etc. In some important aspects the above analyses rest on the study by Enßlin & Heinz (2002), mentioned above, that creates a relatively reliable 2-D–3-D deprojection at the larger cluster centric radii. For further details the reader is commended to the paper by Diehl *et al.* (2008). Clusters of galaxies are discussed further in Chapter 9.

7.5   Some specific ideas on extraction of magnetic energy at the central BH

7.5.1   *General comments*

The *electromagnetic* extraction of energy from the immediate environs of a central galactic black hole is an integral part of the above class of model. The radius of extraction could be (i) close in to the ergosphere of the rotating black hole, or (ii) somewhere not much further out in the accretion disc, in which case space–time distortions might not be of primary importance. A rapidly rotating object requires a Kerr metric, instead of a Schwarzschild metric description of the space–time environment close to the ergosphere. A full treatment of electromagnetic coupling of black hole rotational energy in a Kerr metric environment was published by Blandford & Znajek (1977). A wider discussion and overview of accretion power in astrophysical systems, from stellar to extragalactic, can be found in the CUP book by Frank, King, & Raine (1992).

Since approximately AD 2000, advances in computational power have clarified the advantages and limitations of both hydrodynamic and electromagnetic aspects of jets and lobes from massive black holes. The following subsections focus on *electromagnetic* BH-jet models, since this class of model has made the most impressive recent progress, especially with the help of numerical simulations. These models also lend themselves to electrical circuit analogies. This will become clearer below, and in Chapter 8.

Extragalactic jets feed their energy into the expanding magnetised lobes that ultimately interface with, and "feed" the surrounding intergalactic medium. The density and pressure of the ambient intergalactic medium varies by orders of magnitude, from a galaxy cluster environment having baryonic densities up to $\gtrsim 10^{-3}$ cm^{-3}, out to $\lesssim 10^{-5}$ cm^{-3} in the wider IGM, i.e. more than a factor of 100 contrast. Also, a successful simulation needs to explain the observed properties of a jet-lobe *system* for a range of energy inputs (e.g. BH masses), as well as ambient gas pressures. A self-consistent physical description of an electromagnetic jet-lobe "system" will be determined partly by the environment and, as we discuss below, partly by the jet's current. Improved specification of the immediate IGM environment and the jet and lobe characteristics can provide tighter constraints on the energy transfer from a SMBH to the IGM.

Gravitational infall and angular momentum transfer are required to explain how the accretion energy of a central collapsed object is connected to energy outflow in a collimated beam, or jet. In Chapter 8, observations are described showing that the gravitational-to-magnetic energy conversion can also be very efficient. A physical understanding of these processes, the largest energy transfers in the post-Inflation Universe, is a remarkable phenomenon of Nature. It can also be expected to illuminate other problems in Earthbound and laboratory magneto-plasma phenomena, including fusion energy science.

The following subsections focus on two distinct inner zones of a jet-lobe system. These are: (i) A zone where the initial power flow of the jet is extracted in the highly distorted space-time structure around the BH, and (ii) further out in the BH accretion disc, mostly beyond significant Kerr metric distortions. Both classes of model must successfully explain *both* the effective transfer of angular momentum *and* the relatively high energy conversion efficiency from gravitational to magnetic and particle energy outflow into the lobes.

7.6　Electromagnetic extraction of collimated power flow at the black hole

Various scenarios have been discussed to explain how a rotating black hole can release its rotational energy electromagnetically, and in a very efficient way. Ideas have been explored of a magnetised accretion disc close to a rotating black hole which might be supported by strong magnetic fields and external currents (Penrose 1969, Lynden-Bell 1969 – also containing an historical review, Bisnovatyi-Kogan & Ruzmaikin 1976, Lovelace 1976, Blandford & Znajek 1977). A description of electromagnetic fields *in vacuo* in the vicinity of a Kerr black hole can be found in King, Lasota, & Kundt (1975). The aim of all of these ideas and processes is to understand, as shown by Lovelace (1976), how gravitation energy of infalling material can be removed with high efficiency – e.g. without thermal dissipation in a surrounding disc, in a way that is highly collimated, and in the direction of rotation axis.

An example, following (i) above of a black hole-powered Poynting flux jet comes from the set of simulations by Koide *et al.* (2002). Here, extraction of the black hole's rotational energy occurs well within the rotationally distorted Kerr metric space–time. A rotating black hole is characterised by just two parameters: its mass, M_{BH}, and its angular momentum, J. The maximum value of J is given by

$$J_{\max} = \frac{GM_{\mathrm{BH}}^2}{c}. \tag{7.24}$$

In the model by Koide *et al.*, the rotation parameter $a = J/J_{\max}$ was set at 0.99995, asymptotically close to the Kerr-space limit. The grey "volume" in Fig. 7.4 approximates a spheroid. As a decreases below ~0.8 it asymptotes to a sphere in the slow rotation limit ($a \rightarrow 0$). At a higher rate of rotation (e.g. $a \gtrsim 0.8$), the ergosphere develops surface cusps with relatively smaller radius at the poles, reminiscent of the surface of an apple.

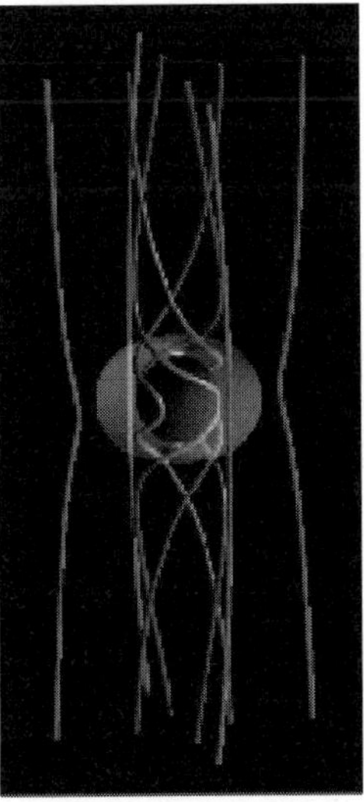

Figure 7.4. A model of the launching of an electromagnetic jet, in which the magnetic field connects both sides of the black hole's ergosphere (Koide *et al.* 2002).

The initial plasma density (ρ_0) within the horizon is low, and the magnetic field strength B_0 is $\sim 10^4$ G. With this dominant magnetic field, set by $B_0{}^2 = 0.10\rho_0\, c^2$, the Alfvén velocity v_A, is 0.593 c in the model in Fig. 7.4. In the frame-dragging of B_φ within the ergosphere, magnetic field lines transfer angular momentum into a torsional Alfvén wave via magnetic perturbations. These propagate against the infalling gas coming down the field lines, as shown in Fig. 7.4. The outflowing free energy from the rotating black hole is of comparable order to that of the infalling plasma. This appears required to explain the high efficiency of energy conversion from gravitational infall energy to the magnetic energy that is ultimately conveyed to the outer lobes.

Koide *et al.*'s simulation focuses on the initial stage of energy extraction from the rotating black hole. In subsequent phases, at larger distances along the jet, and into the lobes; additional processes come into play.

This process of angular momentum extraction from the black hole works by essentially decreasing the energy inside the ergosphere, depicted by the central zone in the figure. As the negative energy plasma within the ergosphere reaches the horizon at the ergosphere boundary, the rotational energy of the black hole decreases, while the energy of angular motion is transferred into Alfvén wave perturbations that propagate away along the field lines shown in lighter grey in the model. Further details are described in Koide *et al.* (2002) and references therein.

7.7   Another concept: Extraction of BH energy from the inner accretion disc, outside the ergosphere

In a quite different and composite model by Colgate, Li, & Pariev (2001), infall of material into an inner $\sim$100 AU accretion disc (Fig. 7.5) occurs through transfer of angular momentum by large scale hydrodynamic shocks that occur in Rossby waves. A key point is that the coherence length of these shocks is comparable with the size of the outer disc in Fig. 7.5. This "Rossby vortex disc" (Lovelace *et al.* 1999, Li *et al.* 2001) is a further development of the α-viscosity disc proposed by Shakura & Sunyaev (1973). Given the right physical conditions, it can be close to 100% efficient in feeding the inner disc through transfer of angular momentum (Colgate & Li 1999).

The conditions for a Shakura–Sunyaev or Rossby vortex disc are that the entropy or heat content must stay approximately constant in the disc for several rotation periods. That means that radiative cooling must be suppressed over this time. A necessary condition for this is a mass surface density in the outer disc (o.d.) in Fig. 7.5, $\Sigma_{\mathrm{o.d.}}$ which is $\sim$100 g cm^{-2}. This condition can also be derived from *protogalactic* conditions, beginning with the collapse of a Lyman α cloud until it reaches the first stage of Keplerian support. At this point, $\Sigma_{\mathrm{initial}}$ $\sim$0.001 g cm^{-2} and the proto-disc diameter is $\sim$100 kpc. At the point of collapse all the way down to the outer disc in Fig. 7.7 where $\Sigma_{\mathrm{o.d.}} \sim 100$ g cm^{-2}, the radius is of order 10 pc, and mass $\sim 10^8$ M$_\odot$ (Colgate & Li 1999).

This simplified *ab initio* calculation appears naturally able to produce the mass of a typical central galactic black hole – that is, provided that the angular momentum is very efficiently removed, as proposed in the Rossby vortex model above. In this model, $M_{\mathrm{o.d.}}$ is fed by efficient angular momentum transfer into the inner 100 AU of the accretion disc (i.d.) over the activity lifetime of the AGN, which is $\sim 10^8$ year. This scenario is the basis of the accretion disc dynamo described by Pariev, Colgate, & Finn (2007).

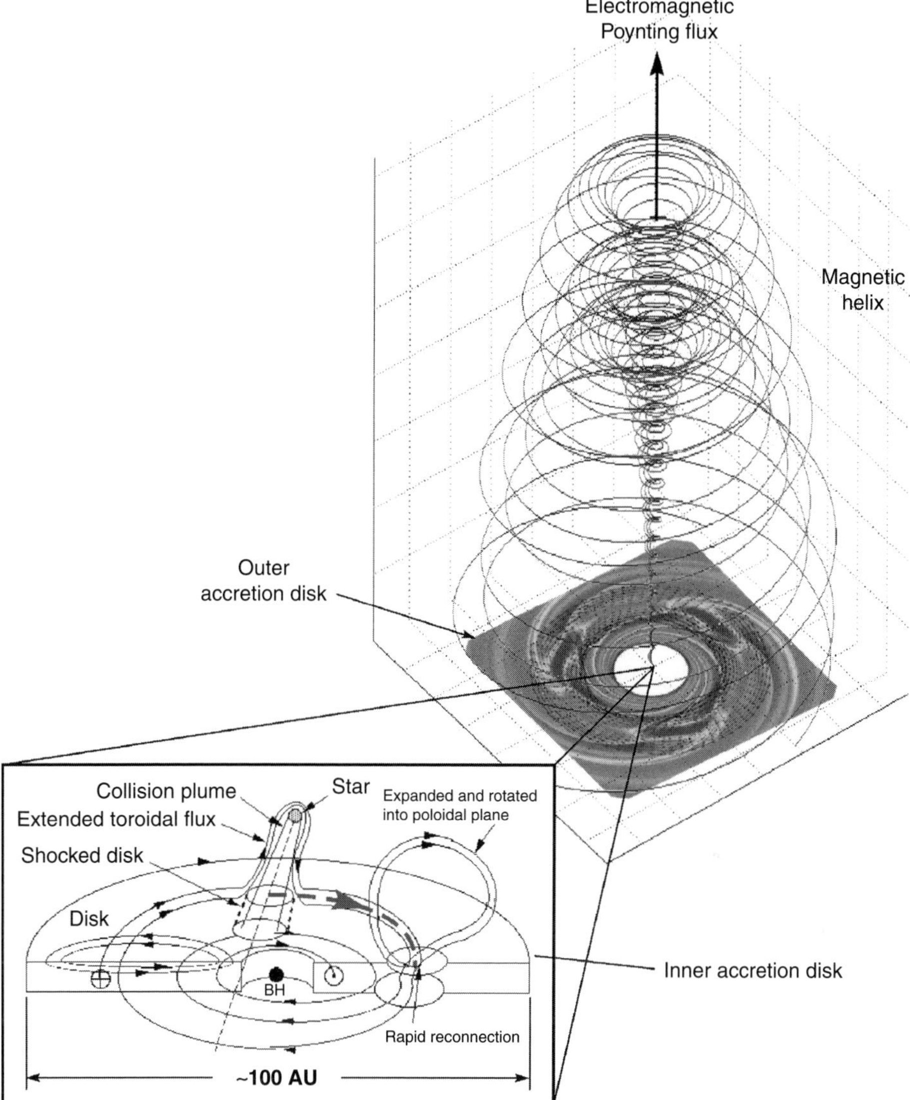

Figure 7.5. Simplified illustration of three interconnected processes to explain how the gravitational infall energy to a massive galactic black hole is converted to a highly collimated electromagnetic Poynting flux energy flow along the rotation axis of the black hole's accretion disc. (1) Angular momentum is transferred outward by hydrodynamic vortices – a system-scale non-linear hydrodynamical instability. (2) The shock zones of stars colliding with the BH accretion disc drag magnetic field lines out of the disc. These are distorted and stretched, and reconnected to form merging flux loops that act as a dynamo, and in the process converting the disc's rotational energy to magnetic field energy. (3) A helical magnetic field structure carries about 10^{19} amperes along the rotation axis. This forms the base of the jet which, in this early phase, is almost entirely electromagnetic. Subsequently, and on much larger scales, the magnetic energy is partially converted to particle energy as the energy flow proceeds out of the parent galaxy and into intergalactic space. This latter phase occurs beyond the scale of the figure. (Source: Pariev, Colgate, & Finn 2007. Figure courtesy of Stirling Colgate.)

Given the gravitational infall energy deposited initially into the "reservoir" of the outer ~10 pc of the hydrodynamic accretion disc, the "inner" accretion disc dynamo further reprocesses the infall energy by an efficient mechanism into a collimated Poynting flux jet. This next phase of the composite model happens within ~ 100 AU in the cartoon image in Fig. 7.5, and is described next.

In Fig. 7.5, the black hole's mass $M_{\rm BH} = 10^8\,{\rm M_\odot}$ and the inner disc innermost radius is ≈ 2 Schwarzschild radii ($R_{\rm s} = 2GM/c^2$) or ~6 × 10^{13} cm, which is the radius of the innermost stable orbit. In contrast to the previous set of models (Section 7.5), Kerr metric terms are not significant over the radius range of this model. The inner disc surface density $\Sigma_{\rm id}$ is ≈ 70 g cm^{-2}, and its black body temperature is ≈ 20 eV which peaks at EUV wavelengths. The differential Keplerian rotation of the disc winds up the magnetic field B_ϕ, to $\approx 10^4$ G. The inner disc will execute $\approx 10^{12}$ turns over the 10^8 year formation lifetime of the black hole. B_ϕ is determined by pressure balance between the local confining gravitational pressure and the magnetic field in the disc.

$$B_\phi \sim \frac{\sqrt{8\pi\Sigma g(r) <h>}}{r^{1/2}} \sim 10^4\ {\rm G}, \tag{7.25}$$

where $<h>$ is the average height of the accretion disc = a few percent of its maximum radius, $g(r)$ the local gravity at distance r from the black hole, and Σ the mass per unit area in the disc (~70 g cm^{-2}).

A key element of the model of Fig. 7.5 is that *stars* will collide with the inner accretion disc. In each collision, the shock front of collision pulls a magnetic loop out of the disc. Conservation of momentum immediately twists this loop (by the Coriolis "force") relative to the inertial frame of the Keplerian rotating disc. This creates a dynamo by analogy with the large scale galactic dynamo discussed in Chapter 5. This dynamo and the B_z component ($\approx B_\varphi$) couple the 10^4 G magnetic field in the inner disc to the base of a rotating helix, which propagates the accretion energy away as electromagnetic [Poynting] flux. Some of these basic physical concepts are justified below.

The number of stars at radii <10 pc is independently estimated as ~10^4. The total number of collisions, n, for a stellar density ρ_*, whose average radius R_* is given by

$$n_{\rm collisions} = \rho_* R_* / \Sigma_{\rm disc} \sim 10^9 \times 10^4\,{\rm stars} = 10^{13}, \tag{7.26}$$

within a 10 pc disc, where $\rho_* R_* \approx 10^{11}$ g cm^{-2} (Colgate & Li 1999). Here, $\rho_* R_*$ is the average mass cross section per star. Most star-accretion disc collisions will take place where the orbital period is ~10^4 s, close to the innermost stable orbit. This is 3.6 × 10^{13} cm = 2.4 AU, of similar order to the footprint radius of the magnetic helix. The collision rate in the inner disc is ~10^{13} over 10^8 year, or 1 every ~300 s. This corresponds to 5–10 collisions *per orbit* in the vicinity of the inner disc in Fig. 7.5. The Mach number of the star-disc collisions is ~100, and the velocities are of order 10^4 kms^{-1}. These conditions are broadly consistent with spectroscopic widths of order 10^3–10^4 kms^{-1} observed in the broad emission line regions around central black holes (Zurek *et al.* 1996). The spectroscopic broadening is interpreted as Doppler broadening of motions in the central black hole's potential at $R \lesssim 100 R_{\rm S}$ (e.g. Peterson 1993, Collier *et al.* 1998).

By a process that remains to be clarified, the reconnecting flux loops in Fig. 7.5 combine into an outwardly propagating set of nested magnetic helices. The torque on the helix is due to $J \times B$ forces, and is transmitted by tension in the field lines that connect the disc matter at the disc's outer and inner radii (Pariev *et al.* 2007). The magnetic energy propagates at sub-Alfvénic speed in the form of continuously wound helical lines, for which, at least initially, $B_z \sim B_\varphi$, the azimuthal magnetic field component. Magnetic flux lines are anchored in the disc, and they must eventually return after some large z-distance, as illustrated. The bulk of the gravitational potential energy is released close to the inner accretion disc, corresponding to the innermost of the nested helices.

This situation sets up an α-Ω dynamo analogous to the large-scale galactic dynamo that was originally proposed to explain large scale fields of differentially rotating spiral galaxy discs (Chapter 5). This powerful dynamo generates magnetic loops which, in the model, combine to form a rapidly rotating helical field tied at its base to the rotating accretion disc. Under certain plausible assumptions the field is self-collimating, assisted by a radial ambient pressure in the (x, y) plane as it propagates away, carrying most of the accretion energy (Fig. 7.7).

The magnetic energy released in one turn at the inner radius is given by

$$W_{\text{mag}} \simeq 2\pi^2 r_{\text{inner}}^3 \times \frac{\left(B_z^2 + B_\varphi^2\right)}{8\pi}, \tag{7.27}$$

where the orbital time is

$$\tau_{\text{orbit}} = \frac{2\pi r_{\text{inner}}}{v_\varphi}. \tag{7.28}$$

The ratio of these expressions represents a luminosity

$$L = \frac{dW}{dt} = \frac{\pi v_\varphi}{8} \left(B_z^2 + B_\varphi^2\right). \tag{7.29}$$

This describes a helical force tube propagating away in the z direction at velocity $v_z \approx v_\varphi$. The helix is force-free, and will propagate over the accretion lifetime of the central black hole. It is assumed that the external pressure of the ambient gas in the first phase of propagation plays some role in the initial collimation of the helix (jet). B_φ, which provides a hoop stress to confine the jet laterally, is proportional to $1/r$, whereas $B_z \propto 1/r^2$. Thus, as r increases, $B_\varphi \gg B_z$ and a kink instability would seem inevitable. But jets appear to be self-regulating to a robust extent, as some observations and simulations indicate.

One attractive aspect of the stellar collision-disc dynamo model is that it is, at minimum, qualitatively consistent with ideas on gravitational collapse of galaxies, and it provides a natural explanation for the mass of the outer accretion disc that feeds the black hole. It also proposes a mechanism for the efficient angular momentum transfer that is required to grow, and feed a black hole. The density cusp and velocity dispersion of the dense stellar cluster at $r \lesssim 100$ AU also appear to be explained in the context of dark matter evolution, in that both the stars and the dark matter behave like a collisionless gas. Among the independently observed characteristics is the stellar velocity dispersion around the black hole, consistent with the spectroscopically observed several thousand kms^{-1} emission line widths of broad line regions. These are independently believed to occur at $r \lesssim 100$ AU. Given the inner accretion disc mass, density and field strength, the model approximately accounts for the observed energy output of a central black hole over a fiducial 10^8 year lifetime, along with an estimate of the jet current (see Chapter 8), which is integral to the magnetic field helix model.

However, the model has some gaps in physical process that are beyond our present ability to confirm. These include the ability of the star-disc collision shocks to actually lift the field loops out of the inner disc. Also, fast reconnection at the loop footprints must occur. The merging of the loops to transfer the torque from the inner disc to the magnetic helix is a complex MHD phenomenon that also needs further study and simulation. Finally, details of the initial collimation of the Poynting flux power flow also need further clarification. Despite all these open questions this model is one of the most comprehensive and attractive – partly because it brings independent observations and concepts into concordance.

7.8   Summary of two SMBH jet models

The above two Poynting flux models present quite different ideas on energy extraction from the black hole. One significant difference is that, in the Blandford & Znajek and the Koide *et al.* models in Section 7.6, the endpoint of gravitational infall occurs at $r \lesssim R_S$ in the strongly deformed space–time environment. By contrast, in the Pariev/Colgate/ Li model it occurs higher up in the black hole's gravitational potential, at $r > R_S$ in the inner accretion disc, and the coupling-out of the rotational energy occurs in the BH accretion disc at radii beyond where Kerr metric space–time distortions dominate.

While neither of these two models should be considered final, they are described here for the purpose of illustrating some of the underlying physics, and to demonstrate that that there are at least two fundamentally different zones in which energy extraction may take place – either within the distorted space–time structure around the boundary of the ergosphere, or in an accretion disc outside of the Schwarzschild or Kerr radius. When we discuss the energy budget of the jet/lobe system at the extragalactic "energy sink" stage (next chapter), this effective end point (inner) radius for gravitational infall will modify the fraction, or efficiency of gravitational energy extraction, by a factor of perhaps a few – e.g. by the ratio of $\approx 0.8 R_S / 4 R_S$ (Section 8.7).

These jet systems can be aptly called Nature's ultimate electricity generator. In the initial phases illustrated here, the energy appears to be carried away almost entirely by magnetic (and electric) fields and electric currents. The current along the inner jet axis is of order 10^{18-19} A. Close to the jet axis it flows mostly parallel to the local magnetic field ($J_\parallel$). The details of magnetic-to-CR energy conversion have, as stated, yet to be well understood and an attempt is not made to conceptualise this in Fig. 7.5. The conversion probably involves a combination of magnetic reconnection and acceleration by coherent electric fields and/or shocks.

These processes will eventually convert magnetic energy into CR particle energy as the system gradually relaxes toward a minimum energy configuration in the mature radio lobes (see e.g. Benford & Protheroe 2008). This means that, as time and distance from the compact energy source increase, the energy flow will gradually transfer from Poynting flux to particle flux. The fact that relativistic particles are usually visible even within the first few parsecs shows that the energy conversion/dissipation in the jets and lobes begins early. How quickly it occurs might depend on ambient gas pressure (much higher in galaxy clusters than in the general IGM) and the degree of jet disruption. However, in *some* jets the current and electromagnetic power flow can evidently occur over kpc–Mpc scales without much dissipation or disruption. The relaxation to a minimum energy state, and the dissipation to relativistic, and eventually thermal gas is not as yet well specified or understood at this time of writing. The conversion to particle energy varies widely depending on the environment, extent, and morphology of extragalactic radio sources.

7.9 Simulations of protostellar jets

Simulations of magnetised *stellar* jets use somewhat different model parameters than described above for extragalactic SMBH–powered sources. In one simulation of a protostellar jet (Camenzind 2005), helical magnetic fields propagate into a molecular cloud having $|B| = 0$. In this computational model, turbulence in the molecular cloud can excite internal pinch and kink modes. These would appear in observations as moving "knots" along the jet (Thiele & Camenzind 2002). The relative growth rates of B_ϕ and B_z along the jet axis (z) are qualitatively similar to those in the extragalactic simulations in Fig. 7.1(b). What they have in common is a magnetised rotating system at their base.

Here, the ordered magnetic field lines are anchored in the stellar accretion disc, and the ambient pressure near to the base can act to initially assist the collimated power flow. Further along the z-axis, the jet continues on its own. A simulation by Moll *et al.* (2008) in Fig. 7.6 shows the jet, whose Keplerian rotating disc provides the rotating base, and the initial

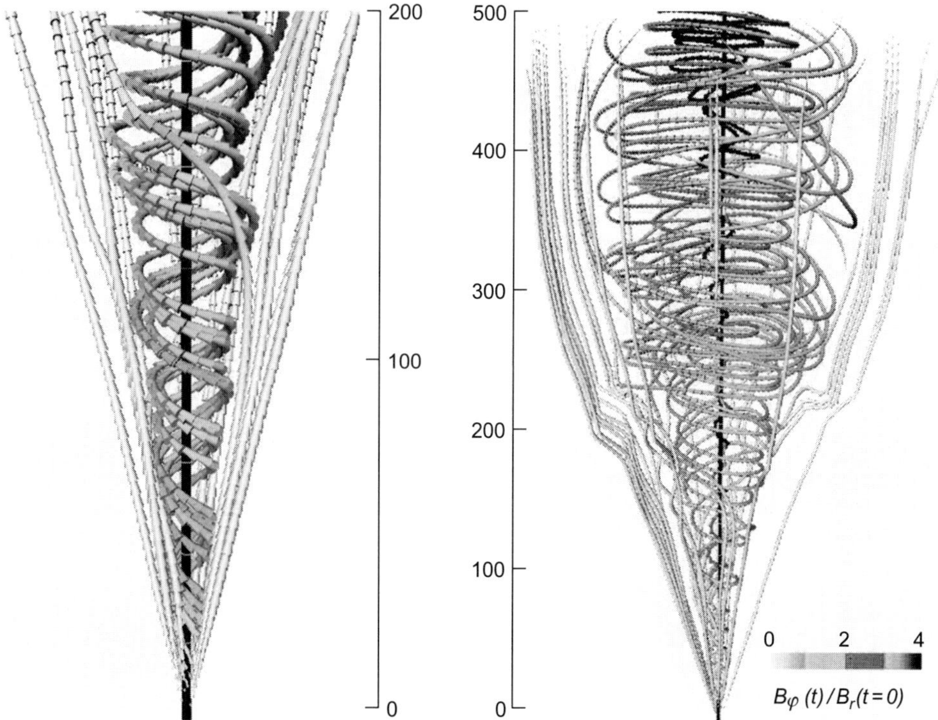

Figure 7.6. Magnetic field line configuration in a simulated Poynting flux-driven *stellar* jet that is driven by a Keplerian rotational profile at its base. Here, the initial acceleration is due to ambient gas pressure. The left side shows the z axis scale (r in the authors' notation) up to 200, and the right side shows the further progression and onset of kink instabilities up to a z-height of 505. (Moll *et al.* 2008 and shown by kind permission of H. C. Spruit.)

angular momentum. These simulations were carried out to $z \sim 1000$ times the initial radius. Further along the jet in Fig. 7.6, the structure of the rotating magnetic flux tubes shows the onset of pinch instabilities as B_φ/B_z evolves. In this model, the magnetic energy, initially 70% of the power flow, becomes gradually dissipated into kinetic and heat energy.

References

Appl, S. & Camenzind, M. 1992, The Stability of Current-Carrying Jets, *Astron. Astrophys.*, 256, 354

Balsara, D. S. & Spicer, D. S. 1999a, Maintaining Pressure Positivity in Magnetohydrodyamic Simulations, *J. Comput. Phys.*, 148, 133

Balsara, D. S. & Spicer, D. S. 1999b, A Staggered Mesh Algorithm Using High Order Godunov Fluxes to Ensure Solenoidal Magnetic Fields in Magnetohydrodyamic Simulations, *J. Comput. Phys.*, 149, 270

Benford, G. 1978, Current-Carrying Beams in Astrophysics – Models for Double Radio Sources and Jets, *MNRAS*, 183, 29

Benford, G. 2006, Stability of Magnetic Equilibria in Radio Bubbles, *MNRAS*, 369, 77

Benford, G. & Protheroe, R. J. 2008, Fossil AGN Jets as Ultrahigh-Energy Particle Accelerators, *MNRAS*, 383, 663

Best, P. N., Eales, S. A., Longair, M. S., Rawlings, S., & Röttgering, H. J. A. 1999, Studies of a Sample of 6C Radio Galaxies at a Redshift of 1 – I. *Deep Multifrequency Radio Observations, MNRAS*, 303, 616

Bisnovatyi-Kogan, G. S. & Ruzmaikin, A. A. 1976, The Accretion of Matter by a Collapsing Star in the Presence of a Magnetic Field. II – Selfconsistent Stationary Picture, *Ast. Space Sci.*, 42, 401

Blandford, R. D. & Rees, M. J. 1974, A 'Twin-Exhaust' Model for Double Radio Sources, *MNRAS*, 169, 395

Blandford, R. D. & Znajek, R. L. 1977, Electromagnetic Extraction of Energy from Kerr Black Holes, *MNRAS*, 179, 433

Bridle, A. H., Hough, D. H., Lonsdale, C. J., Burns, J. O., & Laing, R. A. 1994, Deep VLA Imaging of Twelve Extended 3CR Quasars, *Astron. J.*, 108, 766

Burbidge, G R. 1956, On Synchrotron Radiation from Messier 87, *Astrophys. J.*, 124, 416

Camenzind, M. 2005, Numerical Magnetohydrodynamics in Astrophysics, in *Cosmic Magnetic Fields*, ed. R. Wielebinski & R. Beck, *Lecture Notes in Physics* (Berlin: Springer), 664, 255

Colgate, S. A. & Li, H. 1999, Dynamo Dominated Accretion and Energy Flow: The Mechanism of Active Galactic Nuclei, *Astrophys. Space Sci.*, 264, 357

Colgate, S. A., Li, H., & Pariev, V. I. 2001, The Origin of the Magnetic Fields of the Universe: The Plasma Astrophysics of the Free Energy of the Universe, *Phys. Plas.*, 8, 2425

Collier, S. J., Horne, K., Kaspi, S., Netzer, H., Peterson, B. M., Wanders, I., *et al.* 1998, Steps toward Determination of the Size and Structure of the Broad-Line Region in Active Galactic Nuclei. XIV. Intensive Optical Spectrophotometric Observations of NGC 7469, *Astrophys. J.*, 500, 162

Daly, R. A. & Loeb, A. 1990, A Possible Origin of Galactic Magnetic Fields, *Astrophys. J.*, 364, 451

De Young, D. S. 1971, The Dynamics of Extended Extragalactic Radio Sources, *Astrophys. J.*, 167, 541

De Young, D. S. 2002, *The Physics of Extragalactic Radio Sources*, (Chicago: University of Chicago Press)

Diehl, S., Li, H., Fryer, C. L., & Rafferty, D. 2008, Constraining the Nature of X-Ray Cavities in Clusters and Galaxies, *Astrophys. J.*, 687, 173

Eilek, J. A. & Hughes, P. A. 1991, Particle Acceleration and Magnetic Field Evolution, in *Beams and Jets in Astrophysics*, ed. P. A. Hughes, *Cambridge Astrophysics Series* (Cambridge: Cambridge University Press), 428, 19

Enßlin, T. A. & Heinz, S. 2002, Radio and X-Ray Detectability of Buoyant Radio Plasma Bubbles in Clusters of Galaxies, *Astron. Astrophys.*, 384, L27

Ettori, S. 2000, β-Model and Cooling Flows in X-Ray Clusters of Galaxies, *MNRAS*, 318, 1041

Falke, H., Malkan, M. A., & Biermann, P. L. 1995, The Jet-Disc Symbiosis. II. Interpreting the Radio/UV Correlations in Quasars, *Astron. Astrophys.*, 298, 375

Fermi, E. 1949, On the Origin of Cosmic Radiation, *Phys. Rev.*, 75, 1169

Frank, J., King, A. R., & Raine, D. J., eds. 1992, *Accretion Power in Astrophysics, Cambridge Astrophysics Series 21* (Cambridge: Cambridge University Press)

Frank, J., King, A. R., & Raine, D. J., eds. 1992, *Accretion Power in Astrophysics, Cambridge Astrophysics Series 21* (Cambridge: Cambridge University Press)

Hood, A. W. & Priest, E. R.1979, Kink Instability of Solar Coronal Loops as the Cause of Solar Flares, *Solar Phys.*, 64, 303

Koide, S., Shibata, K., Kudoh, T., & Meier, D. L. 2002, Extraction of Black Hole Rotational Energy by a Magnetic Field and the Formation of Relativistic Jets, *Science*, 295, 1688

Kronberg, P. P., Colgate, S. A., Li, H., & Dufton, Q. W. 2004, Giant Radio Galaxies and Cosmic-Ray Acceleration. *Astrophys. J.*, 604, L77

Kronberg, P. P., Dufton, Q. W., Li, H., & Colgate, S. A. 2001, Magnetic Energy of the Intergalactic Medium from Galactic Black Holes, *Astrophys. J.*, 560, 178

Kruskal, M. D., Johnson, J. L., Gottlieb, M. B., & Goldman, L. M. 1958, Hydromagnetic Instability in a Stellarato, *Phys. Fluids*, 1, 421

Lesch, H., Appl, S., & Camenzind, M. 1989, Collective Plasma Processes in Extragalactic Radio Sources, *Astron. & Astrophys.*, 225, 341

Li, H., Colgate, S. A., Wendroff, B., & Liska, R. 2001, Rossby Wave Instability of Thin Accretion Disks. III. Nonlinear Simulations, *Astrophys. J.*, 551, 874

Li, H., Lagenta, G., Finn, J. M., Li, S. & Colgate, S. A., 2006, *Astrophys. J.* 643, 92

Lovelace, R. V. E. 1976, Dynamo Model of Double Radio Sources, *Nature*, 262, 649

Lovelace, R. V. E., Li, H., Colgate, S. A., & Nelson, A. F. 1999, Rossby Wave Instability of Keplerian Accretion Disks, *Astrophys. J.*, 513, 805L

Lucek, S. C. & Bell, A. R. 1996, The Stability, During Formation, of Magnetohydrodynamic Jets Collimated by an Azimuthal Magnetic Field, *MNRAS*, 281, 245

Lyubarskii, Yu. E. 1999, Kink Instability of Relativistic Force-Free Jets, *MNRAS*, 308, 1006

Marshall, H. L., Miller, B. P., Davis, D. S., Perlman, E. S., Wise, M., Canizares, C. R., & Harris, D. E. 2002, A High-Resolution X-Ray Image of the Jet in M87, *Astrophys. J.*, 564, 683

Moll, R., Spruit, H. C., & Obergaulinger, M. 2008, Kink Instabilities in Jets from Rotating Magnetic Fields, *Astron. & Astrophys.*, 492, 621

Nakamura, M. & Meier, D. 2004, Poynting Flux-Dominated Jets in Decreasing-Density Atmospheres. I. The Nonrelativistic Current-driven Kink Instability and the Formation of "Wiggled" Structures, *Astrophys. J.*, 617, 123

Nakamura, M., Li, H., & Li, S. 2006, Stability Properties of Magnetic Tower Jets, *Astrophys. J.*, 652, 1059

Nakamura, M., Tregillis, I. L., Li, H., & Li, S. 2008, A Numerical Model of Hercules A by Magnetic Tower: Jet/Lobe Transition, Wiggling, and the Magnetic Field Distribution, *Astrophys. J.*, 686, 843

O'Neill, S. M., Tregillis, I. L., Jones, T. W., & Ryu, D. 2005, Three Dimensional Simulations of MHD Jet Propagation through Uniform and Stratified External Environments, *Astrophys. J.*, 633, 717

Pariev, V. I., Colgate, S. A., & Finn, J. M. 2007, A Magnetic α-ω Dynamo in AGN Disks. II. Magnetic Field Generation, Theories, and Simulations, *Astrophys. J.*, 658, 129

Parker, E. N. 1992, Fast Dynamos, Cosmic Rays, and the Galactic Magnetic Field, *Astrophys. J.*, 401, 137

Penrose, R. 1969, Gravitational Collapse: The Role of General Relativity, *Rev. Nuovo Cimento Soc. Ital. Fis.*, 1, 252

Peterson, B. M. 1993, Reverberation Mapping of Active Galactic Nuclei, *Pub. Astron. Soc. Pacific*, 105, 247

Rees, M. J., 1971, A New Interpretation of Extragalactic Radio Sources, *Nature*, 229, 312

Robinson, K., Dursi, L. J., Ricker, P. M., *et al.* 2004, Morphology of Rising Hydrodynamic and Magnetohydrodynamic Bubbles from Numerical Simulations, *Astrophys. J.*, 601, 621

Ryu, D., Ostriker, J. P., Kang, H., & Cen, R. 1993, A Cosmological Hydrodynamic Code Based on the Total Variation Diminishing Scheme, *Astrophys. J.*, 414, 1

Schlickeiser, R. & Shukla, P. 2003, Cosmological Magnetic Field Generation by the Weibel Instability, *Astrophys. J.*, 599, 57

Shafranov, V. D. 1958, On Magnetohydrodynamical Equilibrium Configurations, Soviet Phys. *JETP*, 6, 545

Shakura, N. I. & Sunyaev, R. A. 1973, Black Holes in Binary Systems, Observational Appearance, *Astron. & Astrophys.*, 24, 337

Strom, R. G. & Willis, A. G. 1980, Multifrequency Observations of Very Large Radio Galaxies. II – 3C236, *Astron. & Astrophys.*, 85, 36

Thiele, M. & Camenzind, M. 2002, Knot Production in Magnetized Herbig-Haro Jets, *Astron. & Astrophys.*, 831, L53

Thompson, C. & Duncan, R. C. 1993, Neutron Star Dynamos and the Origins of Pulsar Magnetism, *Astrophys. J.*, 408, 194

Weibel, E. S. 1959, Spontaneously Growing Transverse Waves in a Plasma Due to an Anisotropic Velocity Distribution, *Phys. Rev. Lett.*, 2, 83

Willis, A.G. & Strom, R. G. 1978, Multifrequency Observations of Very Large Radio Galaxies. I – 3C 326, *Astron. & Astrophys.*, 62, 375

Wise, M. W., McNamara, B. R., Nulsen, P. E. J., Houck, J. C., & David, L. P. 2007, X-Ray Supercavities in the Hydra A Cluster and the Outburst History of the Central Galaxy's Active Nucleus, *Astrophys. J.*, 659, 1153

Zurek, W. H., Siemiginowska, A., & Colgate, S. A. 1994, Star-Disk Collisions and the Origin of the Broad Lines in Quasars, *Astrophys. J.*, 434, 46

Zurek, W. H., Siemiginowska, A., & Colgate, S. A. 1996, Star-Disk Collisions and the Origin of the Broad Lines in Quasars: Addendum, *Astrophys. J.*, 470, 652

8

Extragalactic jets and lobes – II. More on magnetic energy flows into the IGM from galaxy nuclei

8.1 Introduction

In Chapter 7, we outlined the roles of magnetic fields in producing galaxy and supra-galactic radio and X-ray sources. Examples of simulations and observational diagnostics were briefly described.

The first part of this chapter explores further concepts that can add insight into the workings of these uniquely energetic systems. Although verification of some of the concepts is presently beyond the capabilities of *some* instruments, it is hoped that the ideas discussed will stimulate new directions of observation and experiment. In the latter part of the chapter we discuss more global aspects of the cosmic ray (CR) and magnetic energy produced by galaxies. These lead to the effects of jets and lobes on the intergalactic medium, and the cosmic evolution of magnetic fields. These phenomena can supplement the role of SN and star-driven outflows that were discussed in Chapter 6.

8.2 An electric circuit model for energy flow from a supermassive black hole

8.2.1 Analogy of an electrical circuit

Analogies to electrical quantities have appeared in the literature in models of black hole accretion discs that are highly conducting and rotating (e.g. Lovelace 1976, Blandford & Znajek 1977, Lovelace & Ruchti 1983, and other papers mentioned in Chapter 7, Section 7.2). The black hole-accretion disc system launches a collimated jet on an initial scale of a few pc which, on several models, carries a net current of order 10^{18} A when scaled to a 10^8 $M_\odot$ black hole progenitor. This same level of current was recently detected about 50 kpc from the AGN in a kpc-scale jet segment by Kronberg *et al.* (2011). Irrespective of what carries the energy in the jet (whether electromagnetic power, an e^+e^- plasma, or a proton/ electron beam), the morphology and energetics of luminous extragalactic radio sources demonstrate that at least the equivalent of a beam current of the above order must be present in these remarkable systems. In the analogy of an electrical circuit, we have a power beam that, in effect, feeds current into an impedance. A reference impedance is that of free space, Z_0, in the absence of current carriers (i.e. *in vacuo*).

$$Z_0 = \frac{\sqrt{\mu_0 \varepsilon_0}}{4\pi} \quad \text{(MKS)} \quad = 30 \ \Omega. \tag{8.1}$$

Here μ_0 and ε_0 are, respectively, the permeability and permittivity of free space, fundamental constants of electromagnetism. Their existence and their values determine Z_0. They also permit the propagation of electromagnetic waves at the velocity of light, c:

$$c = \frac{1}{\sqrt{\mu_0 \varepsilon_0}} \quad \text{(MKS)}. \tag{8.2}$$

In cgs units, $Z_0 = c^{-1}$ in free space. The actual impedance is normally less by some factor, up to a few, than the *in vacuo* value above due to a finite conductivity in the propagation medium. In the case of a jet, the impedance is $Z = (Z_0)(v_z/c)$ where v_z is the bulk axial velocity (in cgs units).

These facts suggest an analogy with a radio or microwave transmission line, e.g. a waveguide, or coaxial cable – familiar laboratory concepts in which power is transmitted by some conduit into a "load". The load, Z, in general a complex quantity, can be modelled as a circuit having components of inductance L, capacitance C, and resistance, R (i.e. Z if a complex impedance). The active quantities, power flow (P), current (I), and voltage (V), are familiar in this context. In this paradigm, the central supermassive black hole (SMBH) can be thought of as an electrical power generator. The power flow, $I^2 Z$, from the SMBH jet progenitor at the galaxy nucleus is calculated to be in the range of 10^{38} W for a SMBH mass of 10^8 M$_\odot$, and may be affected by other ambient conditions close to the accretion disc. Hypothetically, the power output of a 10^{38} W jet would produce a $\sim 10^{18}$A current if it were fed into a 10 ohm impedance. The potential drop across this jet is $\Delta V \approx 10^{20}$ V, given a poloidal magnetic field strength of 10^4 G near the black hole. Here, ΔV might occur across the inner accretion disc, the dimension of which is a few times the BH's gravitational radius (r_G). Such a system can aptly be called Nature's ultimate power generator.

8.2.2 *Observational manifestations of the energy dissipation*

As the jet progresses, the electromagnetic power is transformed, either gradually or suddenly, into a mix of relativistic particles and gradually weaker magnetic field. At some later stage, the magneto-plasma may be largely thermalised. The detailed chain of processes by which this happens is not yet clear, and it could vary among different systems. It may, for example, happen more quickly in the hot thermal gas environment of a galaxy cluster. Here, the jets and lobes may more quickly entrain some of this gas, but such scenarios are not yet entirely clarified.

An independent and approximate verification of the total energy released, or "captured" in space from these SMBH-powered systems can be made from a simple experimental estimate of the accumulated energy that is deposited into the radio-visible lobes. The former is typically 10^{61}–10^{62} ergs for a large progenitor SMBH (see Fig. 8.1 and the related discussion in Section 8.3). This estimate of accumulated energy is independent of whether the jet power was carried electromagnetically by a Poynting flux or by a beam of charged particles.

Powerful massive black hole-jet/lobe systems occur in a great variety of forms, and only a small number have been individually analysed in great detail and in several wavebands. What they have in common is a luminous, often variable, galaxy nucleus source emitting radio, optical and γ-ray photons, a well-collimated jet, and (usually) an outer lobe structure of diverse morphology. The outer lobe structure is commonly asymmetrical about the host galaxy, with a jet that appears one-sided. A widely used, two-class morphological classification scheme (FR-I and II) was introduced by Fanaroff & Riley (1974), and

(a)

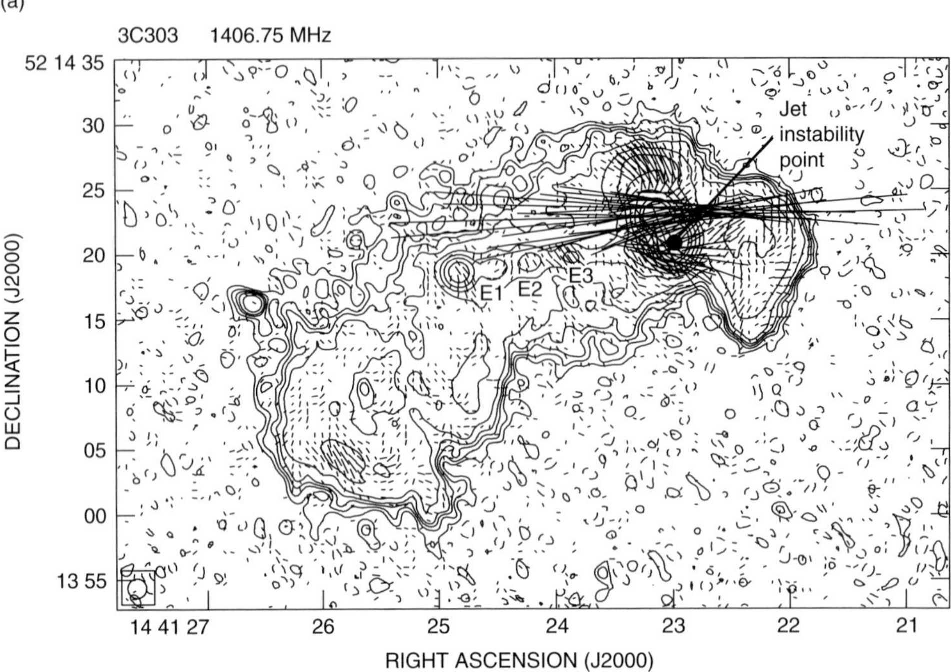

(b)

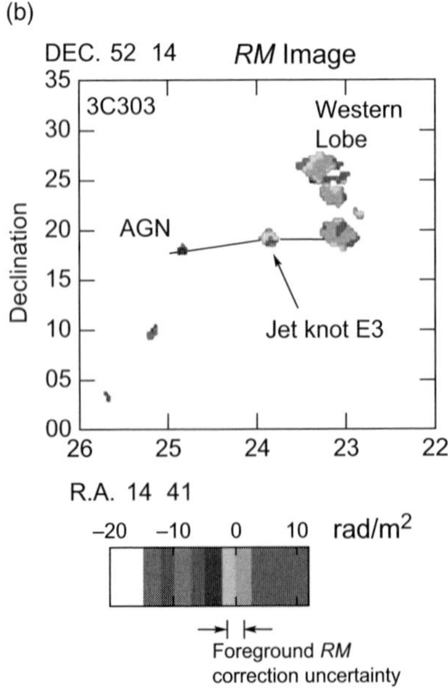

Figure 8.1. (a) The radio galaxy 3C303 at a distance of 600 Mpc. Stokes I contour levels at 1.4 GHz are shown at –0.5, 0.5, 1, 1.5, 2, 3, 6, 2, 24, 50, 100, and 150 mJy/beam, and the angular resolution is 1.5 arcseconds. Linear polarisation intensities are scaled such that a $1''$ equivalent line length = 2.5 mJy/beam. (b) an *RM* image showing the *RM* gradients in knot E3 and in the western lobe. (Source: Kronberg *et al.* 2011.) (c) 4.9GHz. 0.35$''$ resolution VLA image of the jet knots E1, E2, E3 (epoch of observation 1981.3). The highly variable, $< 0.001''$, inverse spectrum AGN source (Preuss *et al.* 1977) is at the left. (Source: Kronberg 1986.)

(c)

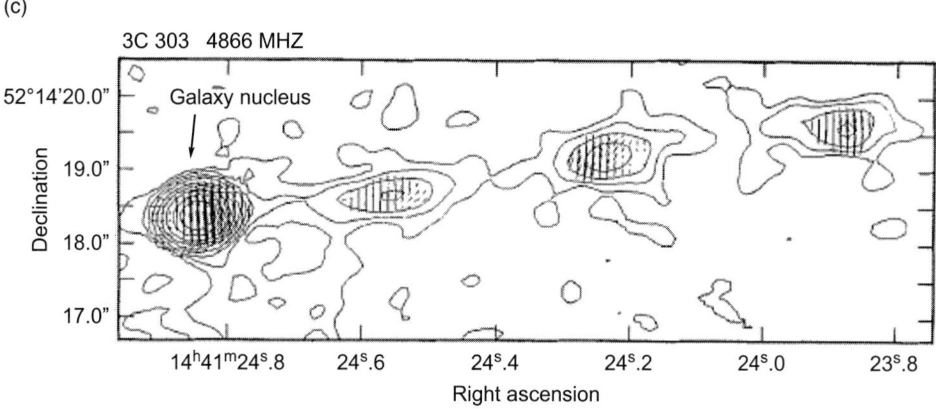

Figure 8.1. (*cont.*)

compilations of images can be found at the website www.jb.man.ac.uk/atlas. Jets in the lower radio luminosity class FR-I often have clear and measurable opening angles, and have been resolved and modelled with data from existing interferometers (e.g. 3C31 – Laing & Bridle 2008).

In higher luminosity FR-II sources the jet widths can remain approximately constant over large distances from the parent AGN (i.e. $\approx 0°$ opening angle); and very few, if any, are well-resolved transversely by existing interferometers. There is a longstanding need here for longer baselines *and* sub-GHz frequencies in order to create detailed Faraday rotation images with good transverse (to the jet axis) angular resolution. Lower frequencies at the highest possible resolution are dictated by the often very small Faraday rotations seen within these systems (see e.g. Section 8.3 below). Future improvements in angular X-ray, optical and i.r. resolution, and photon energy resolution will similarly be important for diagnosing the state of the jet plasma.

8.3   Plasma parameter estimates for a "typical" BH-driven jet-lobe system outside of a large galaxy cluster

8.3.1   *3C303 – a case study of a well-studied, moderately powerful radio galaxy*

The following illustrated discussion shows what has been gleaned from one such jet-lobe system, the 3C303 radio galaxy. It applies basic concepts that are discussed in Chapters 2 and 3 and elsewhere in the book.

3C303 at $z = 0.141(604\ h_{70}^{-1}$ Mpc) is one of many jet-lobe sources at a modestly large distance and luminosity. It is associated with a small group of elliptical galaxies not embedded in a large galaxy cluster. Its interesting one-sided jet, and association with a highly variable, compact ($\lesssim 0.001''$) galaxy nucleus source (e.g. Kronberg 1976 and 1986), initially attracted attention for further study in the radio, optical (Kronberg *et al.* 1977) and X-ray (Kataoka *et al.* 2003) wavelengths, and as a test case for MHD simulations of the jet. The jet knot E3 in Figs. 8.1(a) and (b) has a transverse *RM* gradient. This, combined with an independent estimate of the thermal plasma density around the jet (Lapenta & Kronberg 2005) plus a determination of the *RM* zero level (using independently measured background source *RM*s), indicates that the *RM* changes sign *on* the jet axis. This leads to the closest estimate yet made of an extragalactic jet current – ~$10^{18.5}$A – and its sign which, in this case is directed away from the AGN nucleus (Kronberg *et al.* 2011).

At the relatively high 5 GHz VLA resolution of 0.3″ (Fig. 8.1(c)), each knot from E1 to E3 is elongated by about 2 kpc in the direction of the jet. The knots are also individually visible in X-rays (Kataoka *et al.* 2003). Variations of Faraday rotation over the entire source, including the outer lobe are small (<20 rad m^{-2}) – much less than in a typical galaxy cluster environment or in the Milky Way. Combining the X-ray emission, which in this case is non-thermal, with the low Faraday rotation, the measured synchrotron emissivity and high polarisation degrees down to $v = 1.4$ GHz lead to the conclusion that Faraday rotation within the knots occurs in a very low-β plasma. The average density at $y \approx 0.5$ kpc from the jet's unresolved "spine" is $\lesssim 10^{-5}$ cm^{-3} – an intergalactic density range. These facts, plus the jet's low *RM* from Stokes *I, Q, U* images at five-frequencies, indicate a dominant Poynting flux power flow out to a distance of ~50 kpc from the AGN. The *RM* zero-level was estimated using radio sources in the neighbouring sky field (i.e. removing the Milky Way's and other foreground *RM*s). By applying this *RM* calibration it was found that the *RM* in the best resolved knot (and then only slightly so) passes approximately through zero on the jet axis. The knot also has measurable *RM* gradients, which are: $\nabla RM_z \simeq 0$, and $\nabla RM_y \simeq + 10$ rad m^{-2}/kpc, along and transverse to the jet respectively. The positive sign of the transverse RM_y gradient corresponds to a net positive current directed *away* from the AGN. The magnitude of ∇RM_y across the jet, $\simeq + 10$ rad m^{-2} kpc^{-1}, corresponds to a jet current of $\approx 10^{18.5}$ A.

8.3.2   *What causes a jet's sudden disruption?*

A notable feature of 3C303 in Fig. 8.1 is that it appears to be suddenly disrupted at a point, past which the lobe alignment becomes nearly perpendicular to the jet. Analogous morphological features are seen in many other radio galaxies and quasars. Deep optical images and spectroscopy reveal a faint, 22^m blue, and diffuse, spectrally featureless optical "cloud", or galaxy (Kronberg *et al.* 1977). In projection, at least, it is located exactly where the jet appears to be massively disrupted.

Possible explanations of the disruption seen in Fig. 8.1 are the following: (1) Intersection of the jet with a diffuse cloud, seen in a deep exposure optical field. Another idea (2) comes from soliton model MHD simulations of 3C303's jet. In these simulations, the magnetically dominated jet spontaneously reconfigures itself after ~30–50 Alfvén times (Lapenta & Kronberg 2005), even without ambient pressure variations, or major external disruptions. (3) A simulation by Nakamura *et al.* (2007) shows how a natural gradient in the ambient pressure of the host galaxy's potential well appears to cause a sudden flaring (rather than a deflection) into a radio lobe – see Chapter 7, Fig. 7.2. In that simulation, the jet's (electromagnetic) power flow axis (I_z) is exactly parallel to the ambient pressure gradient (∇P) of the galaxy/cluster halo. If the directions of I_z and ∇P had differed (the more general case), the broadening into a lobe might conceivably have looked more like the kind of disruption in 3C303.

The above short list of ideas is almost certainly incomplete, but exemplifies how internal and external influences may influence jet and lobe structures. Further examples of morphological variations in jets and lobes can be seen in Fig. 1.1 (2147+816), Fig. 8.2 (0634–20), Fig. 9.2 (Hydra A), Fig. 9.4 – three sources around Abell 119, Fig. 9.8 (M87), Fig. 11.2 (Centaurus A), and Fig. 12.4 (3C191 at $z = 1.9$). Further influences of galaxy cluster

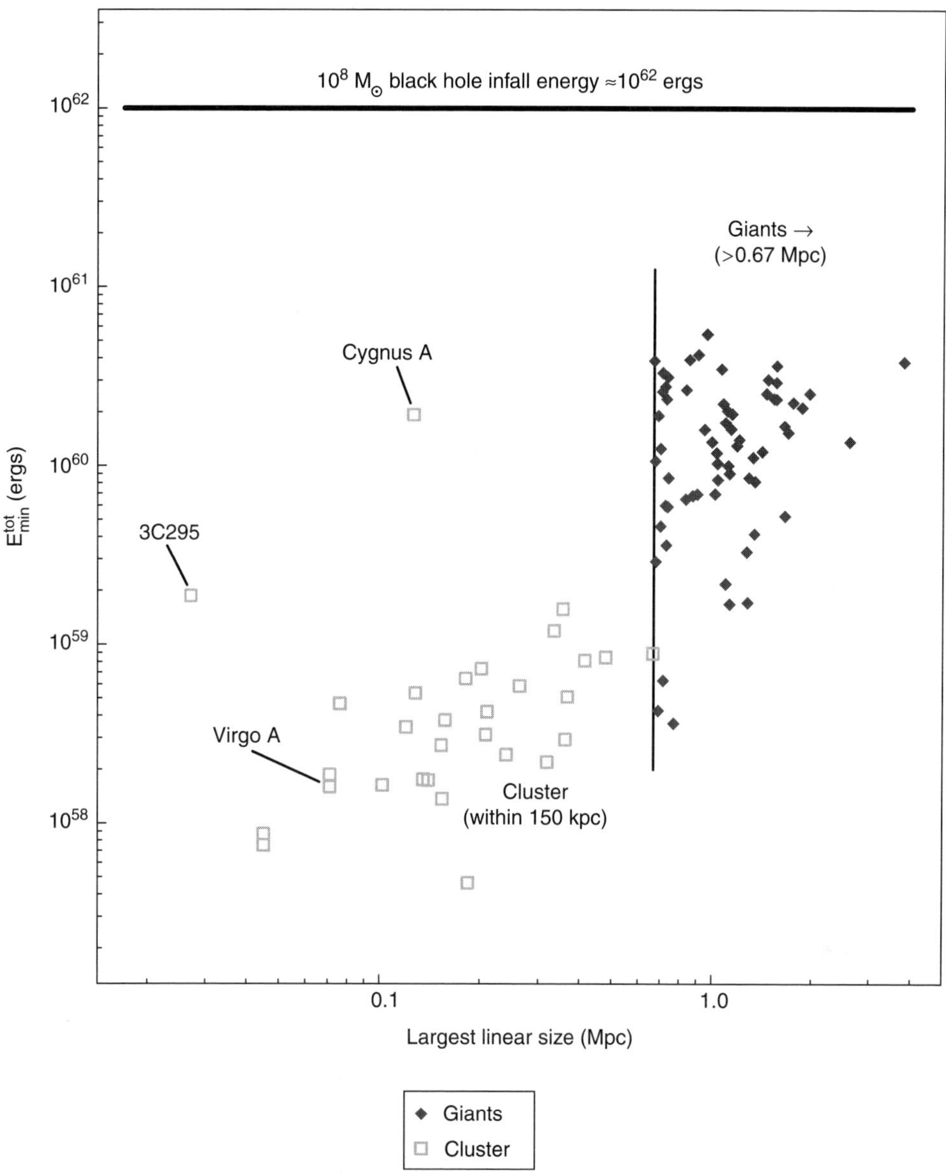

Figure 8.2. The total energy content in relativistic particles and magnetic fields plotted against projected linear size for (i) radio galaxies of projected size $\gtrsim 0.7$ Mpc (solid diamonds). The same quantity is shown for a sample (ii) of galaxy cluster-embedded radio sources that are within ≈ 150 kpc (projected) of the cluster centre. (open squares) (Source: Kronberg, Dufton, Li, & Colgate (2001)).

environments will be mentioned in Chapter 9. One interesting morphological sub-class of sources is characterised by symmetrically opposite jets, *each* of which abruptly changes direction on its respective side of the galaxy nucleus – that is, one side behaves in synchronism with the other.

One general clue might be discerned from the $1''$ resolution Faraday rotation image of the "post-disruption", western lobe of 3C303 (Fig. 8.1(a) and (b)). *RM* variations at higher resolution in this lobe are very small, as they are in the jet, and the magnetic field structure appears to be very regular within each kpc-scale, sub-lobe region. This overall behaviour suggests that the western lobe of 3C303 remains, post-disruption, relatively "starved" of non-relativistic gas. In this case it also appears that the western, post-disruption lobe is still magnetically dominated, like the jet. It gives a hint that, in the disruption, a major magnetic field reconfiguration has occurred. In summary, the substantial absence of non-relativistic gas in the complex western lobe of 3C303, combined with a high degree of kpc-scale field ordering, hints at a major role played by the magnetic field, as was suggested and discussed by Benford (1978). This "plasma-empty" property of 3C303 is common to other extended extragalactic radio Sources: Very low upper limits to the intergalactic plasma density were set from $\Delta RM \lesssim 1$ across the large radio galaxy 3C327 by Kronberg and Crelinsten (1973). The existence of very low intra-lobe gas densities was discovered and described in studies of two giant radio sources, 3C326 and 3C236, respectively by Willis & Strom (1978) and Strom & Willis (1980). It was also demonstrated in the GRG 0634–20 in Fig. 8.3 below.

8.3.3   Milli-arcsecond jet structures close to the black hole progenitor

Jets can also be studied on ultra-compact, parsec-scales, where the collimated energy flow is launched close to the presumed SMBH at the galaxy nucleus. As with kpc-scale jets discussed above, the currently available angular resolutions of 0.1 milliarcsec are not usually sufficient to resolve the jets' structures transversely. Imaging at these resolutions requires long baseline interferometry with baselines of up to several Earth diameters. This involves additional satellite-based antennas participating to create an adequately filled aperture, and extending to tens of thousands of kilometres. A first step in this direction was the Japanese VSOP-1 antenna with an apogee baseline of ca. 22,000km. A VSOP-II is planned at the time of writing, as is an ambitious RADIOASTRON multi-satellite interferometer designed in Russia. Well-calibrated images of *I, Q, U*, and *V* at these resolutions, and at different v_i for *RM* imaging, remain a challenge in the future.

8.4   Extragalactic jets as transmission lines and CR accelerators

8.4.1   Jets as analogues of a transmission line

Transmission lines in the laboratory serve as conduits of propagating electric (*E*) and magnetic (*B*) fields within some conducting boundary. In our astrophysical analogue the conducting boundaries are more diffuse. The idea that astrophysical jets could be thought of as transmission lines was mentioned several years ago by Benford (1978). The axial current is, at least initially, tightly confined, whereas the return current generally flows on a larger transverse ($|x,y|$) scale, and is more diffuse. A more detailed analysis of Poynting flux jets as analogues of transmission lines can be found in Lovelace & Kronberg (2013) and references therein. An important point is that various types of *load impedance* in the jet circuit can cause power reflections from a jet discontinuity – similar to the VSWR measurement in a laboratory transmission line. This leads to another possible explanation of the occasional striking regularity of "knots" seen in many powerful extragalactic jets, and also possibly to clues for mechanisms of particle acceleration in jets. Pursuing this idea further, the magnetic insulation, corresponding to an imbalance in the ratio of $|E|/|B|$, follow from an *inductive* component of the complex impedance of the discontinuity or "load" (Lovelace & Kronberg 2013). Wherever the situation $|E|/|B| \geq 1$ occurs, conditions might also exist in the jet for particle acceleration. The $|E|/|B| \geq 1$ condition is not normally possible in Ohm's law of resistive MHD:

$$E + v \times \frac{B}{c} = 0. \tag{8.3}$$

This can be seen from Equation (8.3) where, dividing by B and applying the inequality $v < c$ gives the condition

$$\frac{|E|}{|B|} < \frac{|v|}{c} < 1. \tag{8.4}$$

A backward reflected wave from an inductive load can cause the magnetic insulation to break down, and at least transiently to produce the situation $|E|/|B| \geq 1$. In an extragalactic context, unlike on Earth, the "transient" can last long enough to permit substantial and efficient particle acceleration. Particle acceleration in a magnetic reconnection layer (see Chapter 3) presents another example of the $|E|/|B| \geq 1$ condition, where $|B|$ can actually go through zero on the reconnection surface "diffusion zone" – Chapter 3.

8.4.2   Particle acceleration in jets

Applying the above ideas to astrophysical jets, the acceleration of particles of both charges can occur in the jet's z-direction. For the electrical parameters of SMBH-powered jets discussed in Section 8.2 ($\mathcal{O}$ 3×10^{18}A and 3×10^{20} V), accelerated particles in the transmission line model above, combined with a 10^4–10^7 year "transient" acceleration time can produce particle energies approaching the régime of ultra high energy cosmic rays (UHECR – Chapter 11). The particle acceleration mechanism is both direct and efficient. It can be distinguished from particle acceleration in shocks, and in a magnetic field reconnection zone.

Magnetised jets may serve in this, and other ways as accelerators of CRs, which, by synchrotron radiation, illuminate both the jets and the large, inflated radio lobes that are fed by the jets. The radio spectra in jets indicate that relativistic electrons must be either accelerated in the jet, or else be transported in a very low B-field environment from the galaxy nucleus, since their synchrotron radiation lifetimes are often much shorter than the transport time from the base of the jet – i.e. the galaxy nucleus, accretion disc, etc. (Willis & Strom 1978, Kronberg *et al.* 2011).

Various possibilities exist for relativistic particle acceleration in jets. As above, it can occur by magnetic reconnection (a collisionless process), in the electric field of the neutral layer close to the jet axis as a result of non-axisymmetric flows, and in turbulence. Reconnection acceleration in this context has been discussed by Romanova & Lovelace (1992). A magnetic reconnection mechanism for accretion discs was also proposed by Lesch (1991). Particle acceleration may be due to a combination of magnetic forces and Kelvin–Helmholtz type velocity shear instabilities (Romanova & Lovelace 1992). See also Königl & Choudhuri (1985), and Eilek & Hughes (1990). In addition, Fermi acceleration due to particle-Alfvén wave scattering in shocks and regions of plasma turbulence might be operative (Axford 1977, Blandford & Ostriker 1978). The proviso is that the particles' average gyroradius r_g is $\ll r_s$, the scale of the shock discontinuity, or that $r_g \ll \lambda_A$, the minimum Alfvén wave wavelength. Fermi-type mechanisms might be effective in accelerating CRs in strong shocks within in radio hotspots (Meisenheimer *et al.* 1989).

8.5   Probes of the internal gas physics in magnetised radio lobes and halos

8.5.1   *Test for the relative lobe-internal energies in relativistic electrons and magnetic fields using Inverse Compton scattered CMB photons*

Inverse Compton (IC) scattering, especially of microwave background photons off mildly relativistic electrons to produce X-rays, has been explored in a variety of astrophysical contexts (Felten & Morrison 1966, Rees 1967). In the commonly used bands of the ROSAT, Chandra, and XMM-Newton telescopes, ≈ 0.1–10 keV, IC scattering of the relativistic electrons (n_r) in radio lobes can compete with synchrotron losses, and with thermal bremsstrahlung as a source of X-ray emission. The relative volume emissivities of synchrotron and IC radiation can be used to estimate, or limit, the magnetic field strength in the lobes. Synchrotron emissivity by itself does not reveal the ratio of the energy densities of the relativistic electrons (ε_{er}^{tot}) to those of magnetic fields ($\varepsilon_B = B^2/8\pi$). The former is given by

$$\varepsilon_{er}^{tot} = \int_{E_{min}}^{E_{max}} N(E)\varepsilon_{er}(E)dE \text{ where } \varepsilon_{er}(E) = CE^{-n}, \tag{8.5}$$

n being the power law index of the relativistic electrons' energy distribution.

Magnetic field estimates derived in this way provide one of the few tests available to separately establish, or limit, the relative energy density of relativistic electrons and magnetic fields and hence the magnetic field strength. Croston *et al.* (2005) and Kataoka & Stawarz (2005) have measured X-ray fluxes for a sample of relatively radio-bright 3C radio source lobes and compared the IC-limited magnetic field strengths with the magnetic field derived for the same lobes on the assumption of energy equipartition between ε_{er}^{tot} and ε_B. (see Chapter 2). They concluded that the IC dominated X-ray emissivities are consistent with approximate energy equipartition between ε_{er}^{tot} and ε_B. This comparison, and conclusion, does not directly account for ε_p, the energy of the protons, or CR nuclei. However, the apparent near-equivalence of ε_{er}^{tot} and ε_B indirectly suggests that ε_{pr}^{tot} is not much larger than ε_{er}^{tot}, and probably not ~ 100 times ε_{er}^{tot} as has been suggested for the local Galactic CR population. Modifications to the ratio of ε_{er}^{tot} to ε_B, and hence ε_B estimates from synchrotron emissivity, may need to take account of a revised distribution in energy of the CR electrons producing the synchrotron radiation. Corrections of this sort, discussed by e.g. Beck & Krause (2005), are influenced by the relative particle energy density at the lowest particle energies. These will be better determined in the future with the help of imaging radio telescopes at frequencies down to a few MHz.

8.5.2   *Magnetic field deduced from self-Compton and synchrotron emission in radio hotspots*

Given a sufficiently high radio surface brightness, the volume density of local synchrotron photons can be comparable to the CMB photon density. Since IC emission can be treated as a classical scattering process with cross section σ_T (provided that $E_e \cdot E_{photon} \ll m^2c^4$), this process of "synchrotron self-Compton scattering" (SSC) produces a slightly different X-ray spectrum from thermal bremsstrahlung. This has also been observed, leading to magnetic field estimates of order 200 μG in some outer hotspots (e.g. Hardcastle *et al.*

2004). We mention incidentally that some hotspots, particularly in less luminous radio sources, have X-ray luminosities that exceed what SSC predicts. A combination of their $\varepsilon_e N(E)$ spectrum with the X-ray spectrum led Hardcastle *et al.* (2004) and others to conclude an additional *synchrotron* X-ray component. This result (see also Röser & Meisenheimer 1987) is of interest in another context, where we consider CR acceleration in locations that are far from the central BH energy source. A constraint is that X-ray emitting synchrotron electrons have a very short lifetime, and hence a correspondingly small transport distance from their acceleration site. That, in turn, puts constraints on the acceleration mechanism.

Optical emission in radio lobes far from the original energy source at the galaxy nucleus has been verified in many extragalactic radio lobes and jets. One of the first for a distant radio galaxy was 3C303 (Lelièvre & Wlérick 1975, Kronberg 1986, Kronberg *et al.* 1977) with its diffuse blue light object that coincides with the end of the jet (Section 8.3). Objects showing optical linear polarisation usually confirm the presence of synchrotron radiation (for the 3C273 jet, see Röser & Meisenheimer 1986, and for Pictor A, Röser & Meisenheimer 1987). More recent large samples of faint optical emission have been analysed and discussed by Gopal-Krishna *et al.* (2001) – and interpreted as IC-scattered photons. In another, more recent analysis, Brunetti *et al.* (2003) discuss and review the evidence for optical emission in outer radio hotspots in this context, and conclude that the evidence favours *in situ* acceleration in the radio hotspots. In addition to the explanation offered by Gopal-Krishna *et al.* (2001), this phenomenon can alternatively be understood in terms of loss-free transport in an environment of force free motion or as a natural consequence of Poynting flux energy transport, as discussed earlier. The latter scenario is consistent with relativistic electrons directly accelerated *in situ* in the strong radiation field of the Poynting flux. This energisation process can also happen along the jet, as mentioned earlier.

A companion phenomenon is the presence of freshly accelerated electrons, that appear throughout the lobes seen in some megaparsec-scale, giant radio sources such as the southern sky radio galaxy, 0634–20 in Fig. 8.3. Freshly accelerated particles in the lobes may well be produced in lobe-internal filaments, so far verified in only a few nearby objects such as Centaurus A and Fornax A. The instrumental "virtue" required to image such faint filaments is a combination of both higher resolution *and* high surface brightness sensitivity. Given this, they may in future be found to be ubiquitous.

The morphology and spectrum of 0.1–2 keV X-ray emission of the outflow gas around starburst galaxies such as M82 (e.g. Schaaf *et al.* 1989) has been analysed to estimate the inverse Compton scattering component of the synchrotron-emitting electron population. In contrast to CMB photon scattering (above), and self-Compton scattering in regions of high synchrotron photon density, these can scatter FIR photons that are generated in the under-lying nuclear starburst region close to the radio synchrotron halo.

8.5.3   *Faraday rotation, depolarisation, ε_B, and n_{th} in radio lobes*

Faraday rotation involves the non-relativistic or thermal component of gas in radio lobes that can also be detected at some level by X-ray thermal bremsstrahlung. This can be done by uniquely identifying the radiation spectrum as inverse Compton, or by identifying X-ray transitions, e.g. emission lines, in the keV band, which would confirm some fraction of thermal emission within the lobes. Definitive specification of *thermal* plasma within radio lobes has been difficult, partly because for most sources the greatest contributions to Faraday

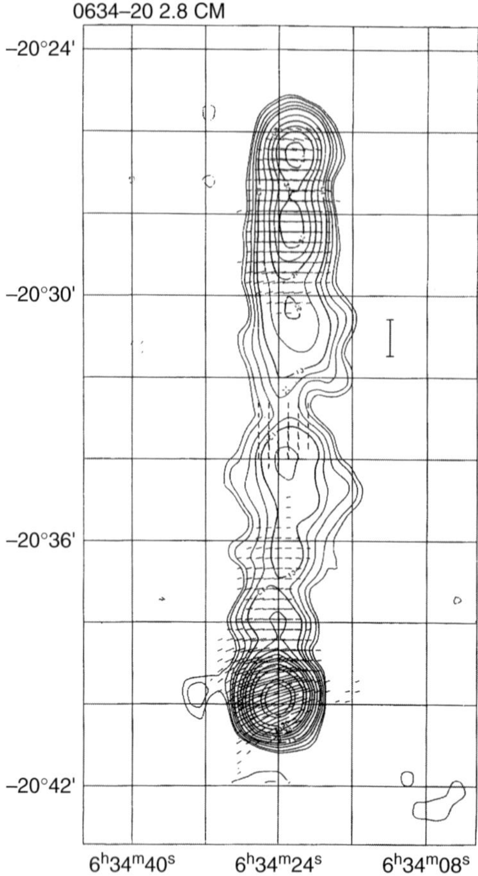

Figure 8.3. A 10.8 GHz image of the giant radio galaxy 0634–20 at $z = 0.056$, and having a projected overall size of ~1 Mpc. It shows highly polarised emission, the orientation of which is normal to the projected magnetic field direction. The total energy calculation for Fig. 8.2 used a higher resolution radio image at 1.4 GHz. The image was made with the Effelsberg 100 m telescope. (Reproduced from Kronberg, Wielebinski, & Graham (1986).)

rotation come either from the Galactic foreground, or from a gaseous halo around the source. Sensitive Faraday rotation images of giant radio sources at sub-GHz radio frequencies can test down to extremely low levels of nonrelativistic gas within the lobes (e.g. Willis & Strom 1978). However some relatively more compact radio lobes, such as in Fornax A reveal a filamentary optical emission line and radio structure. This, as in the Crab Nebula (Chapter 2), might prove much more common, given sufficient angular resolution and surface brightness sensitivity in a Faraday rotation image. Techniques of Faraday rotation synthesis described in Chapter 2 (Section 2.7) can be effective in delineating different plasma density zones within radio lobes.

A companion to the Faraday *RM* test for thermal plasma is measurement of the degree of depolarisation of the lobe in question. Distributed thermal plasma would reduce the polarisation degree most sensitively at the longest wavelengths (see Section 2.7). Furthermore, the depolarisation rate does not normally depend crucially on the field ordering parameter, μ (see Fig. 2.3). Some of the most sensitive limits on thermal plasma in a large radio lobe, ~5 $\times$ 10^{-5} cm^{-3}, were deduced in observations by Willis & Strom (1978) for the GRG 3C326 which they imaged down in frequency to 0.6 GHz. Large source dimensions and long radio wavelengths are favoured for exploring magnetic fields in the lowest plasma density régimes.

8.6 SMBH masses and magnetisation of the IGM

The existence of highly collapsed objects has been independently verified by both imaging and spectroscopic measurements of the orbital speeds around a central super-massive object. For our Galaxy, direct near-in Keplerian velocity measurements give 2.6×10^6 M$_\odot$ (Genzel & Eckart, 1999). For NGC4258, Keplerian velocity curves from orbiting H$_2$O vapour maser line emitters yield 3.6×10^7 M$_\odot$ (Miyoshi *et al.* 1995). Since the 1990s it has been verified that central supermassive objects exist in all moderately large galaxies (e.g. Tremaine *et al.* 2002). They found via correlations between the BH mass and the (easier to measure) bulge luminosity of the galaxy in question. Such observation-based mass estimates range from $\approx 10^{6.5}$ to $\approx 10^9$ M$_\odot$. Observational capabilities are close to verifying the space-time distortions that would uniquely confirm that these are black holes. But we remain aware that full physical description of the interior of black holes is not completely elucidated, and more guiding experimental evidence and theoretical insight are needed. For the discussion in this chapter it will suffice to say that large galaxies contain massive and compact central objects of the mass and approximate dimensions mentioned above, which form the end point of gravitational infall.

At the other end of the SMBH-IGM connection are the extended lobes of radio galaxies and quasars that range from a few kpc to Mpc (see also Chapter 7). It has been demonstrated that strong radio source jets and lobes can magnetise the intergalactic medium, (Rees 1987, Ruzmaikin *et al.* 1989, Jafelice & Opher 1990 & 1992, Daly & Loeb 1990, Kronberg *et al.* 2001, Gopal-Krishna & Wiita 2001).

There is also accumulating evidence to suggest that the magnetic morphologies and related current structures *within the lobes* could be the cause of significant particle acceleration or re-acceleration. (Benford 1978, Benford & Protheroe 2008) – i.e. not just in the jets, hotspots and at the AGN nucleus. This has additional implications: Relativistic electron lifetimes ($\propto B^{-2} E^{-1}$) and their influence on the radiation spectrum have been observed to vary significantly within large lobes. These timescales, including inverse Compton loss times, can "compete" with the intra-lobe transport times of the relevant radiating relativistic particles. Other evidence suggests that the lobes are at least partly governed by magnetic stresses (e.g. Taylor 1986, Benford, 2006, Diehl *et al.* 2008), and by the related current structure, and fed by Poynting flux-dominated jets.

Given our currently incomplete knowledge of the distribution in cosmic time of earlier populations of extended radio sources, it is difficult to accurately calculate the global effect of radio sources on the general IGM field. Daly & Loeb (1990) produced calculations which yield a resulting $B_{\mathrm{IGM}} \sim 10^{-11}$ gauss. If the seed fields of galaxies came from previous generations of extended extragalactic radio sources, then "central BH-seeding" of the IGM must have been efficient at least by redshifts of ≈ 4 and probably earlier. Extended radio sources have been found up to $z \approx 6$, and they could have existed even earlier, where the energies of their synchrotron-emitting electrons were diminished, or "snuffed out" by inverse-Compton scattering off the much denser photon field at higher redshifts (e.g. Rees & Setti 1968).

In the latter case, the magnetic fields would largely remain, still illuminated by CR electrons, and detectable through synchrotron emission at very low frequencies – bearing also in mind that the images we see are received at a frequency that is lower by $(1 + z)$ than that which is emitted. Ideally, old relativistic electrons in the radio lobes, associated with "fossil" magnetic fields, would best be revealed at frequencies below ~10 MHz. The desired detailed images of high-z extended sources between ~10 and ~1 MHz are less likely to be well-defined with Earthbound radio arrays. One reason is because 10 MHz is not far above the typical critical frequency of the Earth's ionospheric F2 layer (f_0F2). Radio imaging down to ~1MHz would be an important scientific mission for a radio array on the far side of the moon. A Lunar far side placement of a low frequency telescope would permit observations away from the Galactic plane at times when the Earth, the Solar corona, and Jupiter (an intermittently strong low frequency radio emitter) are all below the observing horizon.

8.7   Some basic calculations relating to BH-powered outflow

The infall energy onto a black hole can be simply calculated from basic physical principles. We use the Schwarzschild radius, R_S, as a fiducial end-point of infall. That is also where the photons can no longer escape. R_S is given by

$$R_S = \frac{GM_{BH}}{c^2}.$$ (8.6)

The potential (infall) energy, or E_{inf} of any mass that has collapsed to a radius, r, is

$$E_{inf} = \frac{GM^2}{r}.$$ (8.7)

Substituting $r = R_S$, and introducing an output energy conversion efficiency factor η, a very simple expression results for the output energy in terms of M_{BH}:

$$E_{out} = \eta E_{inf} = \eta M_{BH}c^2.$$ (8.8)

The mechanisms of energy conversion and η (see e.g. Sections 7.6 and 7.7) are left to be specified at this point. We see that for a galactic black hole mass of 10^8 M$_\odot$ and the assumption of, say $\eta = 0.1$, then $E_{out} = \eta_{0.1}Mc^2 = 1.8 \times 10^{61}$ ergs. If, as is likely, the BH system has significant angular momentum (Section 7.5), the Schwarzschild space–time metric becomes a more complex Kerr metric which, in the equatorial plane of rotation increases in radius by a modest factor of $\simeq$1–4. It reduces the available gravitational energy reservoir by about this factor. This consideration is relevant only if the effective end point of infall is close to R_S (Figs. 7.4 and 7.5), and not further out. Furthermore, this factor would not apply in the rotation axis direction for a Kerr metric.

Since the radio luminosity is produced by synchrotron emission, knowledge of a system's radio structure, luminosity and redshift permits us to make approximate, very important estimates of the total energy in magnetic fields and CR particles in the radio lobes. With this knowledge, we can attempt to quantify the magnetic energy output into the IGM from populations of central black holes in the Universe. Such a calculation is important for tracing magnetic energy in the mature universe. The basis for this quantification is described next.

8.8 Observational/experimental quantification of BH energy output to the IGM

In connection with the above, it is instructive to compare the minimum total energy, $E_{\min}^{\mathrm{tot}}$, in the radio lobes, quantitatively estimated from observations, with the calculated energy available from BH gravitational infall. Figure 8.2 shows a plot of $E_{\min}^{\mathrm{tot}}$ against the largest radio dimension for two different subclasses of radio source. These are (i) giant radio galaxies (GRGs), "giants" selected with a projected overall dimension ≥ 0.7 Mpc, and a second, control set (ii) of extended radio galaxies *within* a projected 150 kpc of a galaxy cluster's centre. Figure 8.2 shows representative samples for these two categories, for which observation-based energy estimates are plotted against the sources' projected linear size.

The result is striking, in that the upper envelope of a typical GRG's energy content in Fig. 8.2 is only about 1.5 orders of magnitude below the theoretical infall energy of a 10^8 M$_\odot$ BH down to R_{S}. This is remarkable, because it illustrates that more of the BH-released energy is "captured" in GRG lobes than in any other known single-galaxy system. This means that the GRGs are excellent *calorimeters* of central BH energy conversion into CRs and magnetic fields. Consequently, as a class they especially suited for comparison with the independent calculations of energy release from a central black hole.

The factor of $\sim\!10^{1.5}$ gap in Fig. 8.1 is even more noteworthy when we realise that *other* forms of released energy exist which are not represented in the lobe-internal energy calculations. That is, they are not represented in Equation (2.17), and hence not in the plot in Fig. 8.2.

A "budget" of these additional energy components not counted in Fig. 8.1 is as follows:

1. Synchrotron and inverse Compton cooling losses, both $\propto E^2$. Some of the energy injected into the lobes will have been lost by the epoch at which we observe the radio source.
2. Due to projection effects, the average true source volume (V) is inevitably larger than indicated by its projected (observed) dimensions. $E_{\min}^{\mathrm{tot}}$ is proportional to $V^{3/7}$.
3. Particle escape: At all evolutionary stages of a mature radio lobe, some unknown fraction of the CR particles will have escaped from the system.
4. *PdV* work on the ambient medium. This is independently measurable in the denser medium in clusters, but not (yet) in lower density local IGM environments around the GRGs. It is probably the main cause of the cluster/non-cluster member gap in Fig. 8.2.
5. Free expansion energy loss (independent of E_{p}).
6. Photon energy produced at higher energies from infrared to γ-rays (discussed below), although some of this is linked to the inverse-Compton losses cited in item 1.

Another poorly known parameter in this context is the ratio of CR nucleus-to-electron energy k (Sections 2.11 and 8.5.1). It could serve to reduce the total energy requirement of a radio lobe relative to the case where $k = 100$ which was adopted in Fig. 8.1. This is the approximate ratio in local Galactic CRs. For example, adopting $k = 1$, then the lobe energy content, $\propto (1 + k)^{4/7}$, would be reduced by a factor of 9 in Fig. 8.2. Alternatively, if k were $\sim\!10$ in these large lobes, the total energy contents would be $\sim\!3.5$ times lower , i.e. increasing the gap by this factor below the $k = 100$ case represented in Fig. 8.2.

Apart from effects that would increase the measured estimates of the total energy released from the central BH, the gravitational energy reservoir of the central BH is itself subject to a downward revision (i.e. reducing the gap in Fig. 8.2). Recall that E_{inf} was scaled to the final infall radius $r = R_S$ (Equation 8.9). If E_{inf} is extracted at a larger effective radius, $r_{eff} > R_S$, say somewhere further out in the BH's inner accretion disc, then

$$E'_{inf} = E_{inf} \left(\frac{R_S}{r_{eff}} \right) \tag{8.9}$$

where

$$R_S = 0.985 \left(\frac{M_{BH}}{10^8 M_\odot} \right) \text{A.U.} \tag{8.10}$$

A factor that would, on the other hand, reduce the energy content estimates of the lobes is a strong filamentation of the lobes. In Fig. 8.2, this effect was already roughly built-in with the global assumption that the relativistic magneto-plasma fill only 10% of their radiating volume. More recent radio lobe images with high sensitivity and resolution confirm that such filamentation is general. The global effective volume filling factor ϕ (Equation 2.17) may be less than 10%, which would increase the vertical gap in Fig. 8.2. Note, however that this volume filling factor scales the internal energy estimate by only $\phi^{3/7}$, so this effect is not likely to be overwhelming.

A related point of interest is that the CR protons ($\gamma \sim 10^2 - 10^4$) retain their original energy for much longer than the radio emitting relativistic electrons. For the protons, the most likely energy loss will be by p–p collisions with thermal protons having density ρ_p^{th}, producing pions and high energy photons. However outside galaxy clusters, and especially in the rarefied lobes of GRGs, ρ_p^{th} is smaller, $\lesssim 10^{-5}$ cm^{-3}, and the timescale for p–p collisions is very long.

Calculation of the minimum lobe total energy content contains a further uncertainty in our knowledge of the entire CR proton energy spectrum within the radio lobes. This, as mentioned earlier, is difficult to measure directly. However, in future, measurements of γ-ray emission from p–p collisions in denser, e.g. cluster, environments where the proton density is sufficiently high may permit direct measurement – see Chapter 9. This limitation adds to the uncertainty in the effective k-factor. The total CR energy that we infer from the observable radio spectrum may well be somewhat larger, e.g. at γ below ~500. It will depend on the acceleration mechanisms in large radio sources and on the preponderance of old "fossil" electrons in the lower, trans-relativistic energy range. The electron CR energy spectrum that corresponds to the usual upper and lower radio band cutoffs is between ~10^{13} and 10^{14} eV ($\gamma \sim 2 \times 10^4$ and 2×10^5). By contrast, the directly observed spectrum of CR nuclei in the Milky Way covers ~10^{11} to ~3×10^{20} eV (Gaisser & Stanev 2000), a spread of more than nine orders of magnitude!

It is fortunate that one of the "invisible" energy components (No. 4 above) can be independently measured – in galaxy clusters. This is the PdV work that the radio lobes must expend as they expand against any ambient intergalactic gas pressure. At present, PdV energy loss can be measured only within galaxy clusters, because only there can the pressure and density of the confining thermal intracluster gas be independently established via its X-ray-visible thermal emission. This is one reason why a galaxy cluster environment provides an unusually good "laboratory" for studying the physics of radio lobe evolution – as discussed in Section 7.4.

It is for these reasons that cluster-embedded radio sources were chosen as a reference source sample (ii) in Fig. 8.2. Their central galactic BH progenitors (and primary energy source) are presumed to be at least as massive as in the first sample of GRG host galaxies. However, their measured relativistic plasma energy is typically 10^{59}–10^{60} ergs, strikingly lower as a group than that of the GRGs in the first sample. This large difference, seen in Fig. 8.2, can plausibly be attributed to energy loss by PdV work against the dense, hot ICM gas, in addition to some thermal heating of the ICM gas by the expanding and rising lobes. Independent estimates of the PdV work for radio lobes in clusters gives further confidence that we can understand the energy flow paths of BH-generated extragalactic radio sources. With more sensitive X-ray, γ-ray, and low frequency instruments in the future, it should be possible to perform the kind of experiment we have just described for some intergalactic environments that are less dense than in large galaxy clusters, but more dense than LSS filaments and voids.

8.9   Implications of constraints imposed by the energy gap in Fig. 8.2

From the above and Chapter 7 it is clear that the physics that correctly describes the conversion and extraction of gravitational energy is not yet fully understood. One model, the accretion disc dynamo discussed earlier, operates near ≈ 10 AU.

In summary, the upward corrections that we should apply to the GRG energies, if better quantified in Fig. 8.2, would give a better representation of the total BH energy released. Whether, and by how much ηE_{inf} should be downward-corrected, further narrowing the gap in Fig. 8.2, depends on the mechanism by which gravitational infall energy is extracted to produce the output magnetic, particle and photon energy. The interesting fact to note is that both plausible downward corrections to ηE_{inf} and missing energy components not included in the (measured) synchrotron radiation-estimated energies (an upward correction) would combine to *narrow* the gap. If other parameters become well enough known to force k to be small, that in itself could clarify the constraints on the physics and energetics of the radio lobes.

In the absence of a comprehensive model of energy transport from the BH system it is not possible to decide which combination of parameters best accommodates the energy gap in Fig. 8.2. The conversion efficiency from gravitational infall to magnetic energy in lobes is remarkably high, and is on the verge of setting other constraints. Magnetic energy can be expelled from a galaxy into the IGM, and transported to scales of kpc–Mpc within cosmologically short times. This whole chain of processes is a remarkable and still unclarified phenomenon of Nature.

8.10   Additional calculations of global energy release from galactic BHs into the IGM and estimates of the photon energy component

8.10.1   *The average mass density of central galactic black holes*

A simple calculation of the global energisation of the IGM can be made by combining the BHs' dynamically estimated masses from velocity dispersion data in low z bright galaxies with galaxy space density surveys, (Tremaine *et al.* 2002, Pinkney *et al.* 2003), and with the gravitational energy released per galactic BH in Equation (8.7). Recall that Equation (8.7) assumes infall of the entire BH mass to the Schwarzschild radius, R_S. Co-moving values of the mean space density of galactic black holes, ρ_{BH}, have been estimated at

$$\rho_{BH} \approx 2.2 \times 10^5 M_\odot \ \ \mathrm{Mpc}^{-3}. \tag{8.11}$$

Some years before velocity dispersion data were established, Soltan (1982) had pioneered similar estimates of the aggregate QSO photon energy distribution based on quasar populations around the "quasar epoch" at $z \approx 2$. The Soltan method converted the aggregate QSO light output to ρ_{BH} by assuming that the QSO light output represented some fraction, assumed ≈ 0.1, of the gravitational infall energy to the QSOs' central BHs – similarly to what we described for the radio lobes above. These quite independent estimates of ρ_{BH} were subsequently revised and refined by Chokshi & Turner (1992). The latter, in turn, were based in large part on optical QSO luminosity functions and related data compiled by Hartwick & Schade (1990). The results were interestingly close to Soltan's earlier estimate for $<\rho_{BH}>$. Furthermore, the photon efficiency estimate of ~ 0.1 is comparable with the magnetic and CR energies released by galactic BHs. It is also consistent with the constraint that all released energy forms should not, according to this paradigm, exceed E_{inf}. Thus, we see that the average intergalactic space density, for BH masses above $\rho_{BH} \approx 10^{6.5}$ $M_\odot$ is a well-established number, consistent with both light output-based calculations at the QSO epoch and direct galactic BH mass estimates at low redshift.

8.10.2   *Global estimates of magnetic energy density from galaxies*

The information discussed above can be used to arrive at a global estimate of the intergalactic magnetic energy density that is spread into the filaments of large scale structure (LSS) in the present universe. Our calculation needs to include the central BH mass, and the fraction, f_{RG}, of galaxies having central BHs that produced a radio galaxy or radio-luminous (RL) QSO over a Hubble time. Implicit in this reasoning is the assumption that a jet-lobe radio source system is required to inject magnetic fields into the IGM and that, once "planted" in the IGM, the magnetic fields remain there for much of the subsequent cosmic time.

The visible lifetime of a radio source is about 10^8–10^9 years, less than 10% of a Hubble time. η_B is the fraction of BH energy released by the host galaxy as magnetic energy (Equation 8.3). A further global assumption is that the ejected magnetic fields do not fill all of space, but are mostly contained within the LSS filaments and sheets, or "galaxy zones" of the Universe, which fill only a fraction f_{FIL} of all intergalactic space. This latter assumption does not include the possibility that, over cosmic time, magnetic fields and CRs may partially diffuse into the voids on some cosmological timescale. For initial reference, we assume that $\eta_B = f_{RG} = f_{FIL} = 0.1$, and that $<M_{BH}> = 10^8$ $M_\odot$.

Given the above assumptions, the average magnetic energy density, ε_B, that is injected into intergalactic space can be calculated as follows:

$$\varepsilon_B = 1.36 \times 10^{-15} \left(\frac{\eta_B}{0.1}\right) \times \left(\frac{f_{RG}}{0.1}\right) \times \left(\frac{f_{FIL}}{0.1}\right)^{-1} \times \left(\frac{M_{BH}}{10^8 M_\odot}\right) \ \mathrm{erg \ cm}^{-3}. \tag{8.12}$$

Hence, for the scaling numbers discussed above, the smoothed-out, galactic BH-seeded intergalactic magnetic field in LSS filaments is

$$< B_{IG}^{BH} > = \sqrt{8\pi\varepsilon_B} = 1.8 \times 10^{-7} \ \mathrm{G}. \tag{8.13}$$

Despite uncertainties in the scaling factors discussed above, this is an interesting number. It is sufficiently high to be within reach of observational tests. The similarity of ε_B to the density of quasar background light, ε_{QL}, is also striking when we note that Richstone's (2004) estimate of $\varepsilon_{QL} \approx 1.3 \times 10^{-15}$ erg cm^{-3} is quite close to the magnetic energy density in Equation (8.12) above. The *closeness* of agreement between these independently estimated numbers may be deceptive, given the present uncertainties in specifying η_B, f_{RG}, f_{FIL}, and $<M_{BH}>$ above, and additional uncertainties inherent in the estimate of ε_{QL}. But it raises the speculation that intergalactic magnetic fields could be a proxy indicator of the IGM photon density and *vice versa*. If so, does $\varepsilon_{QL}(z) \approx \varepsilon_B(z)$ hold back to the epoch of the BH formation era? In that case, we would expect $<B_{IG}{}^2> \sim (1 + z)^4$ to hold over a wide redshift range if it scales with the intergalactic photon energy density. Does this similarity indicate a more fundamental and/or cosmological significance that has not yet been fully understood? Answers to these questions go beyond the scope of this book. Nevertheless these interestingly coincidental numbers underline the importance of understanding the origins and evolution of cosmic magnetic fields.

The foregoing sections give motivation for discussing intergalactic magnetic fields on large scales, beyond those of galaxy clusters in the wider intergalactic medium. This is also discussed in Chapters 10 and 11. Meanwhile, we explore some of the wider astrophysical ramifications and implications of the conversion, in galaxy nuclei, of large amounts of gravitational energy into magnetic fields and relativistic particles.

8.11   Some consequences of "captured" energy release from galactic BHs

The evidence for an approximate similarity between the photon energy released, and that in magnetic fields and relativistic particles has the following further consequence: The photon energy escapes at $v = c$ as soon as the optical depth around the BH/Quasar becomes small. But the magnetic energy, which we see distributed within the radio source volumes, escapes or diffuses at a much lower speed to form the expanding radio lobes. In a subsequent phase, it will diffuse into the wider IGM, but at a speed $v_{diff} < c$. In this sense, the magnetic energy represents a kind of "captured" energy, though of comparable magnitude to the emitted photon energy.

Thus, on a grand scale and over time, magnetic energy is captured within up to a few megaparsecs for long periods. This fact and the diffusion timescales involved determine the flow of the BH-IGM feedback energy into the wider intergalactic medium. The magnetic energy gets retained within a much smaller IGM volume than what the photons will fill. For further discussion see Kronberg *et al.* (2001), Gopal-Krishna & Wiita (2001), and Furlanetto & Loeb (2001). The latter authors produced calculations to show that the global IGM filling factor of radio lobes may end up as a significant fraction of unity.

This BH-driven energy flow into the surrounding IGM differs in many respects from the star/SN driven galactic winds described in Chapter 6. However, one conclusion is similar, namely that primeval galaxy outflows at $z \gtrsim 10$ could, by star/SN-driven outflows alone, also fill a significant fraction of the IGM filament *volume* by $z \lesssim 7$ (e.g. Kronberg *et al.* 1999). A principal difference, though, is that the energy density from the stellar energy release processes is, when diluted by cosmic expansion to $z \approx 0$, much less than the global energy reservoir available from central BHs of galaxies.

Outwardly transported magnetic fields may remain dynamically important, perhaps for most of a Hubble time. This is because they may interact more strongly with the surrounding IGM than the radiation. For example, direct evidence has been found that star formation is triggered around the overpressured peripheries of FRI radio sources (McNamara & O'Connell 1993). Over-pressured radio lobes, which may have occupied a significant fraction of the early filaments and walls of large scale structure, could have played an important role in triggering star formation, and may also have been a factor in the so-called "quasar epoch".

8.12   Summary of some questions

The connection to energy release from black holes in galaxy nuclei is linked to a number of open questions in physics and cosmology. The timescale over which a jet injects the gravitational-to-magnetic energy into an initial, but significant, volume of intergalactic space is cosmologically very short, 10^6–10^8 years. It is appropriate at this point to list some related puzzles and questions, though our list is by no means complete.

1. Gravitational collapse of baryonic and dark matter to form stars and black holes: How does a self-gravitating, rotating cloud shed its angular momentum efficiently and quickly enough to form supermassive objects within less than a Gyr after the Recombination epoch? The solution(s) to this puzzle may also clarify how stars and planetary systems formed. In both cases angular momentum must be efficiently removed, which implies that the magnetic fields are well coupled to the collapsing matter.

2. What are the processes in the accretion disc and near the BH's ergosphere that so efficiently couple the infall energy into magnetic fields?

3. Related to the above two issues is the question of how the released energy is tightly collimated, both in stellar and AGN systems, and for the latter what makes the collimation persist from dimensions of a few AU to a megaparsec?

4. What is, or what are, the mechanisms of particle acceleration, and how do they produce the very high energy cosmic rays (UHECR), discussed later, that appear to be extragalactic?

5. Given that magnetised CR gas and non-relativistic gas are seeded into the IGM of the filaments and walls of LSS, how does the IGM magnetic field subsequently evolve? Connected with this phenomenon is the AGN-black hole feedback from galaxies, which injects fresh fields and CRs on kpc–Mpc dimensions on timescales of $\sim 10^8$ years while the Universe expands.

6. How do IGM magnetic fields, if in energy balance with thermal, photon and CR pressure, influence the formation of galaxies?

7. What are the consequences for structure evolution, especially on scales less than a few Mpc?

8. What role was played by magnetic fields before the Recombination epoch, and in the very early universe when scales approached the Planck Scale? (Chapter 13)

9. What are the important links to particle physics at these early times and can we find primordial field signatures in the "visible" universe?

The literature of the last 2–3 decades has produced significant progress on several of the above problems; nevertheless virtually all of them remain to be fully clarified.

References

Axford, W. I. 1977, The Three-Dimensional Structure of the Interplanetary Medium, in *Study of Travelling Interplanetary Phenomena 1977; Proceedings of the L. D. de Feiter Memorial Symposium, Tel Aviv, Israel, June 7–10, 1977*, ed. M. A. Shea, D. F. Smart, & S. T. Wu (Dordrecht: D. Reidel), 145

Beck, R. & Krause, M. 2005, Revised Equipartition & Minimum Energy Formula for Magnetic Field Strength Estimates from Radio Synchrotron Observations, *Astron. Nachrichten*, 326, 414

Benford, G. 1978, Current-Carrying Beams in Astrophysics – Models for Double Radio Sources and Jets, *MNRAS*, 183, 29

Benford, G. 2006, Stability of Magnetic Equilibria in Radio Bubbles, *MNRAS*, 369, 77

Benford, G. & Protheroe, R. J. 2008, Fossil AGN Jets as Ultrahigh-Energy Particle Accelerators, *MNRAS*, 383, 663

Blandford, R. D. & Ostriker, J. P. 1978, Particle Acceleration by Astrophysical Shocks, *Astrophys. J. Lett.*, 221, L29

Brunetti, G., Mack, K-H., Prieto, M. A., & Varano, S. 2003, In-Situ Particle Acceleration in Extragalactic Radio Hotspots, *MNRAS*, 345, L40

Cho, J. & Ryu, D. 2009, Characteristic Lengths of Magnetic Field in Magnetohydrodynamic Turbulence, *Astrophys. J.*, 705, 90

Chokshi, A. & Turner, E. L. 1992, Remnants of the Quasars, *MNRAS*, 259, 421

Croston, J. H., Hardcastle, M. J., Harris, D. E., Belsole, E., Birkinshaw, M., & Worrall, D. M. 2005, An X-Ray Study of Magnetic Field Strengths and Particle Content in the Lobes of FR II Radio Sources, *Astrophys. J.*, 626, 733

Daly, R. A. & Loeb, A., 1990, *Astrophys. J.*, 364, 451

Diehl, S., Li, H., Fryer, C. L., & Rafferty, D. 2008, Constraining the Nature of X-Ray Cavities in Clusters and Galaxies, *Astrophys. J.*, 687, 173

Eilek, J. A. & Hughes, P. A. 1991, Particle Acceleration and Magnetic Field Evolution, in *Beams and Jets in Astrophysics*, ed. P. A. Hughes, *Cambridge Astrophysics Series* (Cambridge: Cambridge University Press), 428, 19

Fanaroff, B. L. & Riley, J. M. 1974, The Morphology of Extragalactic Radio Sources of High and Low Luminosity, *MNRAS*, 167P, 31

Felten, J. E. & Morrison, P. 1966, Omnidirectional Inverse Compton and Synchrotron Radiation from Cosmic Distributions of Fast Electrons and Thermal Photons, *Astrophys. J.*, 146, 686

Furlanetto, S. R. & Loeb, A. 2001, Intergalactic Magnetic Fields from Quasar Outflows, *Astrophys. J.*, 556, 619

Gaisser, T. K. & Stanev, T. 2000, Cosmic rays, *Eur. Phys. J. C*, 15, 150

Genzel, R. & Eckart, A. 1999, The Galactic Center Black Hole, in *The Central Parsecs of the Galaxy*, ed. H. Falcke, A. Cotera, W. J. Duschl, F. Melia, & M.J. Rieke, ASP Conf. Ser. (San Francisco: Astronomical Society of the Pacific), 186, 3

Gnedin, Y., Ferarra, A., & Zweibel, E. N. 2000, Generation of the Primordial Magnetic Fields during Cosmological Reionization, *Astrophys. J.*, 539, 505

Gopal-Krishna, Subramanian, P., Wiita, P. J., & Becker, P.A. 2001 Are the Hotspots of Radio Galaxies the Sites of In Situ Acceleration of Relativistic Particles?, *Astron. Astrophys.*, 377, 827

Gopal-Krishna & Wiita, P. J. 2001, Was the Cosmic Web of Protogalactic Material Permeated by Lobes of Radio Galaxies During the Quasar Era? *Astrophys. J. Lett.*, 560, L115

Hardcastle, M. J., Harris, D. E., Worrall, D. M., & Birkinshaw, M. 2004, The Origins of X-Ray Emission from the Hot Spots of FR II Radio Sources, *Astrophys. J.*, 612, 729

Hartwick, F. D. A. & Schade, D.1990, The Space Distribution of Quasars, *Ann. Rev. Astr. Astrophys.*, 28, 437

Jafelice, L. C. & Opher, R. 1990, The Role of Extragalactic Jets in the Magnetization of the Intergalactic Medium, in *Galactic and Intergalactic Magnetic Fields. Proceedings of the 140th. Symposium of the International Astronomical Union, held in Heidelberg, FRG, June 19–23, 1989*, ed. R. Beck, P. P. Kronberg, & R. Wielebinksi, *Int. Astron. Union Symp.* (Dordrecht: Kluwer) 140, 497

Jafelice, L. C. & Opher, R. 1992, The Origin of Intergalactic Magnetic Fields Due to Extragalactic Jets, *MNRAS*, 257, 135

Kataoka, J. & Stawarz, L. 2005, X-Ray Emission Properties of Large-Scale Jets, Hot Spots, and Lobes in Active Galactic Nuclei, *Astrophys. J.*, 622, 797

Koenigl, A. & Choudhuri, A. R. 1985, Force-Free Equilibria of Magnetized Jets, *Astrophys. J.*, 289, 173

Kronberg, P. P. 1976, 3C303: A source with unusual radio and optical properties *Astrophys. J. Lett.* 203, 47

Kronberg, P. P. 1986, 3C 303: A 'Laboratory' Extragalactic Jet Source, *Can. J. Phys.*, 64, 449

Kronberg, P.P., Burbidge, E. M., Smith, H. E., & Strom, R. G. 1977, The Radio Structure and Optical Field of 3C 303, *Astrophys. J.*, 218, 8

Kronberg. P. P., Crelinsten, J. M., 1973, The 49-cm Linear Polarization Distribution in 3C327 and the Density of Intergalactic gas, *Astrophysical Lett.*, 14, 25

Kronberg. P. P., Dufton, Q. W., Li, H., & Colgate, S. A. 2001, Magnetic Energy of the Intergalactic Medium from Galactic Black Holes, *Astrophys. J.*, 560, 178

Kronberg, P. P., Lesch, H., & Hopp, U. 1999, Magnetization of the Intergalactic Medium by Primeval Galaxies, *Astrophys. J.*, 511, 56

Kronberg, P. P., Lovelace, R. V. E., Lapenta, G., & Colgate, S. A. 2011, Measurement of the Electric Current in a kpc-Scale Jet, *Astrophys. J. Lett.*, 741, L15

Kronberg, P. P., Wielebinski, R., & Graham, D. A. 1986, VLA and 100-m Telescope Observations of Two Giant Galaxies 0634 – 20 and 3C445 (2221 – 02). *Astron. Astrophys.*, 169, 63

Laing, R. A. & Bridle, A. H. 2008, Magnetic Field Fluctuations Around the Radio Galaxy 3C 31, in *Extragalactic Jets: Theory and Observation from Radio to Gamma Ray ASP Conference Series, Vol. 386, proceedings of the conference held 21–24 May, 2007 in Girdwood, Alaska, US*, ed. T. A. Rector & D. S. DeYoung, ASP Conf. Ser. (San Francisco: Astronomical Society of the Pacific), 386, 104

Lelièvre, G. & Wlérick, G. 1975, Identification et photométrie des radiosources du catalogue 3CR III. Cas de la source 3 C 303, *Astron. Astrophys.*, 42, 293

Lesch, H. 1991, Electron Heating in Accretion Discs of Active Galactic Nuclei – On the Formation of Ion-Supported Tori, *Astron. Astrophys.*, 245, 48

Lovelace, R. V. E. 1976, Dynamo Model of Radio Sources, *Nature*, 262, 649

Lovelace, R. V. E. & Kronberg, P. P. 2013, Transmission Line Analogy for Relativistic Poynting-Flux Jets, *MNRAS*, 430, 2828

Lovelace, R. V. E. & Ruchti, C. B. 1983, Electron/Positron/Gamma Ray Beams in Cosmic Radio Sources, in *Positron-Electron Pairs in Astrophysics; Proceedings of the Workshop, Greenbelt, MD, January 6–8, 1983*, ed. M. L. Burns, A. K. Harding, & R. Ramaty, AIP Conf. Proc. (New York: American Institute of Physics), 101, 314

McNamara, B. R. & O'Connell, R. W. 1993, Blue Lobe Galaxies in the Cooling Flow Clusters Abell 1795 and Abell 2597, *Astron. J.*, 105, 417

Meisenheimer, K., Roser, H.-J., Hiltner, P. R., Yates, M. G., Longair, M. S., Chini, R., & Perley, R. A. 1989, The Synchrotron Spectra of Radio Hot Spots, *Astron. Astrophys.*, 219, 63

Miyoshi, M., Moran, J., Herrnstein, J., Greenhill, L., Nakai, N., Diamond, P., & Inoue, M. 1995, Evidence for a Black Hole from High Rotation Velocities in a Sub-Parsec Region of NGC4258, *Nature*, 373, 127

Nakamura, M., Li, H., & Li, S. 2007, Stability Properties of Magnetic Tower Jets, *Astrophys. J.*, 656, 721

Pinkney, J., *et al.* 2003, Kinematics of 10 Early-Type Galaxies from Hubble Space Telescope and Ground-Based Spectroscopy, *Astrophys. J.*, 596, 903

Preuss, E., Pauliny-Toth, I. I. K., Witzel, A., Kellermann, K. I., & Shaffer, D. B. 1977, High Resolution Observations of Weak Radio Nuclei in Galaxies and Quasars, *Astron. Astrophys.*, 54, 297

Rees, M. J. 1967, Studies in Radio Source Structure-III. Inverse Compton Radiation from Radio Sources, *MNRAS*, 137,429

Rees, M. J. 1987, The Origin and Cosmogonic Implications of Seed Magnetic Fields, *Q. J. R. Astron. Soc.*, 28, 197

Rees, M. J. & Setti, G. 1968, Model for the Evolution of Extended Radio Sources, *Nature*, 219, 127

Richstone, D. 2004, Supermassive Black Holes: Demographics and Implications, in *Coevolution of Black Holes and Galaxies*, ed. L. C. Ho Carnegie Observatories Astrophys. Ser., ed. L. C. Ho (Cambridge: Cambridge University Press), 1, 281

Röser, H.-J. & Meisenheimer, K. 1986, CCD Photo-Polarimetry of the Jet of 3C273, *Astron. Astrophys.*, 154, 15

Röser, H.-J. & Meisenheimer, K. 1987, A Bright Optical Synchrotron Counterpart of the Western Hot Spot in Pictor A, *Astrophys. J.*, 314, 70

Romanova, M. M. & Lovelace, R. V. E. 1992, Magnetic Field, Reconnection, and Particle Acceleration in Extragalactic Jets, *Astron. Astrophys.*, 262, 26

Ruzmaikin, A. A., Sokoloff, D., & Shukurov, A. 1989, The Dynamo Origin of Magnetic Fields in Galaxy Clusters, *MNRAS*, 241, 1

Ryu, D. S., Kang, H. S., & Biermann, P. L. 1998, Cosmic Magnetic Fields in Large Scale Filaments and Sheets, *Astron. Astrophys.*, 335, 19

Ryu, D. S., Kang, H. S., Hallman, E., & Jones, T. W. 2003, Cosmological Shock Waves and Their Role in the Large-Scale Structure of the Universe, *Astrophys. J.*, 593, 599

Schaaf, R., Biermann, P. L., Kronberg, P. P., & Schmutzler, T. 1989, X-Ray Observations of the Starburst Galaxy M82, *Astrophys. J.*, 336, 722

Soltan, A. 1982, Masses of Quasars, *MNRAS*, 200, 115

Strom, R. G. & Willis, A. G. 1980, Multi-Frequency Observations of Very Large Radio Galaxies. II – 3C236, *Astron. Astrophys.*, 85, 36

Taylor, J. B. 1986, Relaxation and Magnetic Reconnection in Plasmas, *Rev. Mod. Phys.*, 58, 741

Tremaine, S. D. *et al.* 2002, The Slope of the Black Hole Mass Versus Velocity Dispersion Correlation, *Astrophys. J.*, 574, 740

Willis, A. G. & Strom, R. G. 1978, Multi-Frequency Observations of Very Large Radio Galaxies, I – 3C 326, *Astron. Astrophys.*, 62, 375

9

Magnetic fields associated with clusters and groups of galaxies

9.1   Introduction

9.1.1   *Prologue to intracluster gas studies*

Prior to about 1980, studies of galaxy clusters were focussed largely on the member galaxies – their types, stellar populations, and densities as revealed by optical images and spectra (e.g. Dressler 1984). Catalogues of galaxy clusters were compiled by Zwicky (1938) and Abell (1958). The latter constituted a carefully defined, volume-limited sample of 1800 galaxy clusters in the nearby ($z \leq 0.2$) universe, with an empirical scale of richness. Optically gathered evidence for evolution of the member galaxy populations was given important impetus in a study by Butcher & Oemler (1978). None of these studies considered – or arguably needed to consider – magnetic fields, since they were largely hidden to the (then optically invisible) intracluster gas and the galaxies.

9.1.2   *Early radio and optical indications of an ICM*

Evidence for intracluster magnetic fields came first via detection of a radio synchrotron halo, this being a direct tracer of an intracluster magnetic field (see Equation 1.1). This was discovered first in the Coma cluster of galaxies (Willson 1970), the nearest large cluster to us, and subsequently in several other clusters.

Convincing evidence for hot thermal intracluster gas came also many years ago from analyses of the distorted images of "head-tail" radio galaxies that were found to be peculiar to galaxy clusters (Jaffe & Perola 1973). The distortion of the galaxies' morphology and linear polarisation were ascribed to the dynamic and static pressure of the ambient intracluster gas. We recall that these analyses additionally reveal highly organised magnetic fields internal to the radio lobes, as discussed in Chapters 7 and 8. Velocities through the intracluster medium (ICM) of the host galaxies of "head-tail" radio sources can be independently estimated from the optical spectra of their host galaxies.

X-ray observations in the 1980s revealed that the hot, bremsstrahlung-emitting, intracluster gas comprises most (~90%) of the baryonic mass in galaxy clusters. This confirmed a similar, much earlier conclusion by F. Zwicky in the 1930s. Radial velocities of member galaxies (e.g. Rood 1970, Carlberg *et al.* 1997) and gravitational lensing have been used to probe the global masses and mass distributions within galaxy clusters (Schneider *et al.* 2000, Kaiser 2001). These reveal a "missing mass" component, named dark matter, whose mass exceeds, in turn, that of the hot ICM baryonic gas. More recently, researchers have recognized the non-trivial nature of the relationship between the intracluster gas and its associated magnetic fields. Given that the evolution of clusters is bound up and embedded with that of the large-scale structure (LSS) of the Universe, it is compelling to understand the physical role and origin of magnetic fields in these uniquely large aggregates of matter.

9.1.3 Introduction

Principal themes of this chapter are detection methods, origins, and various astrophysical implications of intracluster magnetic fields. The *intergalactic* density of cosmic rays (CRs), magnetic fields, thermal gas, and galaxies make them unique "laboratories" for understanding the magnetisation of the intergalactic medium. Being the largest gravitationally bound structures in the Universe, they are end points of infall of intergalactic gas into the filaments and walls of cosmological LSS. Clusters contain two identifiable types of magnetised galaxy outflows. They are namely, the jet-lobe systems emanating from the central black holes of cluster member galaxies (Chapters 7 and 8), and the stellar and supernova wind–driven outflows. The latter were discussed by Völk & Atoyan (2000) who investigated the consequences of star/SN-driven outflows within clusters.

The infall of gas on supra-cluster scales further affects the complex mix of processes within the cluster. We can imagine that infall happens in different ways, and with different effects. First, some magnetic pre-seeding of the ICM will occur over the cosmological formation timescale of the cluster. The infalling "pre-cluster" intergalactic gas was possibly seeded by weak primordial fields, and even more likely by more cosmologically recent (i.e. since recombination) galaxy outflow processes already discussed. Further, as discussed in the previous chapter, simulations of the (post-recombination) shearing and turbulent motions of large scale flows/inflows in LSS can produce magnetic field amplification prior to infall of gas into a cluster.

Apart from any pre-magnetised, inflowing intergalactic gas, cluster–cluster mergers can cause large scale shocks and compression near the cluster periphery. There is observational evidence for shock- and compression-amplified fields, as infalling intergalactic clouds are decelerated at the cluster periphery. Magnetic fields and CRs energised in this process will eventually get mixed into the ICM. Such processes are difficult to quantify analytically, and are better elucidated by simulations, combined with observational verification. Specification of detailed, observationally verified 3-D magnetic field structures of intracluster gas are only at their beginning. But progress, and prospects for better observations are good, and they are discussed further in Section 9.2.

Outward flows of energy to large cluster radii have also been confirmed in some cases. Some have been identified as buoyantly rising, intact, and probably magnetically dominated "bubble" radio lobes. Being old, they are difficult to detect in radio bands, except at low radio frequencies. A case in point is the outer radio structure of Hydra A, imaged at ~300 MHz and illustrated in Fig. 9.1(a). This raises the general possibility that on a longer timescale, some magnetic flux and energy may be buoyantly *removed* to the periphery of a cluster – that is, working in the opposite direction to the infalling intergalactic gas scenario described above.

Alternative situations to buoyant removal of magnetic energy arise when the lobes become entrained and mixed into the virial velocity field of the ICM *before* they can buoyantly rise to large cluster-centric radii. Diehl *et al.* (2008) describe simulations of this process. An example of entrainment and mixing into the ICM is also specifically indicated in

(a)

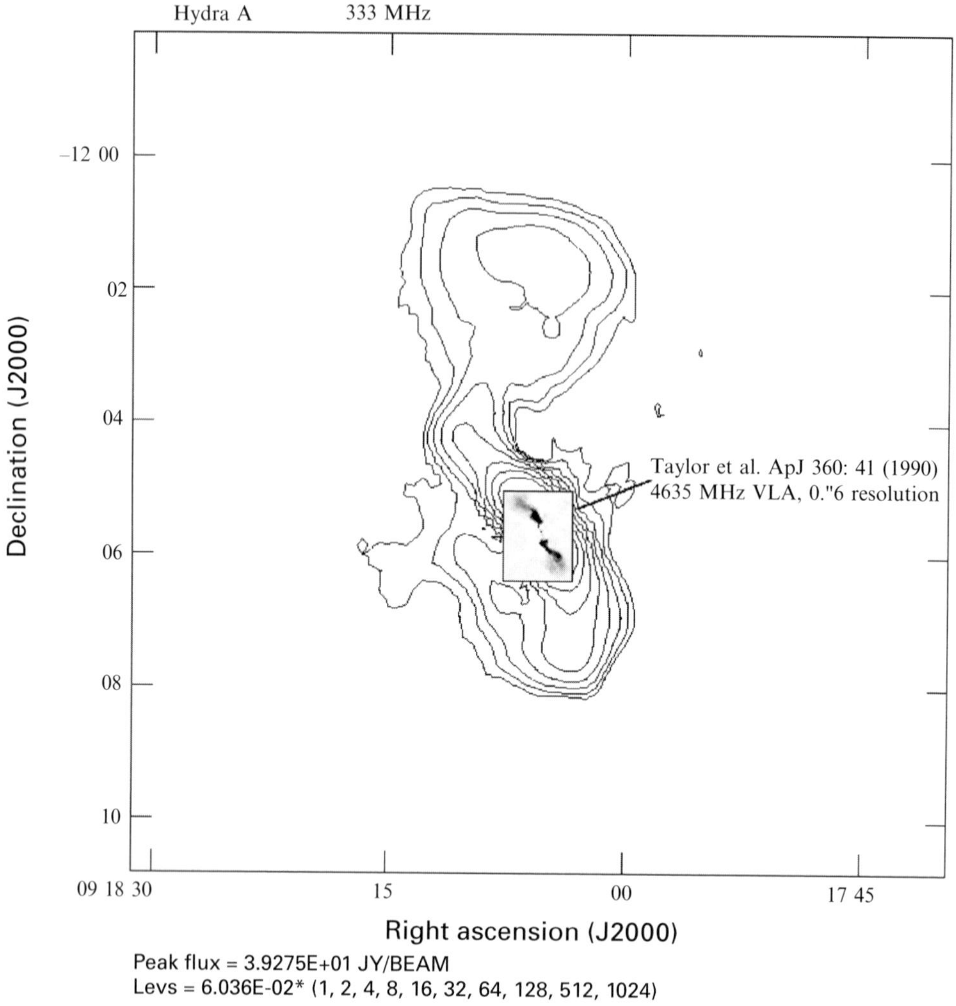

Figure 9.1. (a) (L) The buoyantly rising outermost radio lobes of Hydra A, in the Hydra cluster of galaxies (source: Lane *et al.* 2004). (b) (R) For the Coma cluster, a combined NRC-DRAO Interferometer and NRAO-VLA image at 1.4 GHz of the diffuse synchrotron emission in the central radio halo. Discrete sources have been removed to better isolate the synchrotron radio halo structure. When this is done, the low-level boundary of the extended head-tail radio source 5C4.81, after clipping off the brighter tail regions, suggests an entrainment of the radio tail source into the Coma cluster's large scale velocity structure. This contrasts with the case of Hydra A in (a) (source: Kim *et al.* (1990)).

the instance of the Coma cluster (Kim et al. 1990 and Fig. 9.1(b)). Here, in a combined DRAO-VLA image, very faint extensions of the prominent cluster-internal head-tail radio source, 5C4.81, appear to become entrained and captured within the intracluster velocity field (The large constant-color zone in Fig. 9.1 (b)).

(b)

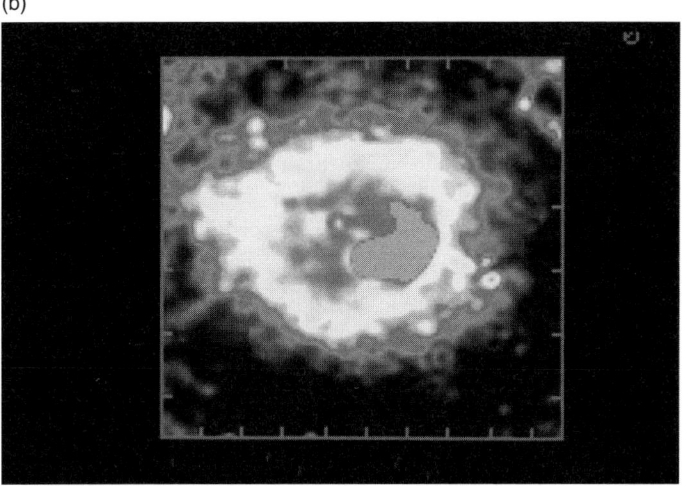

Figure 9.1. (*cont.*)

9.1.4 A single dominant BH-powered source for the Coma Cluster's enhanced radio halo?

The apparent entrainment of the "head-tail" lobe into a significant fraction of the central Coma cluster in Fig. 9.1(b) suggests an intriguing counter-example to the theory that strong cm-λ cluster-centric radio halos are normally caused by cosmologically recent cluster merging activity. Here, at least part of the Coma radio halo appears to be substantially fed by a single *cluster-internal* radio galaxy, whose energy source is the galaxy's central BH, rather than merger/infall energy. It will also be kept in mind that the radiative age of this radio galaxy lobe is at least an order of magnitude less than the likely dynamical age of the Coma cluster itself. The internal energy contents of BH-powered intracluster lobes were discussed in Chapter 8; see in particular Fig. 8.2.

Additionally, over a cluster's lifetime intracluster magnetic fields have been proposed to be seeded in supersonic cluster-internal shock fronts. In any case, they can be amplified in such processes. Field amplification and diffusion might also occur *within* the cluster-internal radio lobes. These scenarios illustrate the complexity and variety of physical processes involved in the magnetisation of a galaxy cluster ICM (see below).

9.1.5 Overview of the causes of the cluster halo emission

To summarise the above, intracluster magnetic fields can be seeded and strengthened in different ways: (1) As above, one, or small number of powerful SMBH-jet-lobe radio sources, if retained within the cluster, may supply enough magnetic energy to "power" a radio halo over ~10^9 year. Figure 9.1(b) suggests this for the Coma Cluster. If this is generally true, then most cluster radio halos may have come from a small number of individual intra-cluster radio sources. (2) B_{ICM} could additionally be enhanced by infalling material in the course of the cluster's past evolution. They could also be generated and augmented by (3) the star- and supernovae-driven outflow winds of member galaxies (Völk & Atoyan 2000), and/or (4) by supersonic shocks associated with the cluster-internal motions, mentioned above. Another source (5), discussed below, could be associated with proton–proton collisions. Such collisions produce CR leptons and γ-rays, however their effect on intracluster magnetic energy is not so straightforwardly calculated. In any case, the CR leptons thus produced will also serve to "illuminate" the cluster fields for radio observations. Explanation (1) above has the advantage that, at least for the Coma cluster, it can be demonstrated by direct observation, and hence relatively free of model assumptions. Faraday rotation probes involving magnetic fields and non-relativistic electrons (Equation 2.7) can also greatly improve our ability to probe intracluster magnetic fields, as discussed in the next section.

The synchrotron-emitting relativistic electrons in clusters can have two basic origins: They can be energised through any combination of processes (1)–(4) above, or they can be produced as a by-product of p–p collisions between CR protons and thermal ICM protons ((5) above). In the latter case, both π^0 (neutral) and $\pi^\pm$ (charged) pions are produced. These, in particular the π_0 particles, decay to produce γ rays, and an additional population of *hadronically induced* relativistic electrons, along with muons and electron and muon neutrinos, and antineutrinos (see also Section 9.8). For further reading see Dennison 1980, Dermer 1986, and Pfrommer & Enßlin 2004a,b,c. The pion decay processes are as follows:

$$\pi_0 \rightarrow 2\gamma \tag{9.1}$$

$$\pi^\pm \rightarrow \mu^\pm + \nu_\mu/\overline{\nu}_\mu \rightarrow e^\pm + \nu_e/\overline{\nu}_e + \nu_\mu + \overline{\nu_\mu}. \tag{9.2}$$

The intergalactic p–p interaction is more or less unique to clusters because of their relatively high intergalactic baryon density. Importantly, it brings cluster γ-ray observations into the mix of diagnostics that can then be connected to intracluster magnetic fields.

9.2   Methods for probing galaxy cluster magnetic fields

9.2.1   *General*
X-ray images of intergalactic gas in galaxy clusters reveal a typical electron density of 10^{-4}–10^{-2} cm^{-3}, a temperature in the range 10^6–10^8 K (=0.9–8.6 keV), and an extent up to ~1 Mpc. Typical X-ray luminosities, L_x, range from 10^{43} to 10^{45} erg s^{-1}. The ion sound speed is comparable to the galaxies' velocity dispersion in the cluster, which is ~400–1200 km s^{-1}.

9.2.2   *Two-dimensional RM* **mapping of a single cluster using background radio sources**
This method consists in performing multi-frequency integrated polarisation observations on polarised radio sources that are behind, inside, and beside the cluster. The *RMs*, combined with the thermal gas X-ray images (exemplified in Fig. 9.2) can reveal the strength of ICM magnetic fields. They range from ~1 to several μG depending on the cluster and location within the cluster. For Coma, the excess cluster Faraday rotation ($\approx$60 rad m^{-2}), together with X-ray data from Abramopoulos & Ku (1983), enabled Kim *et al.* (1990) to quantitatively derive an ICM magnetic field strength for the Coma cluster of galaxies:

$$<|B|_{\mathrm{ICM}}> \simeq 1.9 < \frac{l_0 h_{70}^{-1}}{10\mathbf{kpc}} >^{-0.5} \mu G. \tag{9.3}$$

An elusive parameter is the magnetic turbulence *scale*, l_0. It can be independently estimated from the *RM* fluctuations of the long head-tail source, as described by Kim *et al.* (1990). The

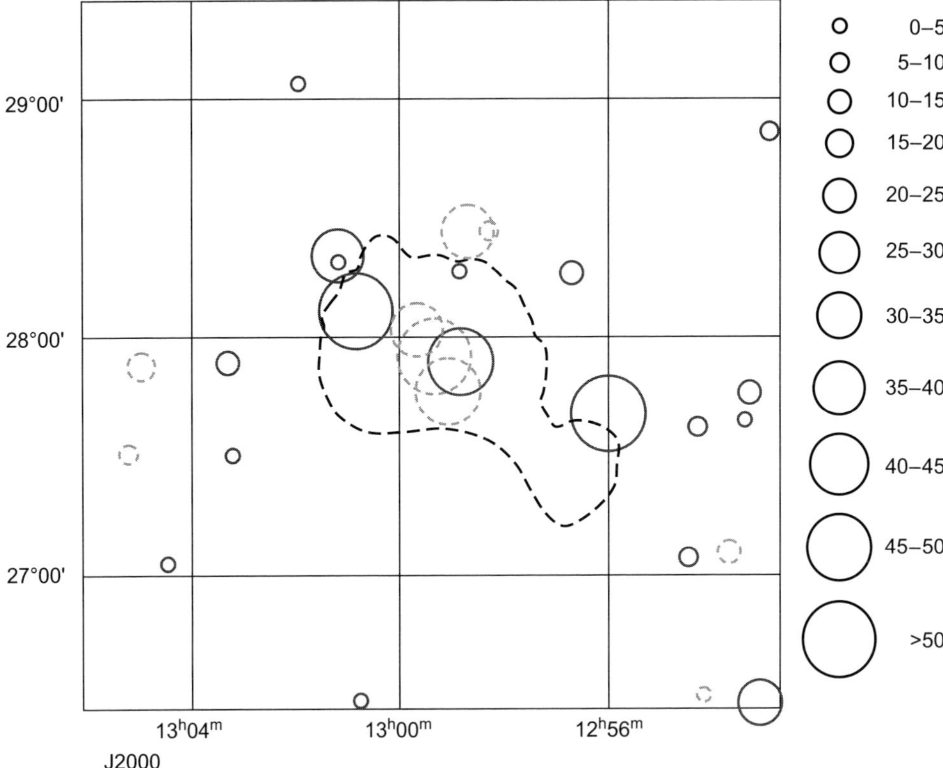

Figure 9.2. *RM* distribution of 29 background and internal radio sources relative to the X-ray boundary of the Coma Cluster of galaxies which shows a clear excess *RM* added by the intracluster medium. Solid and dashed circles show positive and negative *RM*s respectively, where the circle size scales as |*RM*|. The dashed line shows the ROSAT X-ray boundary at approx. 0.12 counts/400 arc s^2 for the ROSAT PSPC detector, in the range 0.5–2.4 keV (from Briel *et al.* 1992). This figure has not previously been published, and updates the original by Kim *et al.* (1990) by adding 11 more *RM*s. It obtains a similar statistical significance (~99%) to the 1990 publication. The additional *RM*s were provided by: the author, R. Kothes, R. Wielebinski, & E. Fürst.

fainter, larger extensions of this "head-tail" source (5C4.81) can be seen in Fig. 9.1(b) – see also Section 9.1.3 and the related discussion in Section 9.2.3 below. A repeat of these same *RM* fluctuation measurements at higher resolution (Feretti *et al.* 1995) has since shown l_0 to be smaller than 10 kpc by a factor of a few. The consequently revised l_0 in Equation (9.3) raises the estimate of $<|\boldsymbol{B}|>_{\mathrm{IGM}}$ from 1.9 to ~7 μG for Coma. This illustrates the influence of just the instrumental resolution, and suggests that smaller-scale and stronger magnetic field components may be discovered as radio telescope sensitivity improves at higher resolutions. This prospect has important implications for the energetics, given that $\varepsilon_{\mathrm{B}} = 9.9 \times 10^{-13} (B/5\ \mu\mathrm{G})^2$ erg cm^{-3}, and it suggests that the magnetic energy density in galaxy clusters is characteristically a few percent of the thermal energy. The hot gas, with $\varepsilon_{\mathrm{TH}} = 2.8 \times 10^{-11} n_{-3}T_8$ erg cm^{-3}, dominates the baryonic mass in clusters and groups. See also Section 9.8.

Initially, there was much surprise at the microgauss-level values in a cluster ICM, given that typical intracluster gas *densities* are approximately two orders of magnitude lower than in the Milky Way disc which has comparable |*B*|! However, basic thermal physics concepts make it evident that the relative *energy* densities (*nkT*) of the ICM and Galactic ISM are not that dissimilar. That is because *T* is typically higher in the ICM by roughly the inverse ratio of the respective ICM/ISM gas densities. Cluster magnetic turbulence is discussed below in Section 9.2.5.

9.2.3   A *statistical RM* probe by "stacking" many clusters

Current radio telescope sensitivities permit only relatively scant 2-D "coverage" of background *RM* sources for individual galaxy clusters. This is illustrated in Fig. 9.2 for the Coma cluster, which, being the closest to us, has a large angular size. However the relatively low angular density of background *RM*s can be compensated by comparing *RM*s for large numbers of background sources at differing impact parameter distances for a "stacked" sample of *many* clusters. Using this concept, Lawler & Dennison (1982) obtained the first tentative, positive statistical signal of cluster ICM rotation measures. This was followed by a similar attempt for a single cluster by Vallée *et al.* (1986). Subsequently, Kim *et al.* (1991) investigated a sample of 53 Abell galaxy clusters, 19 of which had X-ray core size measurements at that time. Clarke *et al.* (2001) investigated a better specified, controlled selection of low redshift Abell clusters that had companion ROSAT X-ray images. Here, *RM*s were determined with the VLA for sources at various impact parameters to several "stacked" clusters. These latter, multi-cluster tests showed that a typical galaxy cluster adds up to ~100 rad m^{-2} for lines of sight near its core. They showed that magnetic fields of a few μG are commonplace in X-ray emitting rich clusters within *ca.* 200 kpc of the clusters' cores. The plot in Fig. 9.3 shows the result of stacking many clusters. It also demonstrates a relative absence of small (otherwise "normal") *RM*s for sightlines at low cluster impact parameter. If borne out in future such probes, it would suggest *prima facie* that, on average, a cluster's ICM *B*-filling factor is relatively high.

A selection criterion for the cluster sample in Fig. 9.3 was a smooth, quasi-symmetrical X-ray image. This selects against clusters with a recent merger history – which also were thought to correlate with strong synchrotron-emitting radio halos. It confirmed the universality of intracluster μG-level fields, in that they exist *independent* of the presence in the ICM of a radio halo of freshly energised, synchrotron-emitting relativistic electrons (as is the case in Coma – Figs. 9.1(b) and 10.1).

The fact that magnetic fields near μG levels are widespread in galaxy clusters is independently consistent with a *lower* limit of 0.3–0.5 μG, which is implied by the observed absence of inverse-Compton-generated X-rays in clusters (*see* Henriksen & Mushotzky 1986, Rephaeli & Gruber 1988). The keV-range X-ray spectra of inverse-Compton and bremsstrahlung emissions are similar, making it difficult to distinguish between the two, solely from their overall X-ray spectral forms. A firm confirmation of the thermal component of the X-rays comes from the distribution, strength, and width of highly ionised *Fe* emission transitions in the keV range. The spectral resolutions of X-ray satellites are now often sufficient to measure the excitation temperatures and densities of transitions in the hot ICM plasma. These independently confirm the density and temperature of the thermally emitting hot gas, and can also be used to set upper limits on any inverse Compton component of keV X-ray emission (Chapter 2, Section 2.12.1).

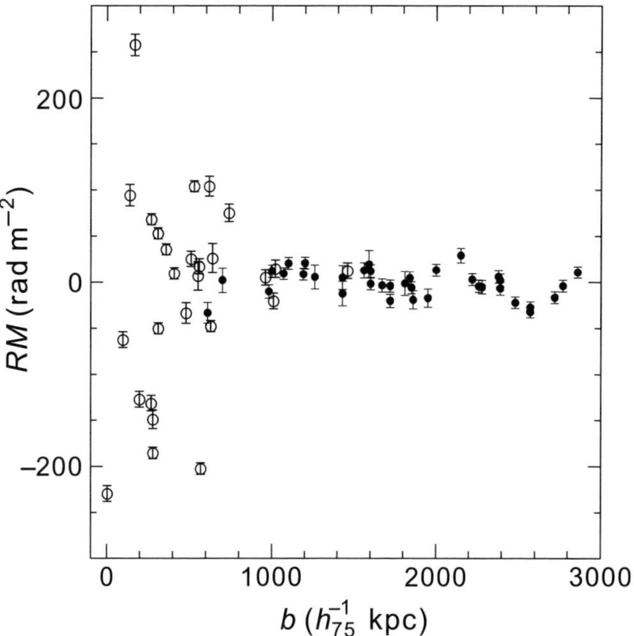

Figure 9.3. Plot of the Faraday rotation measures of 200 quasars and radio galaxies as a function of their impact parameter distance (*b*) from the centre of 16 stacked and scaled galaxy clusters without strong radio halos, and having $z \leq 0.1$ (reproduced from Clarke, Kronberg & Böhringer 2001).

9.2.4 *Multiple source 2-D Faraday rotation mapping over a single cluster*

Extended radio galaxies typically have low lobe-internal *RM*s, $\lesssim 20$ rad m^{-2}, consistent with their usually low internal thermal plasma content. Correspondingly, the low internal depolarisation rate of extended radio source lobes (Chapter 2, Section 2.7) indicates that low densities of thermal plasma that are well below those in the ambient ICM – causing the "holes" discussed earlier. Thus, when such an extended radio galaxy lies either embedded within or behind a galaxy cluster, it acts as a polarised "radiating surface", against which foreground intra-cluster variations of *RM* (Faraday depth, ϕ) can be measured.

Such *RM* (α,δ) variations over extended radio sources (see also Section 9.2.1) provide one of the best direct ways of estimating the field reversal scale structure in a cluster. A good example of this is provided by three extended radio sources found projected onto one cluster, Abell 119 (Feretti *et al.* 1999), shown in Fig. 9.4 below.

The *RM* results for these sources reveal the approximate dependence of the ICM magnetic field on the radio source location: 3C29, with its projected location near the cluster periphery, has small *RM* variations ($\Delta RM_{\mathrm{max}} \sim -30$ to $+30$ rad m^{-2}) and is consistent with little or no cluster ICM effect. As the projected impact parameter distances approach the cluster centre, ΔRM_{max} becomes ~500 rad m^{-2} for 0053–016, and ~800 rad m^{-2} for 0053–015 (Fig. 9.4). The *RM* images show a range of magnetic "turbulence" scales down to the 3.7″ image resolution ($\approx 3.4\ h_{70}^{-1}$ kpc). When combined with deprojected, model 3-D estimates of $n_{\mathrm{e}}(x, y, z)$ from the ROSAT X-ray images, the intracluster field strengths are estimated in the range 5–10 μG (Feretti *et al.* 1999). A119 (like Coma) does not have a cool core.

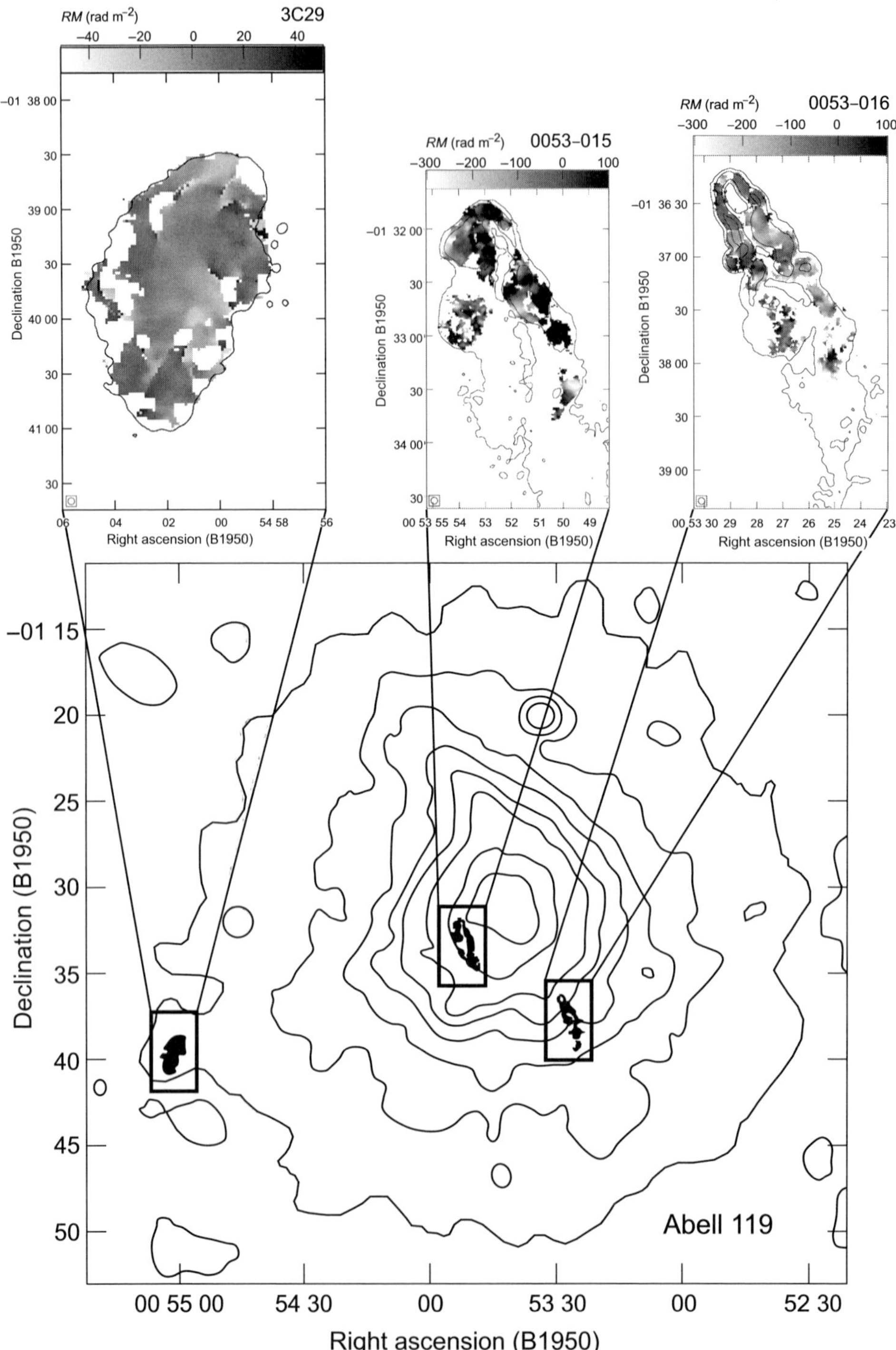

Figure 9.4. $RM(\alpha,\delta)$ variations of the radio structure of three extended radio sources in the cluster Abell 119, shown relative to the cluster's projected hot gas profile (a ROSAT PSPC X-ray image). *Insets: $RM(\alpha,\delta)$* images are shown from left to right for 3C29, 0053–015, and 0053–016. All inset images are shown here with same angular (and approximate linear) scale. Note the very much lower *RM* fluctuation range for 3C29 at the cluster periphery. See text below for further details. (Adapted from Feretti *et al.* (1999).)

Garrington *et al.* (1988) and Garrington & Conway (1991) proposed an extension of the method described in Section 9.2.3 into galaxy group environments. They discovered a systematic side-to-side asymmetry of the *depolarisation ratio* between two frequencies in a large sample of FR II radio sources covering a substantial redshift range. They interpreted this near side/far side difference relative to the radio galaxy nucleus as due to a surrounding 30–100 kpc radius ionised gas halo, having a magnetic field strength of at least 1 μG in low redshift sources, and tangled on a scale of $<$5 kpc. This type of ambient environment seems typical around cD galaxies in galaxy groups and in poor clusters of galaxies. This method extends the rich-cluster findings described above. It is apparent that magnetic fields of at least ~1 μG are widespread in these (cosmologically overdense) galaxy groups.

Before proceeding further, we clarify the definition of intracluster quantities in 3-D, such as T_e (x, y, z) and $n_e(x, y, z)$, by defining the spatial coordinates (x, y, z) as relative to the cluster's physical centre $(0,0,0)$. The unit vector directions for x,y $(\boldsymbol{i,j})$ are in the plane perpendicular to the line of sight to the cluster centre, and z (unit vector $\boldsymbol{k}$) in the line of sight and increasing with increasing distance. Finally, $r = |\boldsymbol{i}x + \boldsymbol{j}y + \boldsymbol{k}z|$ is the radial distance from $(0,0,0)$, and the impact parameter b (linear or angular units) is the line-of-sight projection of $\boldsymbol{r}$ in e.g. (α,δ) coordinates.

Each of the radio methods requires a companion X-ray image, so that estimates of the 3-D distributions of electron temperature, $T_e(x, y, z)$, and density, n_e (x, y, z), within the cluster can be made before a weighted $<|\boldsymbol{B}|>$ is estimated from the projected $\Delta RM(\alpha,\delta)$, as illustrated in Fig. 9.4. The following subsection gives a sample analytical model for doing this.

9.2.5 Deduction of the ICM magnetic field strength from σ_{RM} (α,δ) for varying model cluster parameters

An explicit relation was derived by Felten (1996) between the statistical distribution of background source rotation measures σ_{RM} (α,δ) and a cluster model with a magnetic field strength profile, $|\boldsymbol{B}|(r)$, ionised gas distribution $n_e(r)$, and range of mass distribution profiles. Spherical symmetry is assumed for the cluster, as appears justified by many cluster X-ray images, and a King model was adopted for the electron density.

$$n_e(r) = n_0 \left(1 + \frac{r^2}{r_c^2} \right)^{\frac{-3\beta}{2}} \tag{9.4}$$

(King 1972), where r_c is the core radius, and the parameter β is typically $\approx$0.7. As before, b is the impact parameter of the ray path from a background polarised radio source. As a starting estimate, it seems physically plausible that $\varepsilon_B = B^2/8\pi$ will scale with the thermal energy density of the hot gas, $\varepsilon_{th} = 2n_e kT$. This gives a scaling between the magnetic and thermal energy densities as

$$\varepsilon_B = \alpha(r) \cdot \varepsilon_{th}. \tag{9.5}$$

For the simplest case of a constant α, σ_{RM} (b) is given by

$$\sigma_{RM} = \frac{811.9 \times 10^6 \pi^{\frac{1}{4}}}{\sqrt{3}} \left(1 + \frac{b^2}{r_c^2} \right)^{\frac{1-9\beta}{4}} \cdot \gamma(\beta) \cdot \sqrt{16\pi\alpha n_0^3 kT \cdot l_0 r_c} \tag{9.6}$$

where

$$\gamma(\beta) = \sqrt{\frac{\Gamma\left(\frac{9\beta-1}{2}\right)}{\Gamma\left(\frac{9\beta}{2}\right)}}. \tag{9.7}$$

In Felten's (1996) formulation, B is contained in α ($B \propto \alpha^{0.5}$), n_0 is the electron density at $r = 0$, and l_0 is the reversal scale of the ICM magnetic field. As illustrated in Fig. 9.4, both σ_{RM} (b) and l_0 can be estimated from the $RM(\alpha,\delta)$ image data. Note that for the purpose of estimating l_0 it may not be very crucial whether the extended source is embedded within the cluster or behind it. If the extended source is behind the cluster, which cannot always be verified, then the amplitude of the $RM(\alpha,\delta)$ variations give an alternative, but not greatly different, measure of the cluster $<B(r)>$ at the b location of the source. In any case n_0, r_c and T can be obtained from an X-ray image, which leaves just α to be determined. This permits $<B(r)>$ to be determined if we assume a constant α as above.

For simplicity and illustrative purposes, the above discussion has adopted a single l_0 (or k_0 in inverse units) to represent the magnetic field turbulence scale. To have a more realistic understanding of cluster magnetic turbulence, cluster X-ray images can be combined with $\sigma(RM)$ data as illustrated in Fig. 9.5. Sufficiently good data can provide enough information to model the magnetic power spectrum $\varepsilon(k)$ over a *range* of k by adducing a limited number of assumptions about the inclination angle and location of the extended source, and the Faraday screen in 3-D within the cluster.

Figure 9.5(a) shows a representative plot of $\varepsilon(k)$ by Enßlin & Vogt (2006) that may be typical near the core regions of clusters having cool core zones (in contrast to A119). Figure 9.5(b) shows the data source of the derived RM fluctuation spectrum in Fig. 9.5(a),

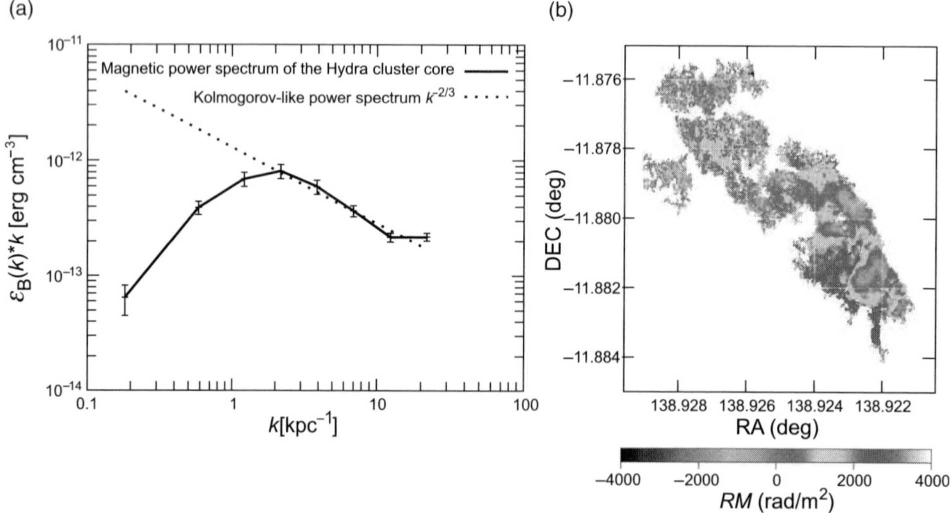

Figure 9.5. (a) An example of the derivation of $\varepsilon_B(k)$ (erg cm^{-3}) vs. k for Hydra A obtained from the observations in (b). A Kolmogorov-like spectrum is shown for comparison (dotted line). (Source: Enßlin & Vogt (2006).) (b) An RM image of the northern, inner lobe of the Hydra A radio galaxy near the core of the Hydra cluster of galaxies. (Source: Vogt & Enßlin (2005).)

calculated by Vogt & Enßlin (2005) using a Bayesian maximum likelihood estimator. This is the inner northern lobe of Hydra A from the inset image of Fig. 9.1(a). It gives a good representation of the observed $RM(\alpha,\delta)$ fluctuations over the surface of the polarised radio lobe.

Being within a cluster central cool gas zone, it is embedded in higher ICM densities and stronger magnetic fields than A119 (or Coma). By way of further comparison, we note that the Hydra cluster's central cool zone is less "extreme" than that of A2163 which is discussed later in connection with the magnetic SZ effect in Section 9.6.2.

9.2.6 Further prospects for 3-D magnetic probes of clusters

The quantity $\alpha(r)$ in Equation (9.5) is of particular interest to specify, in that it represents a non-thermal pressure term in the equation of hydrostatic equilibrium. This, in turn, relates to estimates of the total mass and mass profiles of clusters.

Although the $(1 + z)$ factors in Equation (2.7) are negligible, B determinations in low-z clusters have another dependence on cosmological scaling. This occurs purely through the dependence of their (quasi-Euclidean) distance on the Hubble constant h ($h = H_0/100$ km sec^{-1} Mpc^{-1}). Since l, b, r_c, and n_0 all depend on the cluster's distance, the B value derived from Equation (9.6) will scale as $\sqrt{h}$, so that ε_B scales as h.

A powerful diagnostic would be a combination of Stokes' parameter $I(\alpha,\delta)$, $Q(\alpha,\delta)$, and $U(\alpha,\delta)$ images of cluster synchrotron halos over a suitable range of radio frequencies. With a few exceptions, this method has not yet worked well, but it has good promise with more sensitive future generations of radio (and X-ray) telescopes. Lower radio frequencies, from ~2 GHz down to ~30 MHz are of special interest here. In the radio, a given column through the ICM will, at some (α,δ) location, show different polarisation angles at the different frequencies due to the integrated Faraday rotation along the line of sight through that column (Equation 2.7). Given the finite resolution of a telescope beam, this rotation might also be accompanied by a wavelength-dependent change in the *degree* of depolarisation. This can be a consequence of 3-D variations of $\boldsymbol{B}(x, y, z)$, and $n_\mathrm{e}(x, y, z)$ taking, as before, (0,0,0) as the physical centre of the cluster, plus the fact that the common resolution of multi-wavelength I, Q and U images defines a z column of *finite* cross-section through the cluster.

If the column for a single sightline were infinitely small ($d\alpha = d\delta \cong 0$), the Faraday rotation law (Equation 2.7) would simply sum the Faraday rotations along a unique line of sight, i.e. give a unique Faraday depth, ϕ. If the intrinsic polarisation degree $m(x, y, z)$ and spectral index $\alpha(x, y, z)$ of the synchrotron radiation is constant along the chosen sightline, there will be no wavelength dependence of cluster-modified integrated polarisation degree over the column, just a net Faraday rotation, $RM(\alpha_0,\delta_0)$.

In practice, the column cross section $d\alpha d\delta$ is finite, corresponding to the finite resolution of the radio telescope, so that a single position within a hypothetical telescope raster scan over the cluster samples a *bundle* of sightlines close to some (α_0,δ_0), each of which may have a different Faraday rotation. When the χ differences between sightlines approach the order of a radian, the observed polarisation will be generally be smaller, i.e. depolarisation will occur. The rate at which the observed (α,δ) and orientation $\chi(\alpha,\delta)$ change with λ^2 (or λ) can be compared with model calculations of $\boldsymbol{B}(x, y, z)$, and $n_\mathrm{e}(x, y, z)$ within the cluster halo. The model would ideally contain the complete spectrum of $\boldsymbol{B}(k)$. Basic mathematical frameworks for this modelling are presented in Burn (1966) and by Brentjens & de Bruyn (2006) – see Chapter 2.

Comparisons of $p(\alpha,\delta,\lambda^2)$ and $\chi(\alpha,\delta,\lambda^2)$ with a fully modelled $\boldsymbol{B}(x, y, z)$ for clusters with radio halos will require higher resolution and sensitivity at significantly shorter wavelengths, where the surface brightness of the synchrotron emission has often been too weak to image with present radio instruments. The intra-cluster field strengths, as we shall see below, and electron densities (several μG, and $n_e \gtrsim 10^{-4}$ cm^{-3}) are such that the form of variations in $p(\alpha,\delta, \lambda^2)$ and $\chi(\alpha,\delta, \lambda^2)$ will be verified at shorter radio wavelengths and higher resolutions where there are smaller ray bundles. A set of $m(\alpha,\delta, \lambda^2)$ and $\chi(\alpha,\delta, \lambda^2)$ measurements over a wide range of Faraday depth, together with independent optical or X-ray observations to give electron density, will be necessary to approximate a full 3-D model of $\boldsymbol{B}(x, y, z)$. These measurements require companion imaging at the *longer* radio wavelengths, in the régimes where Faraday rotations $\Delta\chi$ $(\propto\lambda^2)$ are more readily detected. The modelling can be done by an extension of the Faraday depth analysis described in Chapter 2, sometimes known as "Faraday rotation synthesis". We recall that a unique, *assumption-free* 3-D specification of all of the ICM thermal magnetoplasma parameters is not permitted mathematically because λ^2 is not defined in the negative domain (see Chapter 2).

It will be evident at this point that, given enhanced sensitivity and resolution at many adjacent frequencies it will be possible to extend probes of $|\boldsymbol{B}|(k)$ in the ICM over a wide range of k, as shown in Fig. 9.5(a). Similar methods have been used to similarly probe some nearby galaxies. The more regular motions and shapes of individual spiral galaxies make a 3-D Faraday synthesis modelling of the (brighter) diffuse radio emission and magnetic field structure relatively easier than in clusters – see Chapter 5 and Heald *et al.* (2009).

9.3   Magnetic fields and cluster cooling

Previous sections of this chapter have suggested that seeded and amplified magnetic fields exist at all stages of a cluster's evolution. When we combine this with gravitational infall over a cluster's evolution, it is also plausible that intracluster magnetic fields will affect the infall process. Cooling happens when the hot gas falls progressively into the cluster's gravitational potential well, with higher central densities in consequence. The rate of bremsstrahlung cooling increases as $n_e^2 \, T^{0.5} \, Z^2$ (Equation 2.9). The atomic density and the nuclear charge, Z, affect the timescale of this process. In particular, certain emission lines of electronic transitions from higher-Z atoms accelerate the radiative energy loss rate. By the time the consequent cooling reduces the ICM temperature from $\sim\!10^8$ to $\sim\!10^6$ K and less, and n_e increases from $\sim\!10^{-4} \rightarrow 10^{-2}$ cm^{-3}, a catastrophic, runaway gas infall toward the cluster centre will be in progress in the absence of counterforces. It is slowed in part by counterpressure, which can be estimated from the virial theorem.

Several possible sources of energy injection have been thought to balance the accelerated cooling of the dense central gas. Given our improved understanding of the physical processes involved, the term "cool core zone", following Enßlin & Vogt (2006), more appropriately denotes these anomalously dense (and magnetised) cluster centre zones, whose complex flows of energy and mass are as yet incompletely understood. At the n_e and T values above, the Spitzer conductivity may be enough to permit heat conduction between the external ICM and the cool zone. A different heat transfer process has been observed in the transfer of buoyant PdV expansion energy from the BH of the central galaxy toward the outer periphery of the cluster, as illustrated in Fig. 9.1. This clear example of a flow in the *outward* direction has been discussed by Churazov *et al.* (2002), among others.

In the cool core zone this expansion energy can also drive strong turbulent energy – see e.g. Churazov *et al.* (2004).

Magnetic twisting of the strong magnetic flux tubes can also release energy. It has been shown that this energy release might proceed into the innermost zones ($\lesssim 10$ pc), and may be very large. In effect, rotational energy is converted into magnetic field energy at a rate that can be greater than the radiation rate of the thermal gas mentioned above (see Sturrock & Stern 1980, Soker 1997). In conclusion, a range of processes can become important in inner cool core zones.

Clusters have different degrees of central cooling. Those with denser cool zones tend to have significantly stronger magnetic fields. Table 9.1, adapted from Taylor *et al.* (2002), contains an illustrative list of some parameters for 14 low redshift clusters containing extended radio sources. The first two columns list the cluster Abell catalogue number or commonly known constellation designation and/or radio source catalogue number. Column (3) lists the distance/redshift, and (4) a model-calculated "cooling flow rate" from references given in Taylor *et al.* (2002). Columns (5), (6), (7), and (8) list, respectively, the X-ray luminosity, optical classification of the host radio galaxy, monochromatic radio power at 5GHz, and radio source size. Finally, column (9) gives a single Faraday *RM* for the radio source, corrected to the (small) redshift. Further explanation of these quantities can

Table 9.1 *Model calculated "cooling flow rates", dM/dt (Column (4)), compared with other, observation-based, parameters for a sample of 14 galaxy clusters at z $\leq$ 0.11. These parameters are the cluster X-ray luminosity, the cluster-embedded radio source's host galaxy type, monochromatic luminosity, size, and z-corrected RM. Source: Taylor, Fabian, & Allen (2002), where details and background on the calculations and measurements can be found.*

(1)	(2)	(3)	(4)	(5)	(6)	(7)	(8)	(9)
Cluster	Radio source	z	$\dot{M}\,(M_\odot\,\text{year}^{-1})$	$\log L_x$ (erg s^{-1})	Gal. Type	$\log P_{5000}$ (W Hz^{-1})	Size (kpc)	$RM\,(1+z)^2$ (rad m^{-2})
Cygnus A	3C 405	0.057	320^{+71}_{-32}	44.8	E	27.7	180	3000
Virgo	3C 274	18 Mpc	$24^{+10.6}_{-5.7}$	43.1	E	24.4	58	2000
A426	3C 84	0.0183	587^{+86}_{-44}	45.0	pec	25.9	640	N/A
Hydra A	3C 218	0.0522	267^{+48}_{-30}	44.5	cD	26.3	670	-12000
Centaurus	1246–410	0.0099	$39^{+25.5}_{-7.8}$	43.8	E	25.8	15	1500
A2052	3C 317	0.0348	125^{+26}_{-6}	44.1	cD	24.7	39	-800
A2199	3C 338	0.031	197^{+20}_{-41}	44.5	cD	24.3	67	2000
A119	0053–015	0.044	0^{+2}_{-0}	44.4	E	24.5	350	400
	0053–016	0.044	0^{+2}_{-0}	44.4	E	24.4	300	150
0745-191	0745–191	0.1028	1038^{+116}_{-68}	45.4	cD	25.3	25	N/A
A2597	2322–123	0.0824	423^{+91}_{-99}	44.8	cD	25.0	10	4700
A1795	4C 26.42	0.062	462^{+108}_{-56}	44.9	cD	24.6	15	3000
3C 129	3C 129.1	0.0223	$0^{+4.2}_{-0}$	44.3	E	23.4	40	640
	3C 129	0.0208	$0^{-4.2}_{-0}$	44.3	E	24.5	500	260
A4059	2354–350	0.0478	115^{+57}_{-37}	44.3	cD	24.0	70	-1500
A2029	1508+059	0.0767	590^{+27}_{-96}	45.3	cD	24.4	80	-8000

be found in the original paper. Note the correspondence between low *dM/dt* (resulting in a weak or no cool zone) and a relatively low *RM*. A related table with more detailed hydrodynamical and magnetic parameters for 10 clusters can be found in Enßlin & Vogt (2006).

X-ray data show that at least 50% of the brightest clusters have *some* central cool zone (Peres *et al.* 1998). That said, the boundary between a "normal zone" and a "cool zone" or "cooling flow" is becoming more difficult to define. In fact, clusters have a range of cool core intensity. The cool zones are usually seen in X-rays at distances $\lesssim 100$ kpc from the cluster core (Soker 1997). For strong cooling flows, whose putative flow rate, *dM/dt*, ranges from 200 to 1000 $M_\odot$/year, the cool core radius varies from 100 to 200 kpc. In addition to large Faraday rotations in cool zones, evidence has been found for blue light (Heckman *et al.*, 1989), which may be caused by the formation of massive stars from compressed and shocked regions in the dense cool gas zone. In some extreme cases the inferred star formation rate is of order 100 $M_\odot$/year – higher than even the most luminous known individual IR/starburst galaxies (Fabian 1999).

Because the central cool gas zone represents only a small fraction of a cluster's cross section, we can conclude that background *RM* probe observations described in Sections 9.2.2 and 9.2.3 have a relatively small probability of intersecting the central dense cool zones. Thus, the more modest non-cool zone *RM*s of 50–100 rad m^{-2} (Fig. 9.3) and the corresponding magnetic fields of ≤ 10 μG are the more typically registered values in a statistical probe of the type shown in Fig. 9.3.

However, extended radio sources *within* a cool zone radius can exhibit rotation measures of order 500–10000 rad/m^{-2} – as exemplified in the Hydra cluster, in Fig. 9.5(b). Here, magnetic field values as high as 20–90 μG are commonly deduced when the *RM* values are combined with n_e estimates in the denser core. The remarkable strength of such magnetic fields exceeds that of almost any other known intergalactic gaseous system in the Universe! Known exceptions are a few high-z quasar interveners – see Chapter 12. Central cluster environments around the extended radio sources Cygnus A (3C405) (Dreher *et al.* 1987), Hydra A (3C218 – Fig. 9.5) and 3C295 (Perley & Taylor 1991) are further examples of radio systems near, or in, these dense highly magnetised cool cores of clusters. Conceivable modifications to the Sunyaev–Zel'dovich effect by such highly magnetised cool zones are discussed in Section 9.6.2

9.4   Energy components of the intracluster medium (ICM)

We have described how magnetic fields are all-pervasive in the intracluster gas, and typically at levels in above ~1 μgauss. X-ray emission and Faraday rotation studies make it possible to directly estimate a cluster's magnetic energy E_m, independent of its mass, energy in turbulent galaxy motions, and CR energy. Out to projected limits in (x, y) where the cluster's baryonic matter can be detected we can write

$$E_{\mathrm{m}} = \frac{1}{8\pi} \iiint B^2(x, y, z) dx dy dz. \tag{9.8}$$

The strong magnetic fields in galaxy clusters require an explanation of (a) where they were initially "planted" into the ICM and (b) how the field strength evolved, specifically how the fields were regenerated from presumably weaker initial strengths. The source

(a) probably predates the formation of galaxy clusters, as discussed earlier. Question (b) implies that, if magnetic fields have been substantially amplified in the ICM, the gain in magnetic field energy must occur at the expense of other energy source(s), which need to be identified. It is therefore useful at this point to discuss relevant components of energy in clusters.

For fiducial purposes, we adopt typical cluster energy components based on a series of simulations by H. Xu, H. Li, D.C. Collins, S. Li, & M. L. Norman (2009, 2011). Calculated values for the total thermal, kinetic, and magnetic energies are as follows (kindly provided by Hao Xu):

$$E_{\text{TH}} \approx 5 \times 10^{63} \ \text{ergs} \tag{9.9}$$

$$E_{\text{KIN}} = 1.2 \times 10^{63} \ \text{ergs} \tag{9.10}$$

$$E_{\text{m}} = 1.5 \times 10^{61} \ \text{ergs.} \tag{9.11}$$

In each case, the cluster boundaries were defined by including all matter within the cluster having an overdensity $\gtrsim 200$ times the average cosmological matter density at $z = 0$.

The total gravitational "self energy" of the diffuse, including dark, matter in a cluster is

$$E_{\text{GRAV}}^{\text{diff}} = \frac{GM_{\text{cl}}^2}{R_{\text{cl}}} = 1.7 \times 10^{65} \left(\frac{M_{\text{cl}}}{10^{15}M_{\odot}}\right)^2 \left(\frac{R_{\text{cl}}}{500 \ kpc}\right)^{-1} \ \text{erg,} \tag{9.12}$$

where M_{cl} represents all of the cluster mass that is distributed on the scale of the hot thermal gas mass (≈ 10 times that of the galaxies), which is assumed to be the same as R_{cl} for the dark matter. The total kinetic energy, $E_{\text{KIN}}^{\text{gal}}$, of the member *galaxies*, estimated from the masses and redshifts of cluster member galaxies.

A similar number, $E_{\text{KIN}}^{\text{gas}}$, can be calculated for the total turbulent kinetic energy in the intra-cluster thermal gas. Finally, not to be neglected is the cluster's total energy in hadronic and leptonic CRs. Excepting the minority of clusters whose CR electrons are visible via a cluster synchrotron halo, the CR nucleus component is not directly visible to most radio observations, but they could be visible to p–p collision-induced γ-rays. A further, large fraction of clusters may yet reveal their electron CR component at metre to decametre wavelengths by upcoming generations of low frequency radio arrays. The advent of γ-ray telescopes makes it possible to independently measure the CR nucleus (proton) component and its distribution in energy (Section 9.8). A simple quantitative expression for E_{CR} is not given here, but we expect it to be of comparable order to E_{M}, or of order 1–2% of the thermal energy content.

Note also that these energy components are not necessarily independent. Two of the cluster energy categories can be considered as "primary", namely E_{GRAV} and E_{TN}. The remaining sources – E_{M}, E_{TH}, $E_{\text{KIN}}^{\text{gal}}$, $K_{\text{KIN}}^{\text{gas}}$, and E_{CR} – are directly or indirectly derivative from E_{GRAV} and E_{TN}. The similarity of some of them suggests that, over a cluster's lifetime, various physical intercouplings among them have brought them into approximate equipartition.

Can the above global energies help to place constraints on *sources* of magnetic energy in galaxy clusters, or provide other information? For example, the thermonuclear energy and magnetic energies E_{TN} and E_{M} are comparable within a factor less than 10. If, however, the entire source of the magnetic energy were thermonuclear, e.g. in starburst-driven outflows, the conversion efficiency would need to be $\gtrsim 10\%$. It seems unlikely that a chain of thermonuclear-ICM magnetic field processes would have an overall efficiency that high. That leaves us to conclude that much, of cluster magnetic energy likely came from primary gravitational sources. Given the long lifetime of intergalactic fields, the "tapping into" gravitational energy might be expected to occur at pre-cluster epochs, that is, at earlier stages in the formation of universal LSS. These aspects are discussed in the next section.

9.5   Regeneration and amplification of magnetic fields in the intracluster medium

9.5.1   *Scenarios before and after cluster formation that influence the magnetic state of the intracluster medium*

Radio observations have confirmed (Chapter 6) the rapid magnetic field amplification on starburst–driven outflow timescales (e.g. Reuter *et al.* 1994 for M82, and Chyży *et al.* 2006 for NGC4659 in the Virgo cluster). Magnetic field strengths in these outflow halos reach or exceed $\sim 10\ \mu$G, and contain magnetic coherence scales of a kpc or more (see Chapters 5 and 6).

When the first central black hole accretion discs formed in primeval galaxies, we expect that they released large amounts of (gravitational) energy via outflow jets that also magnetised the surrounding IGM in a short space of time (Chapters 7 and 8). Later, in the "post-primeval", pre-cluster LSS, BH-powered energy outflows probably contributed to the magnetisation of the intergalactic gas, as calculated by Daly & Loeb (1990), and Clarke (1993).

Simulations of the development of ("pre-cluster") LSS produce some amplification of pre-existing fields in the large scale shear flows and shocks of the supra-galactic filaments and bubbles. Diffuse matter and galaxies flowed into the deepest potential wells in LSS, which are the formation zones of galaxy clusters and groups. For descriptions of simulations, and comparison with measured fields (up to $\sim \mu$G in the clusters) the reader could consult Kulsrud *et al.* (1997), Ryu *et al.* (1998), H. Xu *et al.* (2009), Cho & Ryu (2009), Section 12.8, and related references. The latter also includes modelling of the magnetic turbulence spectrum in LSS filaments.

After a galaxy cluster has formed, star and SN-driven winds can be expected to contribute to the magnetisation of the ICM. As with the starburst-seeded intracluster fields, BH-powered jets and lobes within clusters also inject magnetic energy into the ICM, as illustrated in Fig. 9.1 and in Chapter 7, Section 7.2.

9.5.2   *Field regeneration in merger-driven shocks and turbulence*

Early ideas on magnetic field amplification in galaxy clusters came from attempts to explain a possibly related phenomenon – the reacceleration of ICM relativistic electrons in clusters like Coma that are pervaded by synchrotron-emitting CR electrons. Some large scale acceleration process is required to explain why the electron CRs, which have finite radiative lifetimes, stay energised at large distances from their original sites of ejection, which are likely radio galaxies in the cluster. Turbulent wakes behind galaxies moving through the ICM have been invoked as a way of re-accelerating relativistic electrons, and to maintain the magnetic field at μG levels (Jaffe 1980). Another idea is that a dynamo mechanism operates in the ICM (e.g. Ruzmaikin *et al.* 1989, Eilek 1993), and has been proposed to explain the strong fields seen in "cool zone flows". These electron acceleration processes are in addition to the hadronically produced $e^{\pm}$ CRs in clusters (Section 9.8).

Another explanation for widespread, energised CR electrons is that cluster mergers, which cause large scale shocks, possess an ample gravitational energy source for this purpose. The electrons are proposed to be reaccelerated in the shocks, and in the post-shock turbulence (Goldman & Rephaeli 1991, De Young 1992, Tribble 1993). Magnetic field amplification in consequence of a cluster merger event has been modelled in conjunction with radio and X-ray observations of the cluster. An illustrative example of the effect of a cluster merger on magnetic field amplification is the comparison of simulations and observations for Abell 3667 in Fig. 9.6, below.

A3667 is one of the clearest examples of a merger of two clusters, in which the merging appears to be in the plane of the sky. This is also the plane of the simulation "slice" in Fig. 9.6(left). At the beginning, the magnetic pressure is enhanced in long filaments. As the simulation proceeds to the post-shock phase, the field pressure becomes much more random and turbulent in nature. The later turbulent stage is where significant magnetic field amplification occurs. In Roettiger *et al.*'s (1999) simulation, B^2 increases of from 3 to 20 μG, where the high end may be artificially capped by the limit of numerical resolution used. A key signature of the cluster merger is diffuse, morphologically coherent, cluster-scale radio-emitting cusps on each side of the merging cluster (Fig. 9.6). Analogues of these large intergalactic scale radio emitting clouds, which have no single galaxy association, have also

(a)   (b)

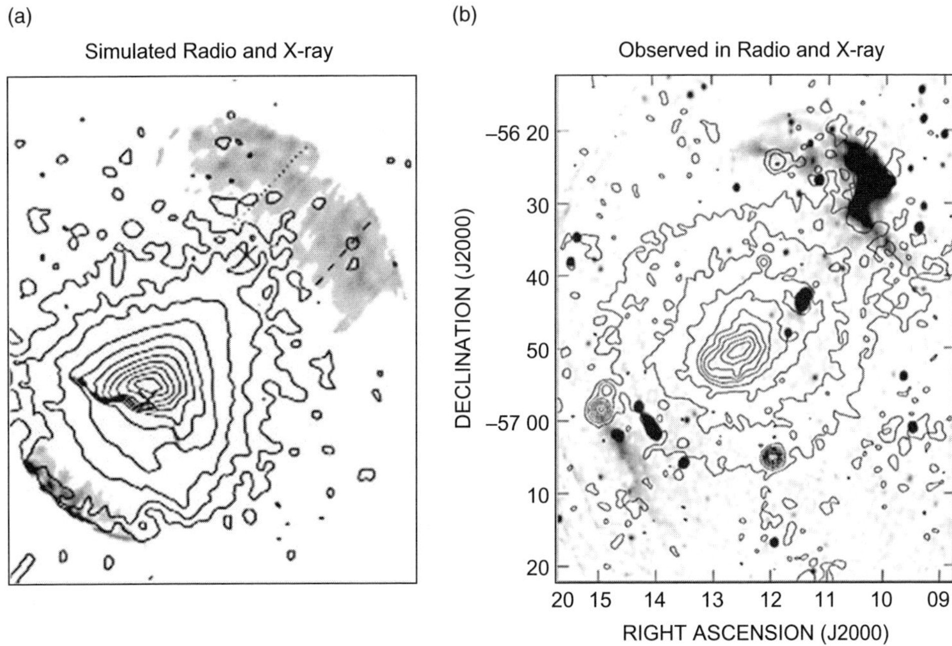

Figure 9.6. Left: Simulated X-ray contours and 1.4 GHz radio (grey scale) emission *after* the cluster merger event in the cluster Abell 3667. Right: Observed X-ray and original 843 MHz radio image from Röttgering *et al.* 1997. Both images are reproduced from Roettiger, Burns, & Stone (1999). The simulation used the ZEUS-3D Eulerian finite difference code of Stone & Norman (1992a, 1992b).

been found in some other extragalactic locations. Diffuse, magnetised intergalactic CR gas is discussed in the next chapter in the context of intergalactic magnetic fields.

Simulations of a cluster that formed out of evolving LSS claim to show how the magnetic field strengths can grow from initial intergalactic values near 10^{-9} G, up to ~1 μG, i.e. by a factor of ~10^6 in energy density. The amplification comes in essence from a combination of two processes: adiabatic compression in the collapse (in which $|B| \propto \rho^{\frac{2}{3}}$), followed in a later phase by stretching and shearing in the infall flows. Dolag, Bartelmann & Lesch (2000) have performed such simulations that begin at $z = 15$, $|\!<\!B_{\mathrm{init}}\!>\!| = 10^{-9}$ G. Among their results was that the final $|B|$ was quite insensitive to the initial degree of tangling. Figure 9.7 shows the result of a cluster evolution simulation that predicts the observed $RM(\alpha,\delta)$ distribution in three orthogonal views.

The simulation shown does not calculate cooling, and so will understate a further stage of magnetic field amplification that could happen at small radii. For a cluster without a central cool zone, the above simulations seem to agree well with observed RM data.

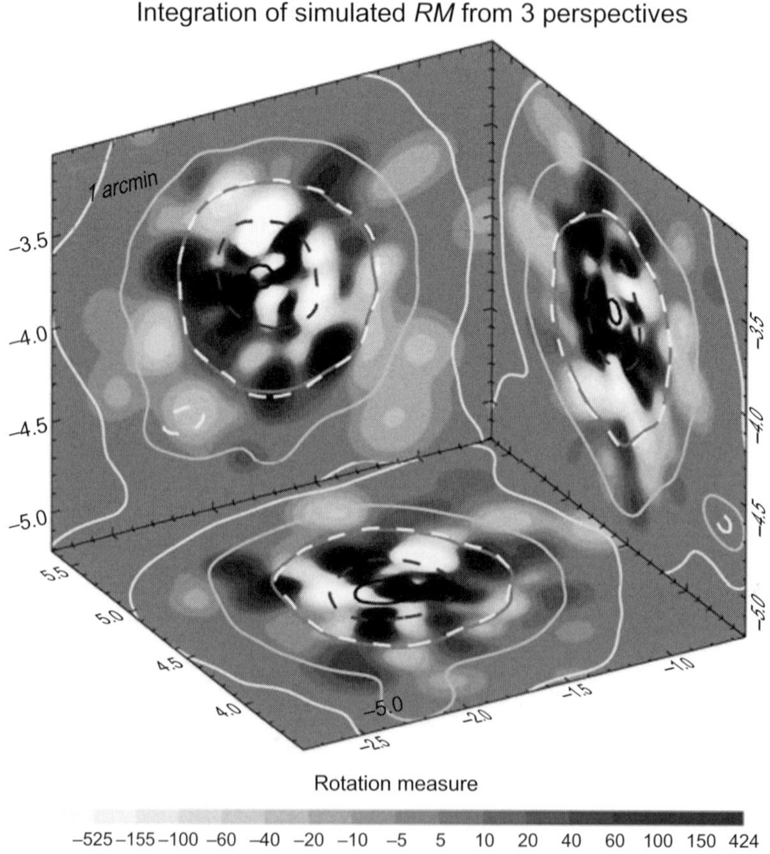

Figure 9.7. Projected 3-D view of a simulated cluster, showing the Faraday rotation distribution that would be seen from three orthogonal viewing directions. Dolag *et al.*'s calculations used the smooth particle hydrodynamics-based GrapeSPH code (Steinmetz 1996). (Reproduced from Fig. 4 of Dolag, Bartelmann, & Lesch (1999).)

9.5.3 Comments on injection of magnetic fields into the ICM by galactic supermassive black holes

The jet-lobe systems have short radiative lifetimes ($\approx 10^8$ years) relative to a cluster's dynamical time ($\approx 10^{10}$ years), and they are "fed" from the "active" nucleus of a galaxy, which harbours a highly collapsed object, presumably a supermassive black hole. Simulations imply that extended radio sources in clusters are transient events in the lifespan of a galaxy cluster. Therefore, we see just a cosmic "snapshot", so that only a fraction of a large cluster sample will contain an active radio galaxy at the cosmological epoch of observation (e.g. Fig. 9.1(b)).

This has several consequences for the physics of the ICM. First, assuming that the energy of the relativistic electrons in the radio lobes is at least matched by that of the non-radiating relativistic protons, large numbers of relativistic protons have also been deposited into the ICM of all clusters. Additionally, as discussed in Chapter 7, AGN-jet radio sources appear capable of regenerating magnetic fields within their lobes, within the short ($\approx 10^8$ year) lobe lifetime. As with the protons, the large magnetic flux, once generated, may not easily be destroyed over the cluster's lifetime, and will remain invisible, except indirectly to Faraday rotation in the presence of thermal electrons, or possibly Zeeman splitting in some special circumstances.

The recent improving resolution, sensitivity, and imaging capability of radio telescopes at lower radio frequencies enables us to detect synchrotron radiation of low energy, "old" relativistic electrons, not only in cluster halos, but also outside clusters, in magnetised clouds of CR electrons up to a few times 10^8 years after active injection phase of the AGN (SMBH) and jet.

Figure 9.8 shows a case in point – an image of 3C274 (Virgo A) produced by the VLA at 327 MHz ($\lambda 92$ cm) and which may be an indicator of multiple production of magnetised

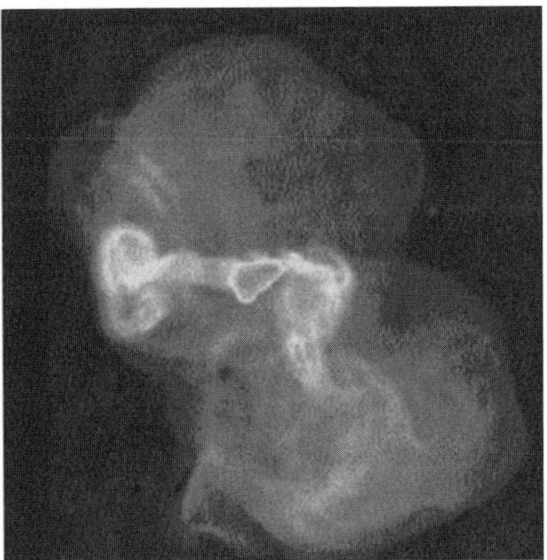

Figure 9.8. A VLA radio image of M87 (3C274) in the Virgo cluster at 327 MHz. At this lower resolution, the east-west oriented jets and two "outer" lobes can still be discerned. The synchrotron-emitting "cloud" extends over 100 kpc, a significant fraction of the cluster's core size. (Reproduced from Owen, Eilek, & Kassim (2000) by kind permission of Frazer Owen.)

lobes. 3C274 is a jet-lobe radio galaxy of modest radio luminosity in the Virgo cluster of galaxies. Virgo is a relatively poor cluster, and at 16 Mpc away the nearest galaxy cluster to us that is visible from Northern latitudes. The east-west orientated jets and lobes can be discerned at the relatively low resolution of this image. The image reveals that the 42 kpc long structure has produced a magnetised zone many times the volume of the original radio source. The overall extent of what we will call the "cosmic ray + magnetic field" zone of this single radio galaxy is a non-negligible fraction of the cluster's core radius. In this sense it is reminiscent of the cluster-size radio extension of 5C4.81 in the Coma cluster – see Section 9.1.3 and Fig. 9.1(R). Its age of a few times 10^8 years is small compared to the Virgo cluster's dynamical age, but the energy loss time of its lower energy CR nuclei and magnetic fields will be many times longer. The M87 and 5C4.81 images thus illustrate how a single supermassive BH can magnetise a large fraction of the entire cluster volume.

Similar filamentary structures of enhanced |B| have been seen in extended radio sources where the instrumental signal-to-noise ratio at high resolution is sufficiently high. The contrast in |B| can be quantitatively estimated directly from the image, and Equation (1.1), since |B|$(x, y, z) \propto \varepsilon_{\mathrm{syn}}^{-2}$, on the assumption that the filaments are fully resolved. Also, given a combination of intra-cluster orbital and turbulent motions, the radio galaxy lobe fields will eventually become distributed and "shredded" within the ICM, as illustrated in the simulations of Diehl *et al.* (2008), discussed in Chapter 7.

Over this multi-Gigayear timespan, at least a few similar such AGN-driven outflow events will have occurred in a cluster. All but the most "recently" accelerated CR electrons will have radiatively cooled, either by synchrotron radiation, or at redshifts $z \gtrsim 1$ by inverse Compton radiation. As discussed earlier, this likely explains why most clusters have no prominent synchrotron radio halos at GHz frequencies. But the magnetic fields from all previous outflow events are not easily destroyed. Destruction of the fields can happen by ohmic diffusion, by some large scale reconnection process, or by removal of magnetic flux from the cluster through buoyancy-driven outflow processes, as illustrated in Fig. 9.1(a). Even if reconnection were able to operate widely in the ICM, turbulence and shocks would act in the other direction – towards energy balance – and re-amplify the magnetic field. Lifetimes of ohmic diffusion losses can be longer than a Hubble time, and so are not significant for most clusters, except in the inner, dense cool zones discussed above. Finally, the numerically large starburst outflows in cluster member galaxies will add to the ICM magnetisation of jet-lobe systems.

9.6   The Sunyaev–Zel'dovich effect as a probe of intracluster gas physics

9.6.1   *The classical thermal and other SZ effects on background CMB photons*

The SZ effect, briefly introduced in Chapter 1, modifies the spectrum of cosmic black body photons after they pass through the hot intracluster gas. More generally, modification of the CMB brightness is possible for any intervening system containing electrons that can Compton scatter the CMB radiation (Zel'dovich & Sunyaev 1969, Sunyaev & Zel'dovich 1970). The modification consists of a decrement in the CMB specific intensity *below* the 218 GHz crossover frequency, $(I(v))$, $\Delta I/I(v) \sim$ a few $\times 10^{-4}$ and a corresponding excess *above* the crossover, perturbing the spectrum in a broad range from approx. 40 to 800 GHz. This small CMB perturbation is readily measurable. It is commonly characterised by the dimensionless Compton y-parameter

$$y = \int \frac{kT_e}{m_e c^2} n_e \sigma_T dl \qquad (9.13)$$

(Kompaneets 1956), where T_e and m_e are the electron temperature and mass, and n_e is the density of the hot gas in a line of sight through the gaseous system, i.e. a galaxy cluster.

In an isothermal cluster, τ_e is the scattering optical depth for a column density, N_e ($\tau_e = \sigma_T N_e$), where σ_T is the Thomson cross-section. The consequent distortion of the CMB spectrum $\Delta T/T(v)$ can be related to a (usually adopted) King model 3-D density distribution of the cluster's hot gas, given in Equation (9.4). Information provided by T_e and N_e from thermal X-ray images of clusters can be combined with $\Delta I/I(v)$ and the cluster's angular size distance. Because $\Delta I/I$ is independent of a cluster's distance, SZ observations over a large redshift range can be used to constrain cosmological parameters such as H_0.

However, other effects can cause slight distortions of the otherwise "straightforward" thermal SZ (SZ_{TH}) effect. These include non-thermal components of N_e, peculiar velocities, *and* intracluster magnetic fields. Thus the SZ decrement/increment, given by the SZ_{TH}, is in addition to several other components discussed below. In the following section we give attention to the *incremental magnetic* SZ component, "δ_B", and also briefly describe the others, but do not treat them in detail. The existence of a "δ_B" is the point at which the SZ effect could be significant, especially in central cool core zones (Section 9.3), where ICM magnetic field strengths are found to be quite high, up to a few *tens* of μG, possibly with an ordered component.

Suprathermal electrons extending in energy up to the trans-relativistic régime further modify the SZ_{TH} (δ_{STH}). Also, a kinetic effect due to the bulk motion of a cluster relative to the CMB photon reference frame will cause a slight modification, δ_{KIN}, to the spectral distortion of the background CMB radiation. Then a relativistic SZ effect (δ_{REL}) can also modify the thermal SZ function in a galaxy cluster. Apart from probing $SZ_{TH} + \delta_{STH}$ in the ICM, measurement of δ_{REL} could be a way of detecting old "fossil" radio lobes in the wider IGM that are otherwise undetectable except at the lowest radio frequencies (Chapter 7). That is, δ_{REL} may indirectly detect magnetic fields, since they are an important constituent of fossil radio lobes, whether in clusters *or* the wider IGM. For these same systems, modified SZ effects might indicate local IGM heating in significant volumes around these old, invisible, still expanding, and magnetised radio lobes that have compressed and heated the surrounding IGM.

The total SZ effect incorporating all the CMB perturbations discussed can be written as

$$\frac{\Delta T_{SZ}}{T} = y \times \left[x \frac{e^x + 1}{e^x - 1} - 4 \right] (1 + \delta_{STH} + \delta_{REL} + \delta_{KIN} + \delta_B), \qquad (9.14)$$

where x is a dimensionless frequency:

$$x = \frac{hv}{kT_{CMB}}. \qquad (9.15)$$

An extensive literature exists on the SZ effect and its variations – see e.g. Rephaeli (1995), Birkinshaw (1999), Enßlin & Kaiser (2000), and Carlstrom *et al.* (2002). Generally, less attention has been paid to the possible effects of magnetic fields on the SZ effect.

9.6.2   *Effect of strong intracluster magnetic fields on the Sunyaev–Zel'dovich effect*

At its microphysical level, the SZ effect is a second order Compton scattering effect caused by the velocity or velocity spread of intervening electrons. These explain the sensitivity to T_e in SZ_{TH}, to the bulk velocity in SZ_{KIN}, and to suprathermal (δ_{STH}) and relativistic electrons, SZ_{REL}, etc. Where a sufficiently strong magnetic field is present, a strong and organised magnetic field will perturb the electron velocity distribution. We denote a magnetic perturbation as δSZ_B (δ_B in Equation 9.14). In the intracluster magneto-plasma, velocities parallel and perpendicular to a segment of ordered $\boldsymbol{B}$ ($v_{e\perp}$ and $v_{e\parallel}$) will differ. This introduces two $\boldsymbol{B}$-orientation-dependent electron "pseudo-temperatures", $T_{e\perp}$ and $T_{e\parallel}$, in the Compton y-parameter above (Equation 9.14).

Thus for a realistic cluster-internal magnetic field fluctuation spectrum with a sufficiently strong $<|\boldsymbol{B}_{ICM}|>$, the SZ spectral distortion can be significantly modified. The effect is illustrated in Fig. 9.9 in a plot that compares the SZ_{TH} differential $\Delta I(v)$ with a B-modified ($SZ_{TH} + \delta SZ_B$)curve (Hu & Lou 2004) for the well-observed cluster A2163. This is a central cool zone cluster with a strong central-zone magnetic field. The magnetic field modifies both the $\Delta I(v)$ decrement and increment regions, and also has a small effect on the crossover frequency near 218 GHz.

Figure 9.9 shows that, independent of other influences on SZ, a cluster core $<|\boldsymbol{B}|>$ in excess of ~10 μG can measurably perturb the SZ effect. It simultaneously illustrates how SZ measurements can be used to set *limits* on cluster magnetic field strengths in cool zones, *independent of Faraday RM measurements*. In Fig. 9.9 this upper limit on $<|\boldsymbol{B}|>$ is ~40 μG for the field turbulence model used in A2163.

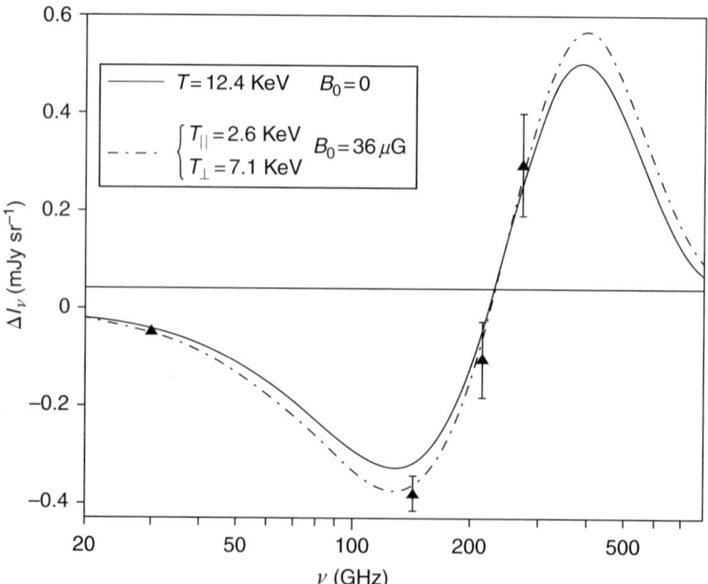

Figure 9.9. A plot of $\Delta I(v)$ for the cluster A2163, showing modeled SZ_{TH} effect, measured data points, and modeled ($SZ_{TH} + SZ_B$) curves for a selected parallel and perpendicular magnetic field strength in the cluster core. From Faraday *RM* and X-ray measurements, $<|\boldsymbol{B}_{ICM}|>$ has been independently estimated at $> 10\,\mu$G in the core cool zone of the cluster. (Source: Hu & Lou (2004).)

All of the above discussion refers to the SZ effect for a given "pointing", *viz.* in the cluster centre direction. The magnetic δSZ effect can also affect the cluster *radial* dependence of the $SZ_{TH} + SZ_B$, thereby in principle providing an extended observational diagnostic of the intracluster magnetic field.

9.7 Inverse-Compton emission and implications for intracluster magnetic field strength

Radiative cooling by inverse Compton (IC) up-scattering of photons by high energy electrons can, in certain circumstances, be a significant energy loss mechanism in galaxy clusters. Normal IC scattering can be treated as a classical scattering phenomenon in which the cross-section is the Thompson cross section ($\sigma_T = 6.6524 \times 10^{-25}$ cm^2), and the number of scattered photons is proportional to $\sigma T \cdot n_e^r \cdot \varepsilon_{ph}^{(i)}$, where n_e^r is the density of relativistic electrons and $\varepsilon_{ph}^{(i)}$ is the energy density of photons to be scattered. Classical Thompson cross section applies as long as $\gamma h v \ll mc^2$. For generality we use the superscript (i) to distinguish 3 types of photon fields in extragalactic situations that can be Compton scattered. The first, $i = 1$, are synchrotron or inverse Compton-generated photons that scatter off the *same* relativistic electron population that produced them. This type of IC radiation is significant in extremely compact AGN sources close to the accretion disc of a QSO or BL Lac object, and will not be considered further here. The second possibility, $i = 2$, is IR photons in a very dense stellar or galactic environment that could scatter off a sufficiently high density of relativistic electrons. For galaxy clusters the most common possibility, $i = 3$, is the up-scattering of microwave photons of the cosmic background radiation off relativistic electrons in the ICM. It is also important to note here that $\varepsilon_{ph}^{(3)} \propto (1 + z)^4$. This means that the IC cooling time is sharply reduced as the cluster redshift approaches and exceeds 1. A single scattering increases the photon frequency by γ^2, where γ is the relativistic boosting factor on the energy γmc^2 of the relativistic electron. The scattered photon, having energy $\gamma^2 h v$, can similarly be upscattered successively n times with σ_T until a limit is reached where

$$\gamma^n \cdot h v \sim mc^2. \tag{9.16}$$

When condition (9.16) is reached – in the relativistic régime – the scattering becomes less effective, i.e. $\sigma_{rel} < \sigma_T$, and it is defined in the relativistic limit by the Klein-Nishina cross section. (For more details the reader could consult Felten & Morrison 1966, Rees 1967, Pacholczyk 1970, Rybicki & Lightman 1979). IC losses for the electrons, like synchrotron losses, increase as γ^2. The IC luminosity spectrum, $L(v)$ can be written, for $z \approx 0$, as

$$L^{IC}(v) = 12\pi\sigma_T \int_0^{\gamma_{max}} n_e^r(\gamma) d\gamma \int_0^1 J\left(\frac{v}{4\gamma^2 x}\right) F(x) dx \tag{9.17}$$

(Sarazin 1999), where

$$F(x) = 1 + x + 2x \ln x - 2x^2 \ \ (x < 1), \text{ and} \tag{9.18}$$

$$x = \frac{1}{4\gamma^2} \times \frac{v_{sc}}{v}. \tag{9.19}$$

$F(x)$ is a geometrical factor that allows for integration over all directions of both the incoming and outgoing photons. $J(v)$ specifies the form of the radiation field to be scattered – in case (3) it is the Planckian cosmic background radiation (CMB). The relativistic electron population has the initial form of a power law, $n_{\mathrm{e}}^{\mathrm{r}}(\gamma) \propto \gamma^{-p}$, giving a synchrotron radiation spectral index $\alpha = -(p - 1)/2$. The IC scattered radiation field is described by Equation (9.17). Since the slope in energy, $p \approx 2.5$, is rather steep, relativistic electrons are relatively much more numerous at lower γ's of about a few hundred. However, electrons having $\gamma \leq 100$ will cause *relatively* fewer scattering events, because in this lowest energy range Coulomb losses will set in and reduce their numbers substantially below the power law values. Over much of its logarithmic range, $L^{\mathrm{IC}}(v)$ will have the same spectral index, α, as the original synchrotron spectrum emitted by the relativistic electron population, $N(\gamma)$ – but translated by γ^2 in frequency relative to pre-scattered photons. Thus, an IC-scattered CMB photon originally near 5×10^9 Hz scattered off an electron having $\gamma = 1000$ will appear at 5×10^{15} Hz. For a wider analysis and derivation of the radiation transfer equations for several mechanisms, a classic paper by Blumenthal & Gould (1970) could be consulted.

The total IC luminosity can be expressed as

$$L_{\mathrm{TOT}}^{\mathrm{IC}} = \frac{32}{9} \frac{\pi c e^4}{\left(mc^2\right)^2} \varepsilon_{\mathrm{rad}} \gamma^2 \tag{9.20}$$

(Pacholczyk 1970), which is emitted in X-ray bands for case (3). As with bremsstrahlung, the inverse Compton emission will be unpolarised, except for special circumstances not discussed here, and its spectral slope α in the range of 1–3 keV is difficult to distinguish from that of thermal bremsstrahlung from thermal electrons in a cluster having both a non-thermal (synchrotron) and thermal (X-ray) halo. The relative dominance of synchrotron (1.1) and inverse Compton loss rates is given by the ratio

$$\frac{L_{\mathrm{TOT}}^{\mathrm{SYNCH}}}{L_{\mathrm{TOT}}^{\mathrm{IC}}} = \frac{\frac{B^2}{8\pi}}{\varepsilon_{\mathrm{rad}}\left(1 + z\right)^4} = \frac{\varepsilon_B}{\varepsilon_{\mathrm{rad}}\left(1 + z\right)^4}, \tag{9.21}$$

which is just that of the magnetic energy density to the radiation energy density. This can also be seen qualitatively from Equation (1.1), in that for a given observed synchrotron luminosity, a higher magnetic field strength means a lower $n_{\mathrm{e}}^{\mathrm{r}}(\gamma)$, hence a lower rate of IC emission. At $z = 0$, $\varepsilon_{\mathrm{rad}} = 4.8 \times 10^{-13}$ erg cm^{-3}, whereas Faraday rotation and X-ray observations (see above) indicate a typical $\varepsilon_B = 6.4 \times 10^{-13}$ erg cm^{-3}. An estimate of $\boldsymbol{B} = 6$ μG for the Coma Cluster corresponds to $\varepsilon_B = 1.4 \times 10^{-12}$ erg cm^{-3}, i.e. three times higher than $\varepsilon_{\mathrm{rad}}$ at $z = 0$. The $\boldsymbol{B}$ estimates for central regions of cool zone clusters, $\varepsilon_B/\varepsilon_{\mathrm{rad}} \gg 1$, would definitely rule out IC radiation here. However, for non-cool core clusters with radio halos, such as Coma, $\varepsilon_B/\varepsilon_{\mathrm{rad}}$ is not much above 1, especially if we allow for spatial variations $|\boldsymbol{B}|(x, y, z)$ and $n_e^r(\gamma)$ within the cluster. A related discussion and instructive plot showing the relation between the high energy photon spectrum and various cluster magnetic field levels for the Coma cluster can be found in Enßlin & Biermann (1998). An IC component of the X-ray luminosities of such clusters is currently difficult to confirm, but it appears at most not overwhelming. This is consistent with cluster magnetic field measurements, and appears to rule out large *intra*cluster zones of weak field strengths below ~0.1 μG.

With the next generation of cluster X-ray imaging and spectroscopy, the ability to identify the relative fractions of bremsstrahlung and inverse Compton components in the X-ray emission will give us another method of placing more precise lower limits to $|\boldsymbol{B}|(x, y, z)$. This is a useful complement to the Faraday rotation data, and illustrates how the ability to isolate

IC emission in cluster X-ray images augments the available diagnostics for constraining and indirectly estimating magnetic field strengths in galaxy clusters.

9.8 Hadronically produced CR electrons, CR protons, and related magnetic field energy

Equations (9.1) and (9.2) showed that collisions between intracluster CR and thermal protons produce γ-rays and CR electrons – in addition to electron and muon neutrinos and antineutrinos. As well, CR electrons may be accelerated by other processes, such as SN-driven winds, BH-driven jets, cluster-internal shocks, etc. In other words, the CR electrons that generate synchrotron radiation in cluster halos can have more than one primary source for their relativistic energy. The significance of p–p collisions for the CR and intracluster magnetic field was first pointed out by Dennison (1980).

Relevant to p–p collision ("hadronic origin") calculations is the question of how CR protons can be accelerated by different processes in clusters. One process that received early attention is acceleration due to resonant pitch angle scattering by Alfvén waves (Schlickeiser *et al.* 1987, Ohno *et al.* 2002, and others). Since high energy CR protons do not lose significant energy through synchrotron or IC losses, they will tend to accumulate in the ICM over the cluster's lifetime. Thus, and also for the reasons mentioned below, constraining CR protons through γ-ray observations is of interest.

Upper limit constraints on the energy spectrum of intracluster TeV protons can be derived by imaging diffuse TeV γ-rays at different energies in and around galaxy clusters. A high energy fall-off around 1–5 TeV has been established from multi-energy TeV photon imaging of the Coma supercluster. Comparison with co-extensive diffuse radio images has produced a global CR proton energy spectrum whose upper cutoff is well-delineated (Aharonian *et al.* – The H.E.S.S. Collaboration 2009).

Alfvén waves involve the ICM magnetic field strength and the thermal ICM density (detected by thermal X-ray bremsstrahlung). It has been argued by Brunetti *et al.* (2004) that, if the energy density ratio $\varepsilon_{CR\ proton}/\varepsilon_{TH}$ is above the threshold level of a few percent, ICM Alfvén waves would be damped out. This underlines the interest in establishing $\varepsilon_{CR\ proton}/\varepsilon_{TH}$. For the Perseus cluster, $\varepsilon_{CR\ proton}/\varepsilon_{TH}$ has been is estimated by Pfrommer & Enßlin (2004) to be in the range of ~2%. This is consistent with $B \sim 8.8\ \mu G$ in the central region of the Perseus cluster on equipartition assumptions, and the detection of a hadronic collision origin of the synchrotron radiating electrons. This strong B is also consistent with negligible IC relative to bremsstrahlung-produced X-rays in the keV range (Section 9.7).

In specifying the physical state of the ICM, including the magnetic field, it is helpful if we can separate the hadron collision-produced CR electrons from other sources. Imaging γ-ray telescopes, can, in principle, isolate the role of π_0 decay (Equation 9.1). Consequently, when data from thermal X-ray imaging of a cluster are combined with gamma ray data, it should also be possible to identify the CR proton population.

9.9 Summary

Widespread, μG-level fields are found to exist in the ICM of galaxy clusters. It was noted that ICM field values are as high as 10–50 μG in some central cool flow zones of clusters. This magnetic energy density is greater than even for spiral galaxy disc fields.

The revelation of widespread cluster *ICM* fields suggests that galaxy fields merge with, or supply those of the ICM. The observational techniques described will help the reader to appreciate how better definition of details will improve in future, given further radio, optical and high resolution X- and γ-ray images.

This chapter, along with Chapters 5, 7, and 8, has presented physical arguments, data and simulations which, combined, suggest more than one origin for the substantial magnetic energy that pervades galaxy clusters. It is shown that significant magnetic energy is injected from black hole/accretion discs of galaxies in the cluster. Nearby cluster radio galaxies show patterns of synchrotron radiation that are consistent with some clear predictions from simulations.

At this point it will be obvious that a complex mix of mechanisms is involved in the origin and evolution of magnetic fields within galaxy clusters. The combination of magneto-plasma processes is complex and we have discussed many mechanisms simultaneously at work in clusters that can affect ICM magnetism.

IC upscattered CMB photons may not, typically, contribute much to the diffuse cluster X-ray emission, because of the typically strong, Galaxy disc-like, $>1\mu G$ ICM magnetic field strengths. Data, as well as various simulations, also show how magnetic field regeneration can take place in the course of formation of the cluster gas halo, whose baryonic mass greatly exceeds that of the galaxies. The primary energy source for this is gravitational, and associated with the large scale collapse of the cluster. A third process, the formation of central cool zones in some clusters, needs further simulation by numerical computation.

A weight of evidence shows that both inflow and outflow from AGNs operate in clusters (e.g. Binney 1998, Owen *et al.* 1999), where the energy input from AGN jet lobe sources is transient and stochastic on a global scale. Large scale inflows should remain relatively constant over a cluster's lifetime. It is reasonable to postulate that, on a timescale that is not too great, *viz.* a fraction of the dynamical time, energy densities approach equilibrium with each other. This chapter does not discuss earlier origins of the magnetic fields that must have pre-existed to provide some of the seed magnetic fields. That topic is briefly covered in Chapters 12 and 13.

We described how strong and ordered cool zone fields can further influence the Sunyaev–Zel'dovich effect and how, correspondingly, the ΔSZ_B effect might in such circumstances be used as an independent upper limit constraint on cluster-central magnetic field strengths. The new era of advanced high energy photon detectors up to EeV energies permits new probes of the ICM not previously possible, and for high energy astroparticle physics phenomena in general. These include measurements of the CR nucleus component and associated high energy particle–photon interactions, hitherto essentially undetectable via synchrotron and inverse Compton emission.

References

Abell, G. O. 1958, The Distribution of Rich Clusters of Galaxies, *Astrophys. J. Suppl.*, 3, 211

Abramopoulos, F. & Ku, W. H.-M. 1983, X-Ray Survey of Clusters of Galaxies with the Einstein Observatory, *Astrophys. J.*, 271, 446

Aharonian, F. A., Akhperjanian, A. D., Anton, G., *et al.* (The H.E.S.S. Collaboration). 2009, Constraints on the Multi-TeV Particle Population in the Coma Galaxy Cluster with HESS Observations, *Astron. Astrophys.*, 502, 437

Binney, J. J. 1998, M 87 and Cooling Flows, in *The Radio Galaxy Messier 87: Proceedings of a Workshop Held at Ringberg Castle, Tegernsee, Germany, 15–19 September 1997*, eds. H. J. Röser & K. Meisenheimer, Lecture Notes in Physics (Berlin: Springer), 530, 116

Birkinshaw, M. 1999, The Sunyaev-Zel'dovich Effect, *Phys. Rep.*, 310, 97

Blumenthal, G. R., & Gould, R. J. 1970, Bremsstrahlung, Synchrotron Radiation, and Compton Scattering of High-Energy Electrons Traversing Dilute Gases, *Rev. Mod. Phys.*, 42, 237

Brentjens, M. & deBruyn, G. 2006, RM-Synthesis of the Perseus Cluster, Astron. *Nachrichten*, 327, 545

Briel, U. G., Henry, J. P., & Böhringer, H. 1992, Observation of the Coma Cluster of Galaxies with ROSAT During the All-Sky Survey, *Astron. Astrophys.*, 259, L31

Brunetti, G., Blasi, P., Cassano, R., & Gabici, S. 2004, Alfvénic Reacceleration of Relativistic Particles in Galaxy Clusters: MHD Waves, *Leptons and Hadrons, MNRAS*, 350, 1174

Burbidge, G. R. 1956, On Synchrotron Radiation from Messier 87, *Astrophys. J.*, 124, 416

Burn, B. J. 1966, On the Depolarization of Discrete Radio Sources by Faraday Dispersion, *MNRAS*, 133, 67

Butcher, H. & Oemler, A. 1978, The Evolution of Galaxies in Clusters. II – The Galaxy Content of Nearby Clusters, *Astrophys. J.*, 226, 559

Carlberg, R. G., Yee, H. Y. C., & Ellingson, E. 1997, The Average Mass and Light Profiles of Galaxy Clusters, *Astrophys. J.*, 478, 462

Cho, J. & Ryu, D. 2009, Characteristic Lengths of Magnetic Field in Magnetohydrodynamic Turbulence, *Astrophys. J.*, 705, L90

Churazov, E., Forman, W., Jones, C., Sunyaev, R., & Böhringer, H. 2004, XMM-Newton Observations of the Perseus Cluster – II. Evidence for Gas Motions in the Core, *MNRAS*, 347, 29

Churazov, E., Sunyaev, R., Forman, W., & Böhringer, H. 2002, Cooling Flows as a Calorimeter of Active Galactic Nucleus Mechanical Power, *MNRAS*, 332, 729

Chyży, K. T., Soida, M., Bomans, D. J., Vollmer, B., Balkowski, Ch., Beck, R., & Urbanik, M. 2006, Large-Scale Magnetized Outflows from the Virgo Cluster Spiral NGC 4569: A Galactic Wind in a Ram Pressure Wind, *Astron. Astrophys.*, 447, 465

Clarke, D. A. 1993, 3-D MHD Simulations of Extragalactic Jets, in *Jets in Extragalactic Radio Sources: Proceedings of a Workshop Held at Ringberg Castle, Tegernsee, FRG September 22–28, 1991*, ed. H.-J. Röser & K. Meisenheimer, Lecture Notes in Physics (Berlin: Springer), 421, 243

Clarke, T. E., Kronberg, P. P., & Böhringer, H. 2001, A New Radio-X-Ray Probe of Galaxy Cluster Magnetic Fields, *Astrophys. J.*, 547, L111

Daly, R. A. & Loeb, A. 1990, A Possible Origin of Galactic Magnetic Fields, *Astrophys. J.*, 364, 451

De Young, D. S. 1992, Galaxy-Driven Turbulence and the Growth of Intracluster Magnetic Fields, *Astrophys. J.*, 386, 464

Dennison, B. 1980, Formation of Radio Halos in Clusters of Galaxies from Cosmic-Ray Protons, *Astrophys. J.*, 239, L93

Dermer, C. D. 1986, Binary Collision Rates of Relativistic Thermal Plasmas. II. Spectra, *Astrophys. J.*, 307, 47

Diehl, S., Li, H., Fryer, C. L., & Rafferty, D. 2008, Constraining the Nature of X-Ray Cavities in Clusters and Galaxies, *Astrophys. J.*, 687, 173

Dolag, K., Bartelmann, M., & Lesch, H. 1999, SPH Simulations of Magnetic Fields in Galaxy Clusters, *Astron. Astrophys.*, 348, 351

Dreher, J. W., Carilli, C. L., & Perley, R. A. 1987, The Faraday Rotation of Cygnus A – Magnetic Fields in Cluster Gas, *Astrophys. J.*, 316, 611

Eilek, J. A. 1993, Dynamo-Supported Magnetic Fields in Radio Jets, in *Jets in Extragalactic Radio Sources, Proceedings of a Workshop Held at Ringberg Castle, Tegernsee, FRG, September 22–28, 1991*, ed. H.-J. Röser & K. Meisenheimer, *Lecture Notes in Physics* (Berlin: Springer*)*, 421, 131

Enßlin, T. A. & Kaiser, C. R. 2000, Comptonization of the Cosmic Microwave Background by Relativistic Plasma, *Astron. Astrophys.*, 360, 417

Enßlin, T. A. & Vogt, C. 2006, Magnetic Turbulence in Cool Cores of Galaxy Clusters, *Astron. Astrophys.*, 453 , 447

Fabian, A. C. 1999, Cooling Flows pre Chandra, XMM and ASTRO-E, in *Diffuse Thermal and Relativistic Plasma in Galaxy Clusters*, ed. H. Böhringer, L. Feretti, & P. Schuecker, *MPE Report* (Garching: Max-Planck-Institut für Extraterrestrische Physik) 271, 123

Felten, J. E. 1996, Mitigating the Baryon Crisis in Clusters: Can Magnetic Pressure be Important?, in *Clusters, Lensing and the Future of the Universe*, ed. V. Trimble & A. Reisenegger, *ASP Conf. Ser.* (San Francisco: Astronomical Society of the Pacific), 88, 271

Felten, J. E. & Morrison, P. 1966, Omnidirectional Inverse Compton and Synchrotron Radiation from Cosmic Distributions of Fast Electrons and Thermal Photons, *Astrophys. J.*, 146, 686

Feretti, L., Dallacasa, D., Giovannini, G., & Tagliani, A. 1995, The Magnetic Field in the Coma Cluster, *Astron. Astrophys.*, 302, 680

Feretti, L., Dallacasa, D., Govoni, F., Giovannini, G., Taylor, G. B., & Klein, U. 1999, The Radio Galaxies and the Magnetic Field in Abell 119, *Astron. Astrophys.*, 344, 472

Garrington, S. T. & Conway, R. G. 1991, The Interpretation of Asymmetric Depolarization in Extragalactic Radio Sources, *MNRAS*, 250, 198

Garrington, S. T., Leahy, J. P., Conway, R. G., & Laing, R. A. 1988, A Systematic Asymmetry in the Polarization Properties of Double Radio Sources with One Jet, *Nature*, 331, 147

Goldman, I. & Rephaeli, Y. 1991, Turbulently Generated Magnetic Fields in Clusters of Galaxies, *Astrophys. J.*, 380, 344

Heald, G., Braun, R., & Edmonds, R. 2009, The Westerbork SINGS Survey. II Polarization, Faraday Rotation, and Magnetic Fields, *Astron. Astrophys.*, 503, 409

Heckman, T. M., Baum, S. A., van Breugel, W. J. M., & McCarthy, P. 1989, Dynamical, Physical, and Chemical Properties of Emission-Line Nebulae in Cooling Flows, *Astrophys. J.*, 338, 48

Henriksen, M. J. & Mushotzky, R. F. 1986, The X-Ray Spectrum of the Coma Cluster of Galaxies, *Astrophys. J.*, 302, 287

Hu, J. & Lou, Y.-Q. 2004, Magnetic Sunyaev-Zel'dovich Effect in Galaxy Clusters, *Astrophys. J.*, 606, L1

Jaffe, W. J. 1980, On the Morphology of the Magnetic Field in Galaxy Clusters, *Astrophys. J.*, 241, 925

Jaffe, W. J. & Perola, G. C. 1973, Dynamical Models of Tailed Radio Sources in Clusters of Galaxies, *Astron. Astrophys.*, 26, 423

Kaiser, N. 2001, Weak Lensing by Galaxy Clusters, in *Gravitational Lensing: Recent Progress and Future Goals*, ed. T. G. Brainerd & C. S. Kochanek, *ASP Conf. Ser.* (San Francisco: Astronomical Society of the Pacific), 237, 269

Kim, K.-T., Kronberg, P. P., Dewdney, P. E., & Landecker, T. L. 1990, The Halo and Magnetic Field of the Coma Cluster of Galaxies, *Astrophys. J.*, 355, 29

Kim, K.-T., Tribble, P. C., & Kronberg, P. P. 1991, Detection of Excess Rotation Measure Due to Intracluster Magnetic Fields in Clusters of Galaxies, *Astrophys. J.*, 379, 80

King, I. R. 1972, Density Data and Emission Measure for a Model of the Coma Cluster, *Astrophys. J.*, 174, L123

Kompaneets, A. S. 1956, The Establishment of Thermal Equilibrium Between Quanta and Electrons, *Zh. Exper. Teor. Fiziki*, 31, 876

Kulsrud, R. M., Cen, R., Ostriker, J. P., Ryu, D. 1997, The Protogalactic Origin for Cosmic Magnetic Fields, *Astrophys. J.*, 480, 481

Lane, W. M., Clarke, T. E., Taylor, G. B., Perley, R. A., & Kassim, N. E. 2004, Hydra A at Low Radio Frequencies, *Astron. J.*, 127, 48

Lawler, J. M. & Dennison, B. 1982, On Intracluster Faraday Rotation. II – Statistical Analysis, *Astrophys. J.*, 252, 81

Ohno, H., Takizawa, M., & Shibata, S. 2002, Radio Halo Formation through Magnetoturbulent Particle Acceleration in Clusters of Galaxies, *Astrophys. J.*, 577, 658

Owen, F., Eilek, J., & Kassim, N. 1999, M87: Outflow Not Cooling Flow?, in *Diffuse Thermal and Relativistic Plasma in Galaxy Clusters: Proceedings of the Workshop...Ringberg Castle, Germany, April 19–23*, 1999, ed. H. Bohringer, L. Feretti, & P. Schuecker, *MPE Report* (Garching, Germany: Max-Planck-Institut für Extraterrestrische Physik), 271, 107

Owen, F. N., Eilek, J. A., & Kassim, N. 2000, M87 at 90 Centimeters: A Different Picture, *Astrophys. J.*, 543, 611

Pacholczyk, A. G. 1970, *Radio Astrophysics* (San Francisco: W. H. Freeman & Co.)

Peres, C. B., Fabian, A. C., Edge, A. C., Allen, S. W., Johnstone, R. M., & White, D. A. 1998, A ROSAT Study of the Cores of Clusters of Galaxies – I. *Cooling Flows in an X-Ray Flux-Limited Sample*, *MNRAS*, 298, 416

Perley, R. A. & Taylor, G. B. 1991, VLA Observations of 3C 295 – A Young Radio Galaxy?, *Astron. J.*, 101, 1623

Pfrommer, C. & Enßlin, T. A. 2004a, Estimating Galaxy Cluster Magnetic Fields by the Classical and Hadronic Minimum Energy Criterion, *MNRAS*, 352, 76

Pfrommer, C. & Enßlin, T. A. 2004b, Constraining the Population of Cosmic Ray Protons in Cooling Flow Clusters with γ-Ray and Radio Observations: Are Radio Mini-Halos of Hadronic Origin?, *Astron. Astrophys.*, 413, 17

Pfrommer, C. & Enßlin, T. A. 2004c, Erratum: Constraining the Population of Cosmic Ray Protons in Cooling Flow Clusters with γ-Ray And Radio Observations: Are Radio Mini-Halos of Hadronic Origin?, *Astron. Astrophys.*, 426, 777

Rephaeli, Y. 1995, Comptonization of the Cosmic Microwave Background: The Sunyaev-Zeldovich Effect, *Ann. Rev. Astr. Astrophys.*, 33, 541

Rephaeli, Y. & Gruber, D. E. 1988, HEAO 1 Hard X-Ray Observation of Clusters of Galaxies and Intracluster Magnetic Fields, *Astrophys. J.*, 333, 133

Reuter, H.-P., Klein, U., Lesch, H., Wielebinski, R., & Kronberg, P. P. 1994, The Magnetic Field in the Halo of M 82. Polarized Radio Emission at Lambda-Lambda(6.2) and 3.6 CM, *Astron. Astrophys.*, 282, 724

Roettiger, K., Burns, J. O., & Stone, J. M. 1999, A Cluster Merger and the Origin of the Extended Radio Emission in Abell 3667, *Astrophys. J.*, 518, 603

Röttgering, H. J. B., Wieringa, M. H., Hunstead, R. W., & Ekers, R. D. 1997, The Extended Radio Emission in the Luminous X-Ray Cluster A3667, *MNRAS*, 290, 577

Rood, H. 1970, Conversion from Measured Line-Of to Space-Velocity Dispersion for a Spherical Cluster of Galaxies: Application to the Mass-Luminosity Ratio of the Coma Cluster, *Astrophys. J.*, 162, 333

Ruzmaikin, A., Sokolov, D., & Shukurov, A. 1989, The Dynamo Origin of Magnetic Fields in Galaxy Clusters, *MNRAS*, 241, 1

Rybicki, G. B. & Lightman A. L. 1979, *Radiative Processes in Astrophysics* (New York: Wiley)

Ryu, D., Kang, H., & Biermann, P. L. 1998, Cosmic Magnetic Fields in Large Scale Filaments and Sheets, *Astron. Astrophys.*, 335, 19

Sarazin, C. L. 1999, The Energy Spectrum of Primary Cosmic-Ray Electrons in Clusters of Galaxies and Inverse Compton Emission, *Astrophys. J.*, 520, 529

Schlickeiser, R., Sievers, A., & Thielmann, H. 1987, The Diffuse Radio Emission from the Coma Cluster, *Astron. Astrophys.*, 182, 21

Schneider, P., King, L., & Erben, Th. 2000, Cluster Mass Profiles from Weak Lensing: Constraints from Shear and Magnification Information, *Astron. Astrophys.*, 353, 41

Soker, N. 1997, Magnetic Fields and Inflow in Cluster Cooling Flows, in *Galactic and Cluster Cooling Flows*, ed. N. Soker, *ASP Conf. Ser.* (San Francisco: Astronomical Society of the Pacific), 115, 182

Stone, J. M. & Norman, M. L. 1992a, ZEUS-2D: A Radiation Magnetohydrodynamics Code for Astrophysical Flows in Two Space Dimensions. I – The Hydrodynamic Algorithms and Tests, *Astrophys. J. Suppl.*, 80, 753

Stone, J. M. & Norman, M. L. 1992b, ZEUS-2D: A Radiation Magnetohydrodynamics Code for Astrophysical Flows in Two Space Dimensions. II. The Magnetohydrodynamic Algorithms and Tests, *Astrophys. J. Suppl.*, 80, 791

Sturrock, P. A. & Stern, R. 1980, Is the Galactic Corona Produced by Galactic Flares?, *Astrophys. J.*, 238, 98

Sunyaev, R. A. & Zel'dovich, Y. B. 1970, The Spectrum of Primordial Radiation, its Distortions and their Significance, *Comments Astrophys. Space Phys.*, 2, 66

Taylor, G. B., Fabian, A. C., & Allen, S. W. 2002, Magnetic Fields in the Centaurus Cluster, *MNRAS*, 334, 769

Tribble, P. C. 1993, Radio Haloes, *Cluster Mergers, and Cooling Flows, MNRAS*, 263, 31

Vallée, J. P., MacLeod, J. M., & Broten, N. W. 1986, A Large-Scale Magnetic Feature in the Galaxy Cluster A 2319, *Astron. Astrophys.*, 156, 386

Vogt, C. & Enßlin, T. A. 2005, A Bayesian View on Faraday Rotation Maps Seeing the Magnetic Power Spectra in Galaxy Clusters, *Astron. Astrophys.*, 434, 67

Völk, H. J. & Atoyan, A. M. 2000, Early Starbursts and Magnetic Field Generation in Galaxy Cluster, *Astrophys. J.*, 541, 88

Willson, M. A. G. 1970, Radio Observations of the Cluster of Galaxies in Coma Berenices – the 5C4 Survey, *MNRAS*, 151, 1

Xu, H., Li, H., Collins, D. C., Li, S., & Norman, M. L. 2009, Turbulence and Dynamo in Galaxy Cluster Medium: Implications on the Origin of Cluster Magnetic Fields, *Astrophys. J.*, 698, L14

Xu, H., Li, H., Collins, D. C., Li, S., & Norman, M. L. 2011, Evolution and Distribution of Magnetic Fields from Active Galactic Nuclei in Galaxy Clusters. II. The Effects of Cluster Size and Dynamical State, *Astrophys. J.*, 739, 77

Xu, Y., Kronberg, P. P., Habib, S., & Dufton, Q. W. 2006, A Faraday Rotation Search for Magnetic Fields in Large-scale Structure, *Astrophys. J.*, 637, 19

Zel'dovich, Y. B. & Sunyaev, R. A. 1969, The Interaction of Matter and Radiation in a Hot-Model Universe, *Astr. Space Sci.*, 4, 301

Zwicky, F. 1938, On the Clustering of Nebulae, *Pub. Ast. Soc. Pacific*, 50, 218

10

Magnetic fields beyond galaxy clusters

10.1 Low frequency radio emission as a tracer of extragalactic magnetic fields

Cosmic ray (CR) electrons are particularly effective "illuminators" of intergalactic magnetic fields at low radio frequencies. This is because the spectral density of synchrotron radiation increases at low frequencies, reflecting values of γ (2.4–3) of the power law distribution of CR electron energies: $N(E) \sim E^{-\gamma}$. The critical frequency, ν_C (near the frequency of highest synchrotron radiation spectral density), is related to E by $\nu_C \simeq (3/4\pi)\omega_B \sin\theta(E_e/mc^2)^3$ Hz (cgs units). An advantage of observing at the lowest possible radio frequencies is that we preferentially detect lower energy CR electrons, which survive the longest to keep "illuminating" the associated magnetic field. This also means that the lower the observing frequency, the further the CR gas will, on average, have diffused from its original acceleration site.

The longest possible loss time for a CR electron can be defined as a "cosmic background-equivalent" magnetic field, B_{bge}, for which a CR electron's energy loss rate by synchrotron radiation ($\propto B^2 E^2$) equals that due to inverse Compton (IC) scattering off the microwave background radiation ($\propto \varepsilon_{\mathrm{bg}} E^2$ – see Equation 9.21) where $\varepsilon_{\mathrm{bg}} = 4.8 \times 10^{-13}(1 + z)^4$ erg cm^{-3}. Thus, for $B \approx B_{\mathrm{bge}}$, the energy density of the magnetic field ($B^2/8\pi$) equals that of the cosmic background radiation ($\varepsilon_{\mathrm{bg}}$). For redshifts $z \ll 1$, to which we restrict the initial discussion, $B_{\mathrm{bge}(0)} \approx 3.3 \times 10^{-6}$ G. Larger magnetic field strengths cause shorter, synchrotron radiation-dominated lifetimes, whereas for $B < B_{\mathrm{bge}}$, a CR electron's loss rate is dominated by IC scattering, and depends only on its energy and $\varepsilon_{\mathrm{bg}}$ (Equation 9.21). A discussion of IC radiation and magnetic fields in the ICM environment appears in Section 9.7, and the basic radiation processes are further described in Felten & Morrison (1966), Rees (1967), and Rybicki & Lightman (1979).

The corresponding expression for maximum radiative lifetime, $\tau_{\mathrm{rad}}^{\mathrm{max}}$, combines radiation losses due to both synchrotron and IC radiation. Expressed in terms of the observing frequency ν, and the magnetic field,

$$\tau_{\mathrm{rad}}^{\mathrm{max}} = \frac{5.4 \times 10^7 \,\text{year}}{\left[B_{\mathrm{bge}(0)}^2 (1 + z)^4 + B^2\right]} \sqrt{\frac{B}{\nu_{10.6\ \mathrm{GHz}}}}, \tag{10.1}$$

where the subscript 10.6 GHz denotes a fiducial observing frequency. $\tau_{\mathrm{rad}}^{\mathrm{max}}$ occurs when the two terms within the square brackets are approximately equal, i.e. at $z = 0$, where $B = B_{\mathrm{bge}(0)} = 3.3\ \mu$G. Thus, the maximum "fossil lifetime" of relativistic electrons scales to first order by $\nu^{-1/2}$ and the maximum radiative lifetime decreases significantly at $z \gtrsim 1$.

192

Practical observing frequencies of interest usually fall between approximately 20 MHz and 800 MHz: Below 20 MHz, refraction and absorption by the ionised component of the interstellar medium and/or the Earth's ionosphere make precise Earth-based imaging more difficult. Above ~1.4 GHz, the most sensitive detection of diffuse intergalactic synchrotron emission becomes increasingly difficult because the emissivity $\varepsilon(v)$ becomes progressively fainter. In addition, the higher energy electrons, which radiate on average at higher frequencies, are less likely to have propagated to large distances from their acceleration sites because of their shorter loss times. Imaging below ~20 MHz is conceivable from space and down to ~1 MHz if the array is placed on the far side of the moon, which has no ionosphere. There, the best observing conditions with minimum interference would occur at times when the Earth, Jupiter, and the Sun are all below the lunar horizon.

For precise imaging at frequencies below ~300 MHz, changes in both in the geomagnetic field and free electron density distribution in the ionosphere need to be monitored and modelled. Ideally this should be done in real time, using data processing techniques that are integrated with the array output. These requirements will become increasingly realizable for low frequency radio telescopes such as LOFAR (EU), the Giant Metre Wave Telescope (GMRT) at Pune, India, and the Long Wavelength Array (LWA) in New Mexico. Such developments will open up new detection possibilities for magnetic fields over supragalactic and intergalactic dimensions.

Relation (10.1) illustrates that, below ~50 MHz, we could search for diffuse radiation from ~10^9 years old intergalactic CRs. Such observations could give clues from the *local* Universe on magnetic field strength evolution over a significant period of cosmic time in systems such as old, extended radio source lobes. An important complement will be sensitive imaging of diffuse X-ray and EUV emission produced by IC radiation.

An early result of this kind using the Westerbork Synthesis Radio Telescope (WSRT) at 327 MHz was produced by Kim *et al.* (1989). The relatively low frequency, in addition to high dynamic range images, revealed diffuse emission extending beyond the Coma cluster, suggesting a magnetised region on a supra-cluster scale. Recall that measurements of this type do not directly yield the field strength (Equation 1.1), but the assumption of approximate equipartition of relativistic particle and magnetic energies permits us to apply a minimum total energy criterion, which gives field strengths of *ca.* $2 \times 10^{-7}(1 + k)^{2/7}$ G (k is the ratio of CR proton to electron energies, usually estimated at between 10 and 100). A combination of the Arecibo 305m radio telescope and the precision imaging capability of the Dominion Radio Astrophysical Observatory (DRAO) Synthesis Telescope interferometer at 0.4 GHz can detect diffuse emission at very low levels over large skyfields. An example shown in Fig. 10.1 is the wide, $9°$ field within the Coma supercluster, and centred on the Coma cluster, near the North Galactic Pole.

The unprecedented sensitivity to diffuse emission in this frequency band is shown in Fig. 10.1. It suggests, as discussed above, that future *lower* frequency images of high sensitivity and comparable resolution will yield new information on the distribution of intergalactic magnetic fields on supra-galaxy and supra-cluster scales. Low frequency multi-channel polarisation imaging will allow the Faraday depth distribution to be modelled at each map pixel point (see Chapter 3). Precision low frequency polarimetric images of the

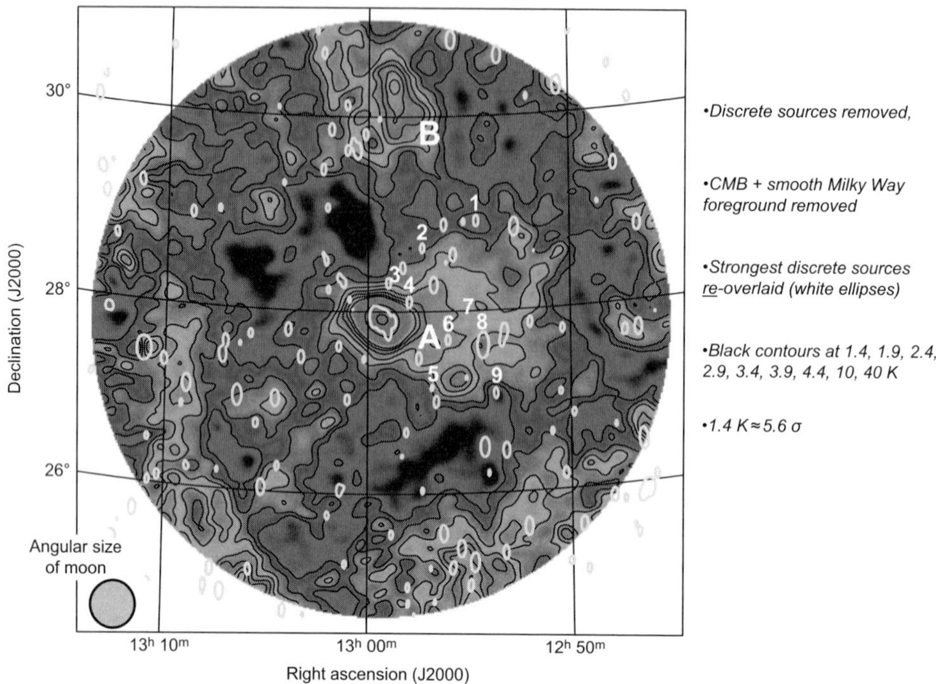

Figure 10.1. Radio image showing 0.4 GHz emission over an area, clipped to 7.5° in diameter, around the Coma cluster of galaxies, whose approximate extent is shown in the centre. Numbers denote radio-loud galaxies having redshifts close to that of the Coma cluster. The lowest contour is at 1.4σ, where σ is 250 mK at 0.4 GHz (Kronberg *et al.* 2007).

extragalactic sky at arcminute resolution will contribute in future to studies of galaxy and LSS evolution, and other aspects of post-Recombination cosmology.

10.2   Large scale intergalactic shocks and "fossil" radio galaxies

As large-scale structure (LSS) developed over cosmic time, intergalactic filaments and bubbles were likely permeated at some level by a magnetic field. Some mechanisms and scenarios for this are discussed in previous chapters. Correspondingly, it is reasonable to assume that, as galaxy clusters and groups formed from large scale inflow in LSS filaments, their intracluster and intragroup gas, and the member galaxies all "inherited" some level of seed magnetic field. This inherited field could have come from various primeval sources such as very early stars, supernova-driven outflows from large numbers of dwarf galaxies, early supermassive massive black holes (Chapters 7 and 8), in addition to possible primordial sources discussed in Chapter 13.

The radiating lobes of the "giant" radio galaxies, usually found at $z < 1$, have measured energy contents in the range 10^{61}–10^{62} ergs as shown in Fig. 8.1 and related discussion. Roughly half of this is magnetic energy, and the largest of these systems extend to dimensions of a typical galaxy–galaxy spacing in LSS. About 100 such Mpc-size radio sources are known at the time of writing, and that number will increase as the capabilities of low frequency telescopes improve. One example of a previously undetected, Mpc-scale fossil radio lobe is the ~1°-long, unidentified radio source "B" in Fig. 10.1. The combination of higher IC loss rates ($\propto E^2$) and the redshift of their photons, $v_{obs} = v_{em}/(1+z)$, "conspire" to make radio images like source B (Fig. 10.1) easiest to detect at lower radio frequencies.

An implication of the above is that hitherto "invisible" fossil remnants of previous generations of such sources may "lurk" undetected in the intergalactic space outside rich clusters. Their associated magnetic fields and CR protons, and hence their intergalactic "energy imprint", can last up to approximately a Hubble time. Apart from radiation losses, they suffer only adiabatic expansion losses, unless some additional energy is subsequently provided to them. The co-moving density of central BH and jet-lobe sources beyond $z \approx 4$ is not well known: It depends on the relative formation rate of galactic black holes at these earlier epochs.

An example of an identified Mpc-scale AGN-fed radio source is shown in Fig. 10.2. The distribution of Faraday rotation over the source is also shown beside the 1.4 and 10.6 GHz radio images.

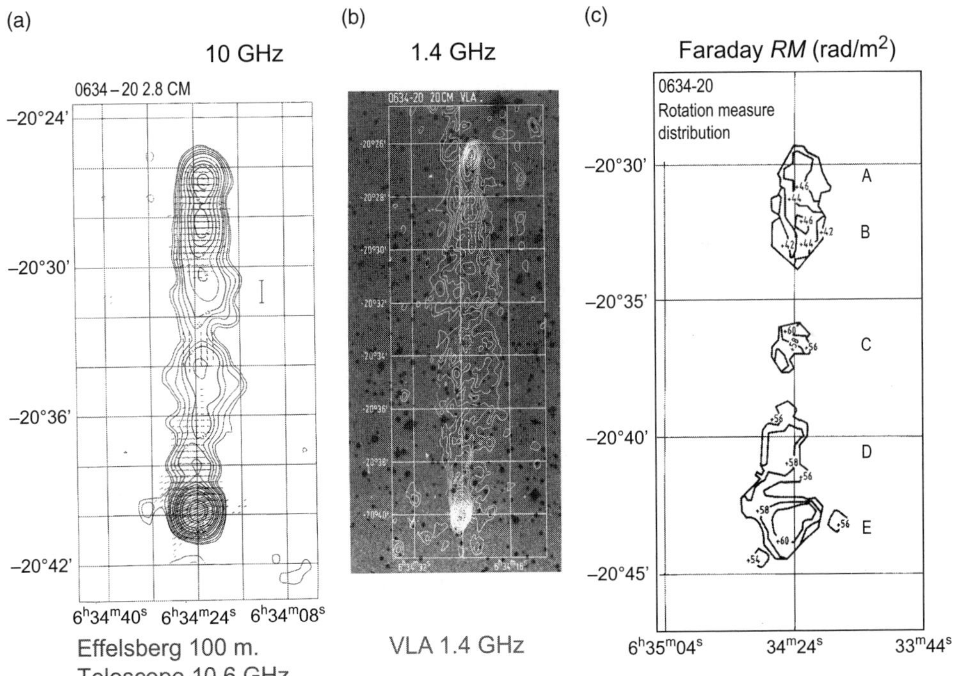

Figure 10.2. (a) 10.6 GHz image of the giant radio source 0634–20 ((l,b) = (230°, −12°)) made with the MPIfR 100 m Effelsberg telescope at a resolution of $75''$ in Stokes parameters *I, Q,* and *U.* (b) A companion 1.4 GHz image made with the NRAO VLA, at the same resolution, but missing some low order spatial frequency harmonius. (c) A 10.4–1.4 GHz Faraday *RM* image made from (a) and (b) at a common resolution of $75''$ (Kronberg, Wielebinski, & Graham 1986).

The radio lobes of 0634–20 appear relatively undisturbed by intergalactic winds, and they contain minimum energy field strengths of $\approx$2–3 μG. The large fractional polarisation and the coherent polarisation orientations over the source indicate that the magnetic fields are highly ordered over much of the large radiating volumes. This is consistent with a lack of disruption by ambient intergalactic "weather". Also, variations of *RM* or "Faraday depth" over the source are small, <10 rad m^{-2}, and possibly even $\lesssim$3 rad m^{-2}. Given that some of this variation could have been increased by galactic and/or intergalactic foreground *RM* fluctuations, the lobes' internal Faraday rotation is very small. This "low *RM* profile", is typical of many large radio galaxies, and implies that the thermal gas density within the large lobes is typically $\lesssim$10^{-5} cm^{-3}. It approaches, and may occasionally be less than, the density of the ambient intergalactic medium.

The radio spectra, α ($S \propto \nu^{\alpha}$), though often difficult to measure precisely, are sometimes (but not always) steeper than in smaller, younger sources. Where this happens, it is normally interpreted as due to particle energy decay at the higher energies because of radiative losses. However, in some giant lobes, such as 0634–20 in Fig. 10.2, fresh relativistic electron acceleration appears to occur *within* the large outer lobe volumes. There appears little evidence for large scale shocks that could cause this reacceleration. An alternative explanation might be a large scale reconnection process within the large lobes such as that proposed for a similarly large radio galaxy, 2147+816 (Fig. 1.1, Kronberg *et al.* 2004). However, the detailed physics of magnetic reconnection on such a large physical scale has not yet been clarified.

Their lobe-internal relativistic and thermal gas densities can be estimated, the latter as an upper limit through measurements of the Faraday *RM* variations, e.g. Fig. 10.2. They can also be used to estimate or limit the local pressure of the ambient intergalactic medium. Furthermore, because of their large dimensions, the morphology of large lobes can in some circumstances serve as a unique kind of intergalactic "weather" probe. They are radio-visible for up to $\approx$1 Gyr (Equation 10.1) within the large scale flows within cosmic bubble walls or sheets that we define as the "galaxy zones" (as distinct from the giant voids).

Such objects might be sensitive to large scale shock fronts in colliding flows of intergalactic gas (and galaxies). It is possible that the relativistic electrons and magnetic fields can, in the right circumstances, be re-energised, either in a visible giant radio source like 0634–20, or in a previously invisible >1 Gyr old fossil – for example, if such a source encounters an external shock or enhanced compression. Unless, or until a fossil giant radio source is broken up by a strong intergalactic shock, or by intergalactic shearing flows, it may confine its relativistic protons and magnetic fields for a very long time.

One candidate for such effects of intergalactic weather is another giant radio galaxy, NGC315 (Enßlin *et al.* 2000). It is located near the Perseus-Pisces supercluster (see Fig. 10.3), in a zone where the galaxy redshifts suggest that just such a collision of large scale flows might be occurring. One of the large (~0.6 Mpc) lobes of NGC315 has a distorted shape, and perhaps more importantly, its radio spectrum appears unusually flat up to ~10 GHz. This also suggests some widespread particle acceleration within the lobe. Because the electrons' radiation age appears less than the energy flow time from the parent galaxy's nucleus, a large scale intergalactic shock, or compression may also explain the re-acceleration of these relativistic electrons over such a large volume of intergalactic space.

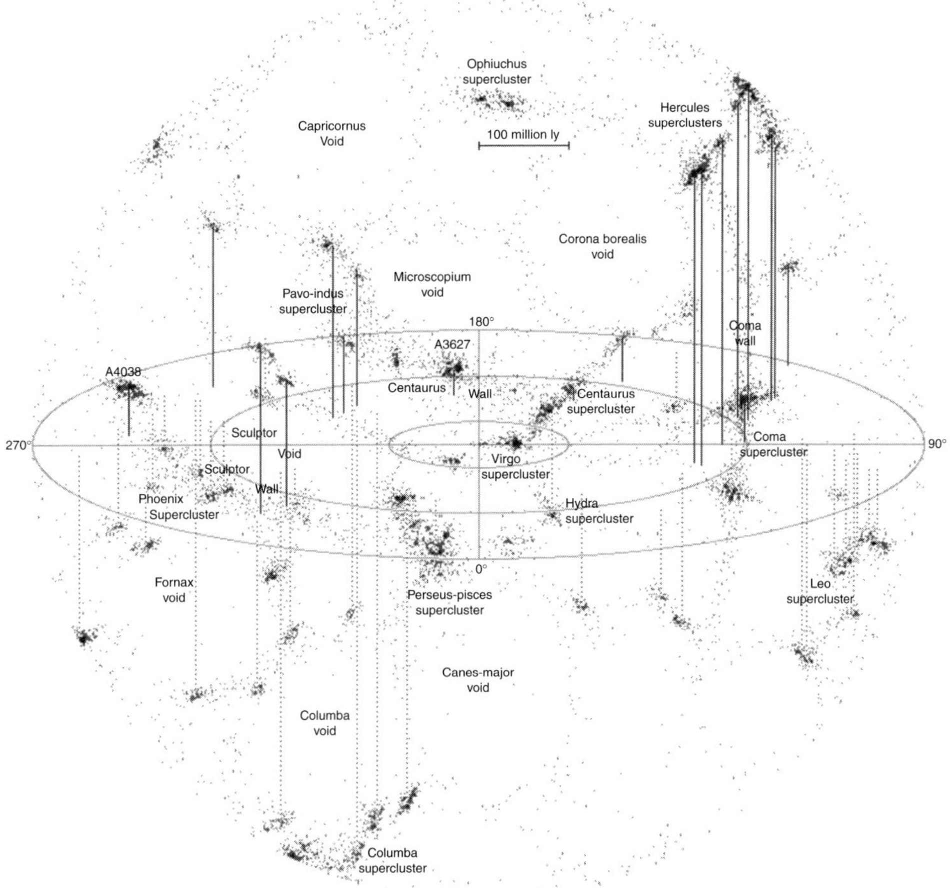

Figure 10.3. A schematic view in quasi-3D, showing clusters, superclusters, groups, and voids of the nearby universe within a radius of *ca.* 155 Mpc. It illustrates the intergalactic régimes where we would like to obtain information on the strength and geometry of magnetic fields, and especially the voids. (Source: Hudson 1993.)

10.3　γ-Ray probes for weak intergalactic fields in cosmic voids

10.3.1　Introduction

We have recently seen an increase in the number and scope of instruments to detect high energy photons, neutrinos, leptons, hadrons and other subatomic particles from space. This has, in effect, turned "empty" space into a newly accessible high energy physics "laboratory" that happens to be permeated with magnetic fields. The word "laboratory" is justified by the fact that photons and subatomic particles at very high energies can be produced through photon–photon interactions – for example, TeV γ-ray photons with the extragalactic background light (EBL). The latter can be *infrared* or optical photons from the aggregate of galaxies, or CMB photons. The existence of these and other interactions gives new opportunities to probe magnetic fields at very low levels in space. At the same time it opens a promising new realm of high energy and particle physics research in an extragalactic context.

One motivation for probing magnetic fields in voids has come from predictions that fields in voids could be relics of important processes and interactions prior to Recombination, beginning at the QCD phase transition, and with nucleogenesis. Further details are discussed in Chapter 13.

In the following pages, we examine interactions among photons and leptons (e^+, e^-) in the presence of an intergalactic magnetic field. These interactions can result in deflections, time delays, and modifications of the original extragalactic γ-ray photon energy spectrum. In general they are energy dependent, and sometimes quite complex. They can be analysed to deduce intergalactic field levels over a wide range of magnetic field strengths, and usually well below those detectable by Faraday rotation, Zeeman splitting, or synchrotron glow measurements (i.e. $\ll 10^{-8}$ G).

Upper and lower limit values of B_{IG} at these weak levels, e.g. in cosmic voids, are becoming progressively tighter with time. It is likely that we will continue to explore intergalactic magnetic fields in this new "magnetic frontier" of intergalactic space well into and beyond the 21st century. Magnetic fields, along with particle and photon fluences through cosmic voids, are also potential tracers of hitherto undetected energy in these largest volumes of the Universe.

The next few subsections and Chapter 11 will illustrate how a myriad of interactions involving particles, photons, and magnetic fields can now be studied and related to B_{IG} at photon and particle energies up to and beyond $\sim 10^{20}$ eV. This is two to three orders of magnitude above the currently highest interaction energies obtainable with the most power-ful Earth-bound particle accelerators, such as the Large Hadron Collider (LHC) at CERN. Chapter 11 continues discussion of (charged) hadrons and secondary particles that interact in the Earth's atmosphere (e^+, e^-, $\mu^{+/-}$) and its interior (neutrinos).

It has been postulated that cosmic voids represent a pristine physical environment that has carried over from primordial epochs before Recombination. To the extent that this may be true, it would provide a rare opportunity to probe, by proxy, magnetic fields prior to Recombination, i.e. further into the plasma epoch, the radiation-dominated era, and possibly even to the first few μs and seconds during the period of nucleogenesis and inflation (Chapter 13).

But this raises other questions: Can we be certain that any void magnetic fields are truly "pristine", in the sense that they derive from primordial conditions? Or could they, over a Hubble Time have been somehow "polluted" by magnetic fields from stars (e.g. SN) and galaxies that have leaked and diffused into the voids? Such questions are not yet definitively answered, though there are reasons to believe that such "polluting" processes have occurred. Figure 10.3 gives a visual overview of the layout of void and galaxy overdensity zones at $z \sim 0$. It shows the actual "3D" locations of galaxy filaments, clusters, voids, etc., within a distance of ~ 155 Mpc.

10.3.2 *General properties of a burst of γ-ray photons from a blazar*

Suppose now that a transient source appears over a very broad band of γ-ray energy, say from 10^9 to 10^{20} eV (0.01 TeV–100 EeV), as illustrated in Fig. 10.4. For a 0.1 TeV γ ray, the mean free path (MFP) for interaction with the first intergalactic photon is comparatively large. We denote it as $\lambda_{\gamma_0\gamma'}$. This means that a short burst in this "lower" energy range would serve as a kind of time marker for higher energy γ rays at $E \sim 1\text{–}100$ TeV. These have shorter $\lambda_{\gamma_0\gamma}$'s for interaction with an intergalactic IR photon.

The primary photon flux, $F(E_{\gamma 0})$, is assumed to have a fluence of the form

$$F\big(E_{\gamma 0}\big) = F_0\big(E_{\gamma 0}\big)e^{-\tau\big(E_{\gamma 0},z\big)}, \tag{10.2}$$

where $F_0(E_{\gamma 0})$ is the initial spectrum emanating from the left side of Fig. 10.4 and τ is the optical depth with respect to pair production on the cosmic infrared background (the CIB component of the EBL). The optical depth (interaction rate) of a primary increases with energy, toward 50 TeV and beyond (i.e. $\lambda_{\gamma_0\gamma}$ decreases).

$$\tau\big(E_{\gamma 0}\big) \propto E_{\gamma 0}, \tag{10.3}$$

where the power law photon energy index Γ of $F_0(E_{\gamma 0})$ is usually taken as $\simeq\!-2$. Although the γ-ray primaries $F(E_{\gamma 0})$ are most attenuated at the highest energies, a significant fraction of this "lost" energy will re-appear in subsequent γ-ray cascades, described in the next section.

High energy $h\nu$ –e^+e^- cascades in the intergalactic medium

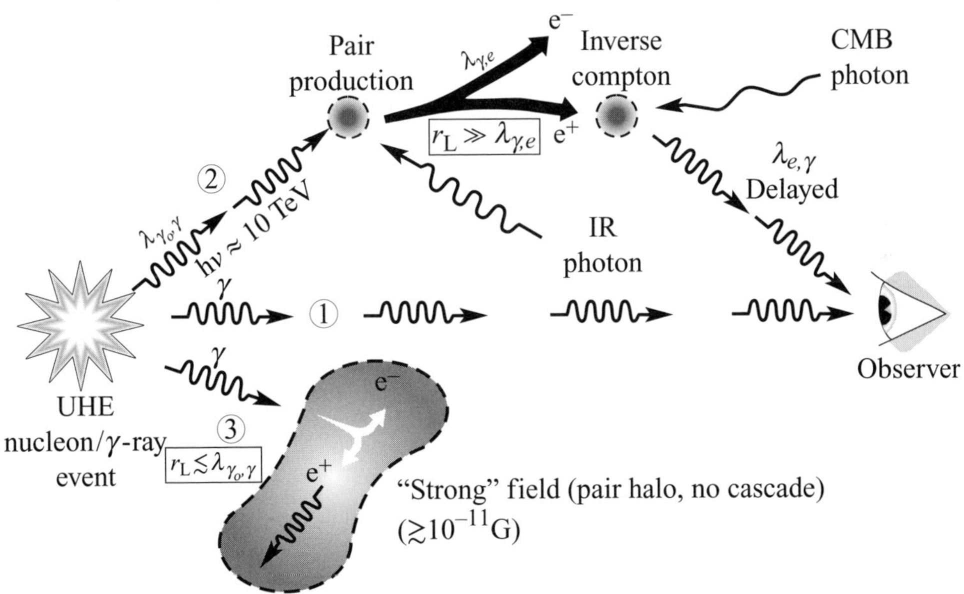

Figure 10.4. Illustration of a primary burst, and γ-ray cascades. The original VHE γ-ray "primary" is produced either directly, or indirectly in the decay of a π^0 meson originating from a UHECR nucleon at the source (Equation 11.7). In propagation path No. 1, the lower energy 0.1 TeV γ-rays can more easily propagate to us directly. In contrast, 100 TeV gamma rays will interact first with IR photons of the EBL (case 2). Case 3 is a variant of case 2 in which the magnetic fields are sufficiently strong that the e^+ and e^- are bent completely out of our line of sight (see Plaga 1995, Waxman & Coppi 1996). (Adapted from Plaga (1995).)

The interaction lengths are energy dependent being, as mentioned above, particularly long in the ~0.1 TeV photon energy range, and comparatively shorter for high energy TeV photon interactions with the IR component of the EBL. Depending on the photon energy, $\lambda_{\gamma_0\gamma}$ can be much larger than, for example the Greisen–Kuzmin–Zatsepin (GZK) p–γ interaction "horizon". This is the x_{loss} distance of ~80 Mpc for UHECR protons, discussed in Chapter 11.

For γ-ray bursts there is an *angular spreading* delay, Δt_{A} (relative to a photon propagating at c) which does not depend on B_{IG}:

$$\Delta t_{\text{A}} \sim (1+z)\frac{\lambda_{\gamma_0\gamma} + \lambda_{\text{CMB}}^{\text{IC}}}{2\gamma_e^2 c} \sim (1+z)\frac{960}{\left(\frac{\gamma}{10^6}\right)^2 \left(\frac{n_{\text{CIB}}}{0.1\text{cm}^{-3}}\right)}\, s. \tag{10.4}$$

For the energies of interest here, the MFP $\lambda_{\gamma_0\gamma}$ for angular broadening ($\mathcal{O}$ 20 Mpc) is much greater than $\lambda_{\text{CMB}}^{\text{IC}}$ for IC scattering (~$0.7(\gamma_e/10^6)$) Mpc, so that in such cases $\lambda_{\text{CMB}}^{\text{IC}}$ is neglected. Note that Δt_{A} does not involve an angular size measurement.

Angular broadening is discussed below, and in Section 10.4. Due to the highly relativistic bulk flow associated with the initial γ-ray flare, the fluence will have a jet opening angle θ_{j}, of order Γ^{-1}. As with Δt_{A}, θ_{j} it is independent of an ambient B_{IG}. Scaling to a bulk relativistic flow of $\Gamma = 10$, the angular spreading of the jet is given by

$$\theta_{\text{j}} \approx \frac{1}{\Gamma_{\text{j},10}} \simeq 6^{\circ}. \tag{10.5}$$

10.3.3 *Effects of B_{IG} on photon interactions and γ-ray $\leftrightarrow e^+e^-$ cascades*

Figure 10.4 illustrates some basic aspects of γ-ray cascade propagation from an extragalactic high energy source in the presence of an intergalactic magnetic field, B_{IG}. The original source could be either transient or steady. It could emit either an ultra high energy cosmic ray (UHECR) nucleus or comparably energetic γ-rays. As before, we denote $\lambda_{\gamma 0,\gamma}$ as the MFP for the first interaction between the primary TeV photon and the EBL where γ_0 and γ refer to the primary and IR background-photon, respectively. Analogously, $\lambda_{\gamma,e}$ is the MFP of the post-interaction e^+e^- in the second stage, and $\lambda_{e,\gamma}$ for the third stage – etc. The cascade is formed by repeats of the second two interactions in Fig. 10.4, producing successively lower energy γ-ray photons and e^+e^- pairs.

Photons of the CIB come from an aggregate of galaxies, quasars, and blazars. The total space *density* of *all* photons, n_{ph}, is, however, dominated by the CMB photons ($\simeq$411 cm^{-3}), these being more dense by an uncertain factor of ~10 – >100 than IR and optical photons of the EBL (Aharonian *et al.* 2006). In any case, the e^+ and e^- leptons will Compton scatter lower energy photons to another E_γ photon and thereby produce a partially self-perpetuating cascade of γ-ray $\leftrightarrow e^+e^-$ pairs, as illustrated in Fig. 10.4.

In a *non*-zero intergalactic magnetic field the subsequent charged e^+e^- pairs, whose energy is still a substantial fraction of the pre-interaction $E_{\gamma 0}$ or E_γ, immediately propagate in curved orbits characterised by a Larmor radius r_{L}. They will emit synchrotron or curvature radiation, and r_{L} is related to B_{IG} (in units of 10^{-15} G) by

$$r_{\rm L} \simeq 1.1 \left(\frac{E_{e^+e^-}}{\rm TeV}\right)\left(\frac{B_{\rm IG}}{10^{-15}\rm G}\right)^{-1} \rm Mpc. \tag{10.6}$$

After some subsequent MFP distance $(\lambda_{\rm e,\gamma})$ the leptons will preferentially IC scatter intergalactic CMB photons, because the latter have the highest photon density, $n_{\rm ph}$. At any stage in the subsequent cascade, a lepton produced after $\lambda_{\gamma,\rm e}$ will be deflected by

$$\theta \simeq \frac{\lambda_{\gamma,\rm e}}{r_{\rm L}}. \tag{10.7}$$

There are different ways in which $B_{\rm IG}$ can be related to the secondary leptons illustrated in Fig. 10.4. In the extreme case (path 3) where $r_{\rm L} \lesssim \lambda_{\rm \gamma,e}$), the cascade in the observer's direction is terminated, and we would observe a faint local halo due to the e^+e^- pairs that are not collectively beamed to us. An isolation of this local cloud of e^+e^- would enable measurement of the magnetic field strength close to the source – but not in the intervening IGM.

We focus here on what is commonly called a "pair halo", or cascade halo which is seen in projection *along the propagation path* to us – path 2 in Fig. 10.4. In this case, where $r_{\rm L} \gg \lambda_{\rm e,\gamma}$ (i.e. a small angle deflection), the cascade of γ-rays is initially close to the propagation direction and it is spread out in energy, time, and (eventually) in angle. For a finite $r_{\rm L}$ and a sufficient number of successive scatterings, the received cascade γ-ray signal will progressively wander away from the propagation direction of the primary γ-ray beam. This gives rise to a pair halo "cone" (Fig. 10.5 (left, upper half)), which will grow in size as the energy of the leptons and scattered photons decreases, commensurate with more segments of $\lambda_{\rm e,\gamma}$ and $\lambda_{\gamma,\rm e}$. Similarly, it will grow in proportion to the strength of $B_{\rm IG}$. When this happens, some of the downstream gamma ray photons in the cascade will leak out of the detector's field of view (FOV, Fig. 10.5 (right)).

Figure 10.5 illustrates these processes and shows how observations can relate to estimates or limits on $B_{\rm IG}$. For the purpose of interpreting the cascade parameters to deduce $B_{\rm IG}$, we refer (Section 10.3.2) to an optical depth term, τ, corresponding to $\lambda_{\gamma_0\gamma}$ in the first phase of the cascade, prior to the production of the first e^+e^- pair in path 2 of Fig. 10.4. '

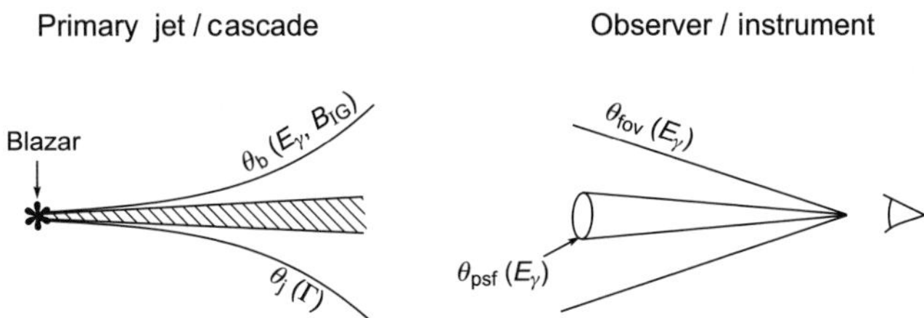

Figure 10.5. Illustration of the angular widths of a TeV blazar's primary and cascade emission beam, (left), compared with the observer- and detector-related beam sizes (right). In the circumstances considered here $\theta_j(\Gamma)$ will be less than $\theta_{\rm b}(E_\gamma, B_{\rm IG})$ – a prerequisite for enabling a measurement of $B_{\rm IG}$. See text for further explanations.

If the original γ-rays at $E_{\gamma 0}$ were emitted in a series of bursts, as distinct from continuous γ-ray emission, how would this affect the analysis of the cascade? The answer lies in a comparison with τ over the cumulative length of a few $\lambda_{e,\gamma} + \lambda_{\gamma,e}$ segments in a subsequent well-developed cascade. For a typical multi-TeV primary photon burst, $\tau \ll$ the optical depth to IC scattering after several $\lambda_{e,\gamma} + \lambda_{\gamma,e}$ segments in the cascade. Where this happens, it will normally be difficult to distinguish a bursting fluence from a steady fluence of primary γ-ray photons, since the latter is an average over a large cascade delay time. The result is that the bursting has little effect on the detected "downstream" γ-ray cascade emission, i.e. pair halo – and consequently on the estimate of B_{IG}.

As the cascade develops, the emission cone is broadened, and then delayed by another effect, namely the *magnetic deflection delay*, Δt_B, due to r_L. This is sensitive to B_{IG}, as can be seen in Fig. 10.4. It is given by

$$\Delta t_B \approx (1 + z)\left(\lambda_{e,\gamma} + \lambda_{\gamma 0,\gamma}\right)\left(\theta_B^2/2c\right), \tag{10.8}$$

where $\lambda_{e,\gamma}$ is the MFP of an cascade IC photon (or lepton) from a scattering off the CMB. $\lambda_{\gamma 0\gamma}$ as before is the MFP to $\gamma\gamma$ scattering of a CIB photon off a high energy γ-ray from the source, and θ_b is the angular broadening due to a B_{IG} (Fig. 10.5). (Murase *et al.* 2008). $\theta_B = \lambda_{\gamma,e}/r_L$ is the magnetic deflection mentioned above. A basic requirement for using cascade pair echo data to detect a weak B_{IG} is that $\Delta t_B \gg \Delta t_A$. That is, the angular broadening due to B_{IG} (upper half of left side of Fig. 10.5) *must be greater* than $\theta_j(\Gamma)$ in the lower half of the left side of Fig. 10.5. Dynamical analyses of pair-creation echoes have been described by Razzaque, Mézáros & Zhang (2004), Casanova, Dingus, & Zhang (2007), Takahashi *et al.* (2008), Ichiki *et al.* (2008), Murase *et al.* (2008), Dolag *et al.* (2011), Taylor, Vovk, & Neronov (2011) and others.

10.3.4 *Example of a fluence delay from a 100 TeV primary photon flux for a range of B_{IG} values*

The average Δt_B, for intervening B_{IG} values ranging from 10^{-15} to 10^{-18}G is shown below in Fig. 10.6, calculated by Taylor *et al.* (2011). The source, at $z = 0.13$, emits a 100 TeV primary photon, and it is assumed that $\theta_j = 0°$ (see Fig. 10.5). The time delay is seen to be sensitive to E_γ to the power of E_γ^{-2} or steeper.

10.4 B_{IG} from the angular broadening of a γ-ray cascade

We discuss different angular size parameters that are important for interpreting a broadening of the γ-ray cascade beam. Consider a blazar jet at distance D from the observer, which emits γ-rays toward us with an opening angle θ_j. As before, an electron and positron are produced after a distance $\lambda_{\gamma 0\gamma}$, and these travel a distance $\lambda_{e,\gamma}$. Provided $\lambda_{e,\gamma} \ll r_L$, they will be deflected in an intergalactic magnetic field by a small angle $\theta \simeq \lambda_{e,\gamma} / r_L$, as above. Normalising the deflection for a 10 TeV electron/positron in a typical $e^+e^- \leftrightarrow \gamma$ cascade (Fig. 10.4), θ can be written as follows (adapted from Neronov & Vovk 2010):

$$\theta \sim 3 \times 10^{-4}\left(\frac{E_e}{10\text{ TeV}}\right)^{-2}\left(\frac{B_{IG}}{10^{-16}\text{G}}\right). \tag{10.9}$$

Here, the magnetic fluctuation scale, λ_B, is much greater than the distance scale for lepton IC scattering: $\lambda_B \gg \lambda^e_{IC}$, where

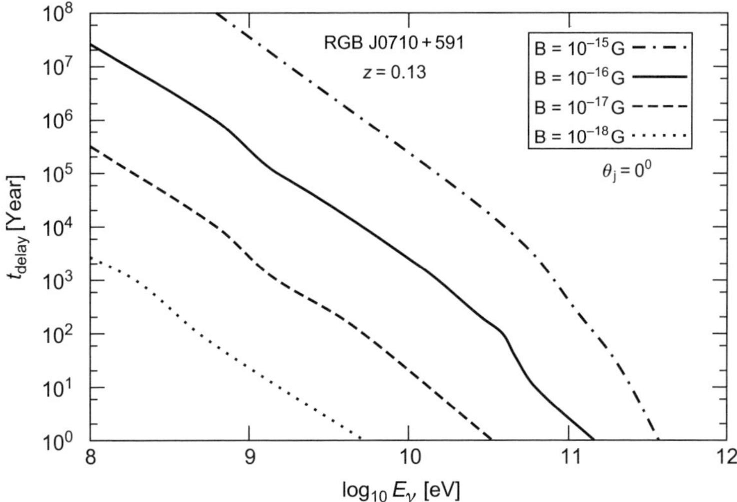

Figure 10.6 The mean time delay of cascade emission from a 100 TeV *primary* photon flux originating at $z = 0.13$. The delay curves, top to bottom, correspond to intervening B_{IG} values of 10^{-15}, 10^{-16}, 10^{-17}, and 10^{-18} G, respectively. (Adapted from Taylor, Vovk, & Neronov 2011.)

$$\lambda_{IC}^{e} \approx 3 \left(\frac{E_e}{10\,\text{TeV}} \right)^{-1} \text{kpc}. \tag{10.10}$$

The result is an angular broadening, θ_B, of an ensemble of cascade photons seen by an observer,

$$\theta_B \simeq \left(\frac{\lambda_{\gamma 0 \gamma}}{D} \right) \delta = \frac{0.3}{\tau(E_{\gamma 0}, z)} \left(\frac{E_\gamma}{\text{TeV}} \right)^{-1/2} \left(\frac{B_{IG}}{10^{-13}\text{G}} \right) \text{degrees} \tag{10.11}$$

(Neronov & Semikoz 2007). Here τ, related to the original energy $E_{\gamma 0}$ and z, is given by $D/\lambda_{\gamma 0 \gamma}$ where $\lambda_{\gamma 0 \gamma}$ is redshift-dependent due to the z-dependence of $\sigma_T n_{EBL}$. Here, n_{EBL} and D are, as before, the EBL photon density and distance to the source, respectively.

The angular broadening steadily increases with successive scatterings i.e. after the initial $\lambda_{\gamma 0 \gamma}$, and several steps of $\lambda_{\gamma,e}$, $\lambda_{e,\gamma}$, each involving successively lower E_γ and $E_{e+,e-}$. As $E_{e+,e-}$ becomes successively lower, the γ-ray scattering angle increases (and r_L decreases) until the photons become increasingly deflected from the original beam direction. Eventually they are lost to the line of sight in the detector's field of view (θ_{FOV}). This depresses the observed fluence spectrum of the observed γ cascade photons below a certain E_γ in the observed pair halo (Fig. 10.5). Thus, $\theta_b(E_\gamma)$, can be linked to an estimate or limit of B_{IG}. Figures 10.5–10.7 illustrate these effects and their relationship to the propagation path to us from the blazar that produced the γ-ray cascade.

At this point the instrument-specific resolution, or point spread function, θ_{PSF}, is important to consider. It depends on both the photon energy and the size and design of the reflector/detector. For the Fermi LAT telescope, it has been empirically estimated as

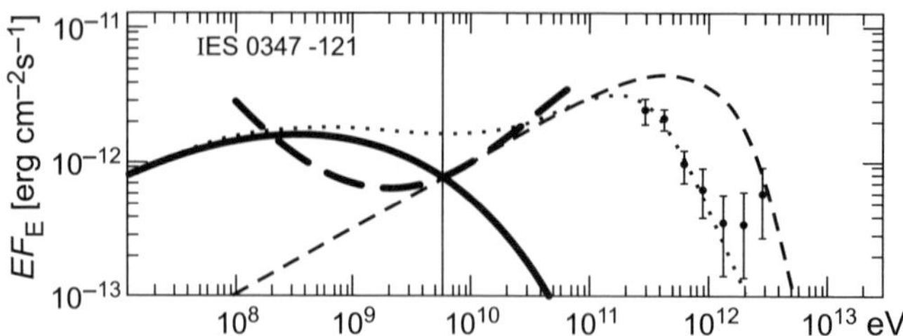

Figure 10.7. The total energy spectrum of E_γ, based on H.E.S.S.-detected TeV measurements of the blazar 0347-121 at $z = 0.188$. It shows the effect of weak intergalactic magnetic fields on the received γ-ray spectrum. The thinner dashed line shows the primary, unabsorbed, spectrum and a thick solid line depicts a model of the cascade emission. The dotted curve shows the initial cascade emission after the first stage of $\lambda_{\gamma,e}$. A vertical line near 6 GeV shows the energy below which a depression in the cascade emission is expected. The thick dashed line shows the approximate E_γ-dependent sensitivity limit of the Fermi LAT. (After Neronov & Vovk 2010, Fig. 1.)

$$\theta_{\mathrm{PSF}} \approx 1.7^\circ \left(\frac{E_\gamma}{\mathrm{Gev}}\right)^{-0.74} \left[1 + \left(\frac{E_\gamma}{15\mathrm{Gev}}\right)^2\right]^{0.37}. \tag{10.12}$$

(Taylor, Vovk, & Neronov (2011). For this detector, θ_{PSF} decreases by nearly an order of magnitude from 1.7° for a 1Gev photon, to $\simeq 0.2^\circ$ for a 1 Tev photon. For us to make a positive measurement of B_{IG} or a limit value, θ_{b} must be greater than the instrumental point spread function of the γ-ray detector $\theta_{\mathrm{b}} > \theta_{\mathrm{PSF}}$ (Fig. 10.5).

θ_b must also, as before, be compared with the primary jet's opening angle, θ_{j}, which is independent of B_{IG}. A calibration of the detailed shape and energy dependence of the detector's θ_{PSF} involves interactions within the 3-D volume of the γ-ray detector, and this is not a trivial task (see Neronov *et al.* 2011).

A further significant instrumental parameter is the γ-ray telescope's angular field of view, θ_{FOV}. For example, an optimally large θ_{FOV}, together with sufficient surface brightness sensitivity (large collecting area) and a smaller and better calibrated θ_{PSF}, would permit detection of B_{IG} at higher levels – associated with the largest angular deflections (θ_{B}) shown in Fig. 10.5.

With coming generations of instruments, B_{IG} could conceivably be detected at levels as high as $\sim 10^{-11}$ G. This would come within 2–3 orders of magnitude of B_{IG} probes using multi-frequency Faraday depth (*RM*) measurements of diffuse emission at low radio frequencies (Chapter 2, and Fig. 10.1), and low-level radiometry of synchrotron radiation at sub-GHz frequencies, as illustrated in Fig. 10.1. To further optimise and extend the *low* B_{IG} régime (e.g. around $B_{\mathrm{IG}} \approx 10^{-16}$ G and below), improved sensitivity in the range of 1–100 GeV will be important, when additionally combined with an optimally small θ_{PSF}.

Summarising the foregoing discussion, some key improvements that can be anticipated in the coming generations of TeV and GeV γ-ray detectors are the following:

(1)   A smaller θ_{PSF} which is optimally calibrated over the entire received range of E_γ. Note that θ_{PSF} applies to the received E_γ. For the Fermi LAT telescope, θ_{PSF} goes from $\sim 0.2^\circ$ to $\sim 1.7^\circ$ over its $E\gamma$ range of 1 TeV to 1 GeV. These θ_{PSF} numbers may change quickly with time as instruments develop.

(2) Larger collecting area for improved sensitivity over the entire (larger) FOV and range of $E\gamma$.

(3) An optimally large FOV, commensurate with improved sensitivity to probe for larger B_{IG}, and lower $E\gamma$ (and pair energies).

(4) An optimal degree of simultaneity between space-based GeV observations and multi-TeV ground based Air Cerenkov Telescopes (ACT). The nature of this improvement would be operational rather than instrumental, and it would extend the range of B_{IG} that can be probed.

We choose sample source parameters $t_d \approx 10^5$ year for $z_{\text{source}} = 0.03$, $\theta(B_{IG}) \sim 5°$, and $E_{\gamma 0} = 40$ TeV. In this case the observer will see a well developed and steady cascade source, broadened by θ_b whether the blazar is bursting or steady. After many $\lambda_{e,\gamma}$ and $\lambda_{\gamma,e}$ along the cascade, the total observed γ-ray flux will eventually be only a few percent of that emitted into the opening angle of the jet at $E_{\gamma 0}$.

The parameters associated with upper and lower limits, B_{IG}^{max} and B_{IG}^{min}, are illustrated by the following equations (Neronov & Semikoz 2007, Taylor, Vovk, & Neronov 2011). Recall that the B_{IG} determination here from the cascade angular broadening is a different measurement from the time delay profile.

$$B_{IG}^{\text{max}} \simeq 3 \times 10^{-12} \left[\frac{\tau(E_{\gamma 0}, z)}{\tau(E_{\gamma 0}, z) - 1} \right] \left[\frac{E_{\text{max}}}{10\text{TeV}} \right]^{1/2} \left[\frac{\theta_B}{5°} \right] \text{ G} \qquad (10.13)$$

and

$$B_{IG}^{\text{min}} \simeq 10^{-16} \tau(E_{\gamma 0}, z) \left[\frac{E_{\text{min}}}{0.1\text{TeV}} \right] \left[\frac{\theta_{\text{PSF}}}{0.1°} \right] \text{ G}. \qquad (10.14)$$

The normalisations are in the range of current angular resolution capability and detection limits for the range 0.1–10 TeV. Future instrumental improvements indicate that firmer values of B_{IG}^{min} (Equation 10.14) will be achievable as both the sensitivity below 100 GeV and θ_{PSF} improve further. Current methods for measuring the angular size of a γ-ray cascade have the potential to verify B_{IG} in the range from $\sim 10^{-12}$ to $\sim 10^{-17}$ G. Lower B_{IG}^{min} down to 10^{-20} G could potentially be probed if we include the TeV flare fluence-monitoring technique described below. For further background reading and details see e.g. Aharonian, Coppi & Völk (1994), Dolag *et al.* (2004), and Elyiv, Neronov, & Semikoz (2009).

10.5 Intergalactic magnetic field γ-ray flux probes down to 10^{-20} G

In Fig. 10.8, we examine the phenomenon of a single and strong, short term 10 TeV γ-ray flare, and the strength and delay of the photon cascade that it produces. For this, or a similar combination of parameters, an observed fluence will develop on time scales of minutes to years. For a sufficiently strong and time-limited burst, the immediately following cascade stage can, at least in principle, be studied to explore intergalactic magnetic fields as weak as $\sim 10^{-20}$ G. Whether such weak magnetic fields actually exist in cosmic void regions, and can be observationally isolated, is not clear at this time.

It is notable that both the *form* and *time delay* of the secondary cascade can be very sensitive to variations in B_{IG} around 10^{-20} G and above. This is illustrated in Fig. 10.8 in a

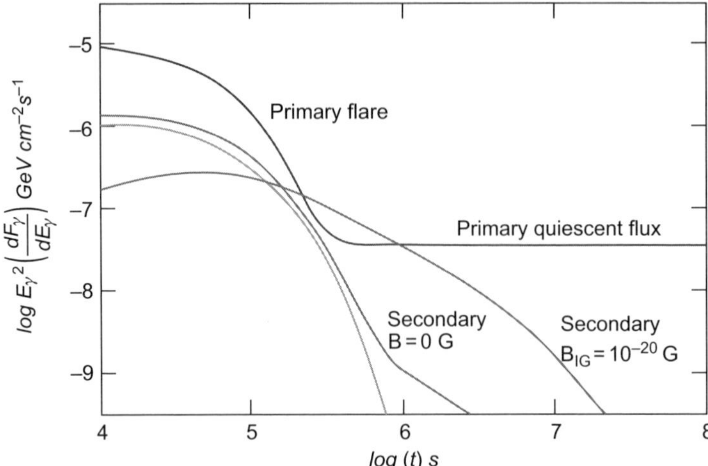

Figure 10.8. The pair cascade halo flux of a hypothetical $10\times$ stronger flare than that in PKS 2155-304 at $z = 0.116$. (Adapted from Murase *et al.* 2008). The secondary cascade photon flux inceases with energy above ~1 GeV. It will also increase as the B_{IG} coherence length, λ_B, decreases from ~ 1 to ~ 0.1 Mpc. λ_B is normally not directly measurable, and was assumed to be 1 Mpc.

model calculation of the flare and cascade components for PKS 2155-304 at $z = 0.116$. A γ-ray flare of this source was actually observed in 2006. The figure shows how, given the sensitivity of the GLAST γ-ray telescope, a hypothetical, sufficiently strong flare could reveal the form and intensity of the fluence $S(t)$ both for the primary flare and for the secondary cascade emission. A key result here is that the secondary pair echo emission is very sensitive to a weak B_{IG} near 10^{-20} G.

The energy dependence of the secondary (cascade) photon flux is also sensitive to the *coherence scale,* λ_B, near 10^{-20} G. For a 1.5-day flare, λ_B changes the E_γ-dependence of the cascade fluence by a factor of order 10. This is the case for the model flare of Fig. 10.8. It happens as λ_B changes between 0.1 and 1 Mpc in an energy range below ~10 GeV. The change can be positive or negative, depending on E_γ. Graphical illustrations and further discussion of these effects can be found in Murase *et al.* (2008).

The long horizontal "tail" of quiescent emission following the flare in Fig. 10.8 is an accompanying feature of a single flare, and it could mask the interesting secondary cascade emissions. However, a flare and its initial cascade at this level should be observable by some instruments. The curves in Fig. 10.8 incorporate models for SSC emission in the flare, and also for the CIB component of the EBL, which is not directly measured. The flare + echo emission contains other subtleties in the determination or limitation of B_{IG}. These are too extensive to cover in detail here, but they involve interesting physics. The interested reader is referred to, among others, the articles by Aharonian *et al.* (2006), Murase *et al.* (2008), Takahashi *et al.* (2008), Dolag *et al.* (2009), Neronov & Vovk (2010).

For a B_{IG} probe near 10^{-20} G to detect the pair echo emission down to a few GeV, the primary burst photons need to be at least a few TeV, and the detector must be sufficiently sensitive. As mentioned, the first interaction in the cascade chain, illustrated in Fig. 10.4, involves the infrared background photon density of the EBL. Unlike the case for lower energy photons scattering off the CMB, this requires modelling of the CIB, which must be independently characterised from other astrophysical information (see e.g. Kneiske *et al.* 2004 and Franceschini, Rodighiero & Vaccari 2008). The CIB models require observations that characterise the aggregate infrared emission from galaxies, AGNs, and stars over a range of redshift. For further reading on this somewhat complex subject, the reader could also consult Primack, Bullock, & Somerville (2005) and Stecker, Baring, & Summerlin (2007).

A possible primordial lower B_{IG} limit in voids might be raised if, at some epoch soon after the first stars were formed, the co-moving void volumes were "polluted" by magnetic energy emanating or "leaking" from very early black hole-jet/lobe systems, Population III stars, etc.

10.6 Summary remarks

Magnetic field probes in cosmic voids are one of the few ways by which we might estimate the primordial magnetic fields created before the current Universe of stars and galaxies. Most of the γ-ray methods outlined here can probe B_{IG} to large extragalactic distances – out to and beyond $z \sim 0.2$, i.e. a few Gpc. Ultimately, they can be compared with much stronger, μG-level magnetic fields that can be probed in QSO Faraday rotating absorption line systems out to $z \sim 2$. These are either more local to the QSO, or in cosmologically intervening galaxy halos. All of these fields or their limits, may, in turn, be linked to particle and quantum physics processes of the primordial Universe. This refers to proper times (**T**) beginning just after recombination at (**T** $\sim 10^5$ year), through the plasma epoch, and back to when the first baryons were created.

We have described three ways by which B_{IG} can be probed using extragalactic γ-ray sources; (1) angular broadening of pair halo emission in a cascade (2) measurement of the fluence vs. time behaviour of a strong primary flare and its cascade emission, and (3) the total energy spectrum of a γ-ray blazar and its development with time. Gamma ray probes cover a very wide range of magnetic field strength, and this is the only known band available to probe the very weakest levels of B_{IG}. Given improvements in γ-ray resolution, sensitivity, and wider field of view of post-Fermi generation of γ-ray telescopes, it will be possible to probe wider areas of cascade deflection. This should also enable energy-dependent pair halo size measurement. Such improvements will permit the types of γ-ray measurements described in this chapter also to probe for magnetic fields at higher B_{IG} strengths, above 10^{-13} G. Better specification of the higher end of B_{IG} in voids will begin to close the measurement gap between γ-ray techniques and radio probes by Faraday *RM* and faint synchrotron "glow". An informative discussion of future prospects can also be found in Elyiv, Neronov, & Semikoz (2009).

A realistic modelling of cascade propagation from an original "event", or source involves propagation of the γ-rays, leptonic pairs, etc., through a *combination* of magnetic field environments. Lines of sight to distant blazars can traverse a *mixture* of voids, filaments, and even outer halos of intervening galaxies. Thus, a "pure" measurement of B_{IG} in a void needs to isolate the void, whose typical dimension at $z \approx 0$ is of order 20–50 Mpc. In an example, the first few kpc might be traversed in the high (μG level) field of the host galaxy; then a cluster, group or filament environment; and finally our Local Group and Milky Way halo/disc environment, with IGM voids in between. This illustrates the complex range of

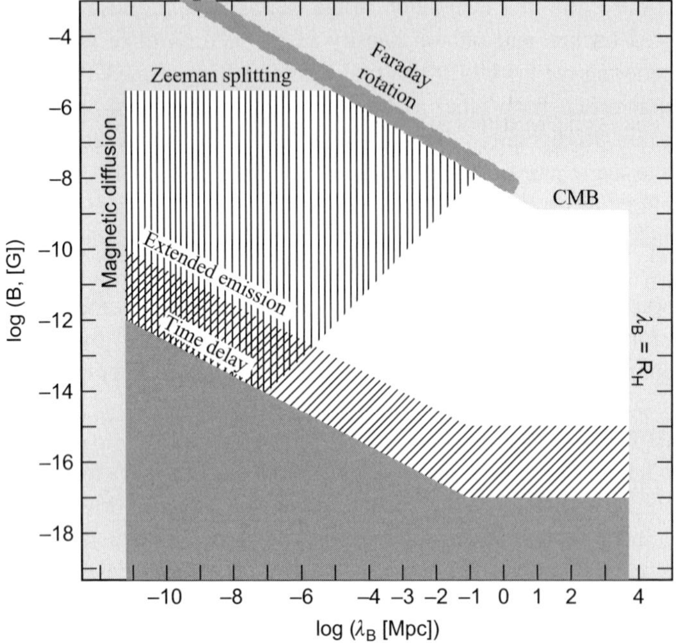

Figure 10.9. Illustration of the bounds of B_{IG} and λ_{BIG} that have been derived from previous simultaneous GeV and TeV data and from other known limits. The $\log(\lambda_B)$ scale ranges from the magnetic diffusion limit at the lower end, to the upper limit of a Hubble radius (R_H). (Modified from Taylor, Vovk, & Neronov (2011).)

possibilities for interpreting cascade propagation to obtain the relevant intergalactic field strengths.

For a primary photon, large void segments are of comparable order to $\lambda_{\gamma 0,\gamma}$ (Fig. 10.4), and dimensions of the voids in redshift space are of order $\Delta z \sim 0.01$. In the case of the γ-ray blazar IES 0347-121 at $z = 0.188$ ($\simeq 0.8$ Gpc), the ray path to us will intersect more than one galaxy filament and/or void. While the intersection path through a LSS sheet or filament may be only $\sim$5% of the neighbouring void pathlength, the strength of B_{IG} in parts of filaments might be up to $10^{-8 \to -7}$ G. This represents a contrast in B_{IG} of up to $\sim$12 orders of magnitude (24 orders in magnetic energy density!). It means that the simple "pathlength weighted" B_{IG} to a typical blazar at $z \sim 1$ could overwhelm a B_{VOID} along the way. Increasingly precise galaxy redshift surveys will be an important complement to γ-ray cascade analyses. This will help to match B_{IG} probes to individual isolated voids. This also means that test sources at lower redshifts are favoured for this purpose, as can be visually gleaned as in Fig. 10.3. It is separate from the issue, mentioned above, that void regions themselves could be at least partly polluted by astrophysical sources that might "leak" magnetic fields and CRs into the voids as LSS formation develops. In spite of all the above uncertainties it is impressive that γ-ray detector capabilities can potentially probe approximate B_{IG} strengths between $\sim 10^{-12}$ and 10^{-20} G – an unprecedented measurement range of 10^8:1.

References

Aharonian, F. A. *et al.* (The H.E.S.S. Collaboration). 2006, A Low Level of Extragalactic Background Light as Revealed by γ-Rays from Blazars, *Nature*, 440, 1018

Aharonian, F. A., Coppi, P. S., & Völk, H. J. 1994, Very High Energy Gamma Rays from Active Galactic Nuclei: Cascading on the Cosmic Background Radiation Fields and the Formation of Pair Halos, *Astrophys. J.*, 423, L5

Casanova, S., Dingus, B. L., & Zhang, B. 2007, Contribution of GRB Emission to the GeV Extragalactic Diffuse Gamma-Ray Flux, *Astrophys. J.*, 656, 306

Dolag, K., Grasso, D., Springel, V., & Tkachev, I. 2004, Mapping Deflections of Ultrahigh Energy Cosmic Rays in Constrained Simulations of Extragalactic Magnetic Fields, *JETP Lett.*, 79, 583

Dolag, K., Kachelrieß, M., Ostapchenko, S., & Tomàs, R. 2009, Blazar Halos as Probe for Extragalactic Magnetic Fields and Maximal Acceleration Energy, *Astrophys. J.*, 703, 1078

Dolag, K., Kachelriess, M., Ostapchenko, S., & Tomàs, R. 2011, Lower Limit on the Strength and Filling Factor of Extragalactic Magnetic Fields, *Astrophys. J.*, 727, L4

Elyiv, A., Neronov, A., & Semikoz, D. V. 2009, Gamma-Ray Induced Cascades and Magnetic Fields in the Intergalactic Medium, *Phys. Rev. D*, 80, 023010

Ensslin, T. E., Simon, P., Biermann, P. L., Klein, U., Kohle, S., Kronberg, P. P., & Mack, K.-H. 2001, Signatures in a Giant Radio Galaxy of a Cosmological Shock Wave at Intersecting Filaments of Galaxies, *Astrophys. J.*, 549, L39

Felten, J. E. & Morrison, P. 1966, Omnidirectional Inverse Compton and Synchrotron Radiation from Cosmic Distributions of Fast Electrons and Thermal Photons, *Astrophys. J.*, 146, 686

Franceschini, A., Rodighiero, G., & Vaccari, M. 2008, Extragalactic Optical-Infrared Background Radiation, Its Time Evolution and the Cosmic Photon-Photon Opacity, *Astron. Astrophys.*, 487, 837

Hudson, M. 1993, Optical Galaxies within 8000 km s^{-1} - I. *The Density Field, MNRAS*, 265, 43 (http://mhvm.uwaterloo.ca/research/large-scale-structure/)

Kim, K.-T., Kronberg, P. P., Giovannini, G., & Venturi, T. 1989, Discovery of Intergalactic Radio Emission in the Coma-A1367 Supercluster, *Nature*, 341, 720

Kneiske, T. M., Bretz, T., Mannheim, K., & Hartmann, D. H. 2004, Implications of Cosmological Gamma-Ray Absorption. II. Modification of Gamma-Ray Spectra, *Astron. Astrophys.*, 413, 807

Kronberg, P. P. 1994, Extragalactic Magnetic Fields, *Rep. Prog. Phys.*, 57, 325

Kronberg, P. P., Wielebinski, R., & Graham, D. A. 1986, VLA and 100-m Telescope Observations of Two Giant Galaxies 0634–20 and 3C445 (2221–02), *Astron. Astrophys.*, 169, 63

Murase, K., Takahashi, K., Inoue, S., Ichiki, K., & Nagataki, S. 2008, Probing Intergalactic Magnetic Fields in the GLAST Era Through Pair Echo Emission from TeV Blazars, *Astrophys. J.*, 686, L67

Neronov, A. & Semikoz, D. V. 2007, A Method of Measurement of Extragalactic Magnetic Fields by TeV Gamma Ray Telescopes, *JETP Lett.*, 85, 473

Neronov, A., Semikoz, D. V., Tinyakov, P. G. & Tkachev, I. I. 2011, No Evidence for Gamma-Ray Halos around Active Galactic Nuclei Resulting from Intergalactic Magnetic Fields, *Astron. Astrophys.*, 526, 90

Neronov. A. & Vovk, Ie. 2010, Evidence for Strong Extragalactic Magnetic Fields from Fermi Observations of TeV Blazars, *Science*, 328, 73

Plaga, R. 1995, Detecting Intergalactic Magnetic Fields Using Time Delays in Pulses of γ-Rays, *Nature*, 374, 430

Primack, J. R., Bullock, J. S., & Somerville, R. S. 2005 Observational Gamma-ray Cosmology, in *High Energy Gamma-Ray Astronomy: 2nd International Symposium*, ed. F. A. Aharonian, H. J. Volk, & D. Horns, AIP Conf. Proc. (New York: American Institute of Physics) 745, 23

Razzaque, S., Mészáros, P., & Zhang, B. 2004, GeV and Higher Energy Photon Interactions in Gamma-Ray Burst Fireballs and Surroundings, *Astrophys. J.*, 613, 1072

Rees, M. J. 1967, Studies in Radio Source Structure-III. *Inverse Compton Radiation from Radio Sources, MNRAS*, 137, 429

Rybicki, G. B. & Lightman, A. L. 1979, *Radiative Processes in Astrophysics* (New York: Wiley)

Stecker, F., Baring, M. G., & Summerlin, E. J. 2007, Blazar γ-Rays, Shock Acceleration, and the Extragalactic Background Light, *Astrophys. J.*, 667, L29

Takahashi, K., Murase, K., Ichiki, K., Inoue, S., & Nagataki, S. 2008, Detectability of Pair Echoes from Gamma-Ray Bursts and Intergalactic Magnetic Fields, *Astrophys. J.*, 687, L5

Taylor, A. M., Vovk, I., & Neronov, A. 2011, Extragalactic Magnetic Fields Constraints from Simultaneous GeV-TeV Observations of Blazars, *Astron. Astrophys.*, 529, A144

Waxman, E. & Coppi, P. 1996, Delayed GeV-TeV Photons from Gamma-Ray Bursts Producing High-Energy Cosmic Rays, *Astrophys. J.*, 464, L75

11

Intergalactic cosmic rays, gamma rays, and magnetic fields

11.1 A brief introduction

In 1962 a surprising discovery was made of the arrival of a cosmic ray (CR) with energy estimated at $\sim 10^{20}$ eV (Linsley 1963). In the intervening years, the CR spectrum has been increasingly well specified, especially at "ultra high" energies above $\sim 10^{18}$ eV (UHECRs) thanks to the The Akeno Giant Air Shower Array (AGASA) (Nagano *et al.* 1984, Yoshida *et al.* 1995), HiRes (Bird *et al.* 1994, Boyer *et al.* 2002) and AUGER (Cronin 1992) detectors, among others. Two key features of the overall CR spectrum shown in Fig. 11.1 are the "knee" at about 10^{16} eV and the "ankle" near $10^{18.5}$ eV. A gradual phase-over from galactic to extragalactic CR origin is thought to occur between $10^{16.5}$ and 10^{18} eV.

Over this same energy range there is also an apparent "chemical switchover", in the sense that UHECRs above $\sim 10^{18}$ eV have been widely believed to be mostly protons ($Z = 1$), whereas at the lower CR energies heavier nuclei are seen. However, some recent analyses have suggested that Fe nuclei might form a significant CR component above $\sim 10^{18}$ eV. This question is not completely resolved, and much of the following discussion in this chapter is based on the assumption of $Z \sim 1$ for the majority of CRs above $\sim 10^{18.5}$ eV. Questions and theories on the composition of CRs have received considerable attention in the literature, much of which goes beyond the scope of this book. Further reading on CR composition could begin with the results of the KASCADE project (Antoni *et al.* 2004) and the KASCADE Grande CR "spectrometer" instrument (Navarra *et al.* 2004) and their predecessors. For a wider background on CRs, Thomas Gaisser's book (1990) is recommended.

It was described earlier how the observed CR isotropy up to $\sim 10^{16}$ eV (measured as of *ca.* 1949) was used to infer an interstellar magnetic field strength of $\sim 5\,\mu G$, in order to isotropise CR trajectories that originate in the very anisotropic distribution of supernova remnants in the Galactic disk. More recent CR data reveal the additional fact that this isotropy continues well past 10^{17} eV. Our puzzlement at this result may be rooted in the fact that the structure, strength, and extent of the Galactic halo magnetic field – hence the propagation and confinement environment there – require better specification. The component above $\sim 10^{18.5}$ eV is subject to extragalactic magnetic deflection, and to energy losses from interactions with intergalactic photons, particularly the Greisen-Kuzmin-Zatsepin (GZK) effect and Bethe–Heitler losses. In addition to line-of-flight effects, there could be high energy cutoffs that are intrinsic to the sources. Such effects, not precisely specified at this point, might possibly "pre-empt" the GZK effect.

Above $\sim 10^{19}$ eV, apparent *ani*sotropies have emerged in the all-sky UHECR arrival data from the Southern hemisphere data (e.g. The Pierre Auger Collaboration, Abreu *et al.* 2010). However with increasing AUGER events and instrumental enhancements, better calibration,

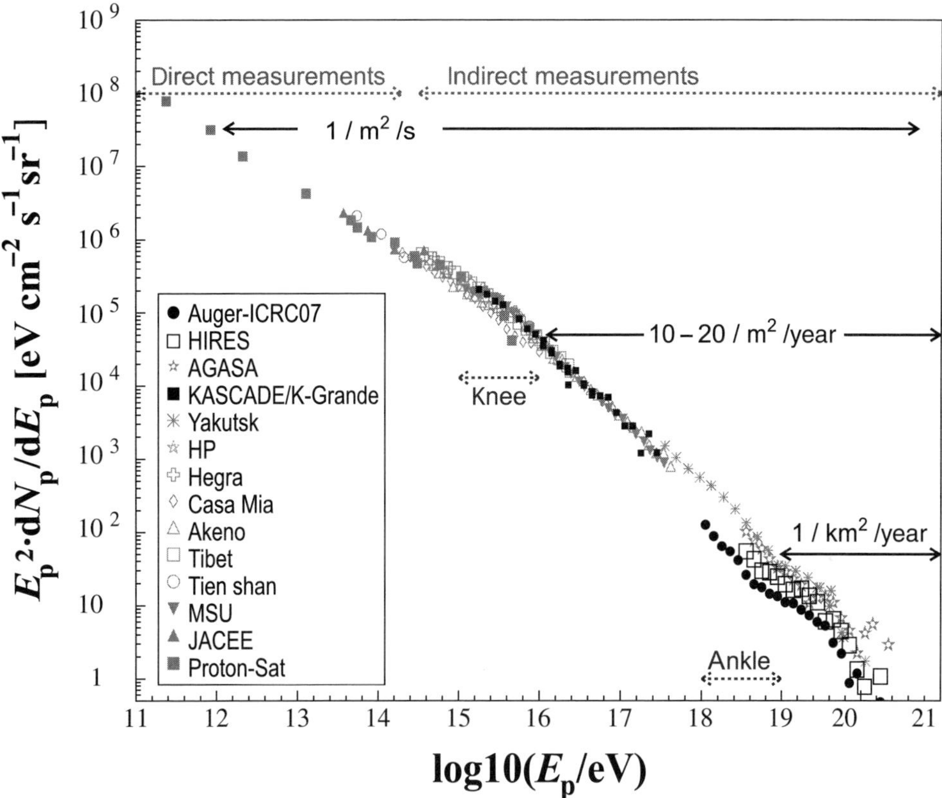

Figure 11.1. The CR spectrum from 10^{11} to $10^{20.5}$ eV, compiled from various instruments. It shows the cumulative event rates above three different threshold energies, and illustrates the value of large collecting areas and long experiment times in order to obtain good event statistics in the interesting UHECR range. Clearly visible are the "knee", the "ankle", and the recently observed "cutoff" above $10^{19.5}$ eV. Refer to the text for further explanation. (Figure after Becker 2008, and updates kindly provided by J. Becker Tjus 2013, private communication.) These include data from Auger 2011, KASCADE 2011, and IceTop 2013.

etc., these now appear less striking (Letessier-Salvon 2013). In the Northern hemisphere (e.g. the HiRes Collaboration), UHECR anisotropies have always been less apparent. They are of particular interest for understanding the extragalactic acceleration sites of UHECRs and the intergalactic magnetic fields through which CRs have propagated to reach us. CR detectors can measure the arrival directions with ever increasing precision, as a function of CR energy and composition. In this sense they have advanced to "telescopes" that can image the sky. The complex and many-faceted UHECR measurements will become an increasingly important tool for studying intergalactic magnetic fields.

11.2   Magnetic effects on UHECR propagation in intergalactic space

A UHECR nucleus of nuclear charge eZ and energy E has a gyroradius

$$r_g \approx 100 \frac{E_{20}}{B_{-9} eZ} \quad \text{Mpc,} \tag{11.1}$$

where E and B are in units of 10^{20} eV and 10^{-9} G, respectively. If typical B_{IG} values are in the range 10^{-7}–10^{-8} G, then propagation trajectories originating from 1 to 100 Mpc away (the distance to the Coma supercluster) will be bent; for example, $r_g \approx 1$ Mpc for $E = 10^{20}$, $Z = 1$ and $|B_{IG}| = 10^{-7}$ G. $|B_{IG}|$ must be $\ll 10^{-8}$ G to avoid a significant deflection (θ) for proton UHECR sources out to 100 Mpc – unless, of course the UHECR sources are much closer. We will learn much more if a UHECR detector can actually "point" to individual sources. At the same time, the above numbers tell us that we might simply be unlucky if B_{IG} within the nearest few Mpc were relatively high, say 10^{-7}–10^{-6} G. In that case the original direction of the CR source would likely be lost. More generally, the structure of $|B_{IG}|$ can, in certain circumstances interact with the UHECR to create detection "blind spots", a phenomenon discussed further in Section 11.11.

Apart from $|B_{IG}|$, the other "given" is the typical distance from us of objects that accelerate UHECRs – they must not be too far away! To minimise the impact of deflections on our observations, we would ideally prefer to detect CRs at $E \gg 10^{20}$ eV. But these are relatively rare, and energy-dependent CR-photon interactions increasingly take effect at higher energies, to the point of drastically reducing the effective loss length, x_{loss}. This reduces their arrival energies at Earth. Potentially, it would place us in a box of parameter (and physical) space, beyond which we cannot directly "see" to an UHECR source.

Before discussing UHECR energy loss mechanisms and then considering candidate sources, we can extend Equation (11.1) to consider UHECR protons propagating from distance D through an IGM with a fluctuating magnetic field having correlation length λ_B.

$$\theta(E,D) \simeq 0.025° \left(\frac{D}{\lambda_B}\right)^{1/2} \left(\frac{\lambda_B}{10 \text{ Mpc}}\right) \left(\frac{B}{10^{-11}\text{G}}\right) \left(\frac{E}{10^{20} \text{ eV}}\right) \tag{11.2}$$

(Waxman & Miralde-Escudé 1996) where D and λ_B are in Mpc. Using characteristic normalisations and a random walk over an intergalactic pathlength $D(>\lambda_B)$, the numerical value of the observed deflection is given by

$$\theta(E,D) \simeq 8°Z \left(\frac{D}{10 \text{ Mpc}}\right)^{0.5} \left(\frac{\lambda_B}{1 \text{ Mpc}}\right)^{0.5} \left(\frac{E}{10^{20} \text{ eV}}\right)^{-1} \left(\frac{B}{10^{-9} \text{ G}}\right) \text{deg} \tag{11.3}$$

(Sigl, Miniati, & Enßlin 2003).

Analogously to the γ-ray pair cascades described in Chapter 10, the spread in angle deflections in the course of propagation has a corresponding time of arrival delay (τ), given by

$$\tau(E,D) \simeq 1.5 \times 10^3 Z^2 \left(\frac{B}{10^{-9} \text{ G}}\right)^2 \left(\frac{D}{10 \text{ Mpc}}\right)^2 \left(\frac{\lambda_B}{1 \text{ Mpc}}\right) \left(\frac{E}{10^{20} \text{ eV}}\right)^{-2} \text{ year} \tag{11.4}$$

(Sigl, Miniati, & Enßlin 2003). These arrival delays can range from hundreds of thousands up to millions of years (Fig. 11.4), depending on the above normalizations, and on whether the propagation path traverses a dense intergalactic filament, a galaxy halo where B might be $\gtrsim 10^{-7}$ G, or a cosmic void where B is probably $\ll 10^{-9}$ G. At present λ_B is imprecisely known, but note that it affects θ only by the 0.5 power and τ linearly. Also, if $Z \gg 1$ (e.g. 26 for Fe), a much larger delay occurs.

The angular deflection could occur in different locations along the line of sight. These could be: (1) An extended and turbulent, magnetised Milky Way halo where $|B|$ could be locally up to several μG over 5–20 kpc; (2) In nearby intergalactic zones of enhanced magnetic field where B could be near 0.1 μG (this could include extended halo zones of galaxy clusters); (3) In a zone local to the UHECR source; (4) In galaxy filaments of cosmological LSS; and finally (5) In the large voids of LSS (Fig. 10.3). CR nucleon deflection in (5) seems less likely, but would depend on the poorly-known range of B values within voids.

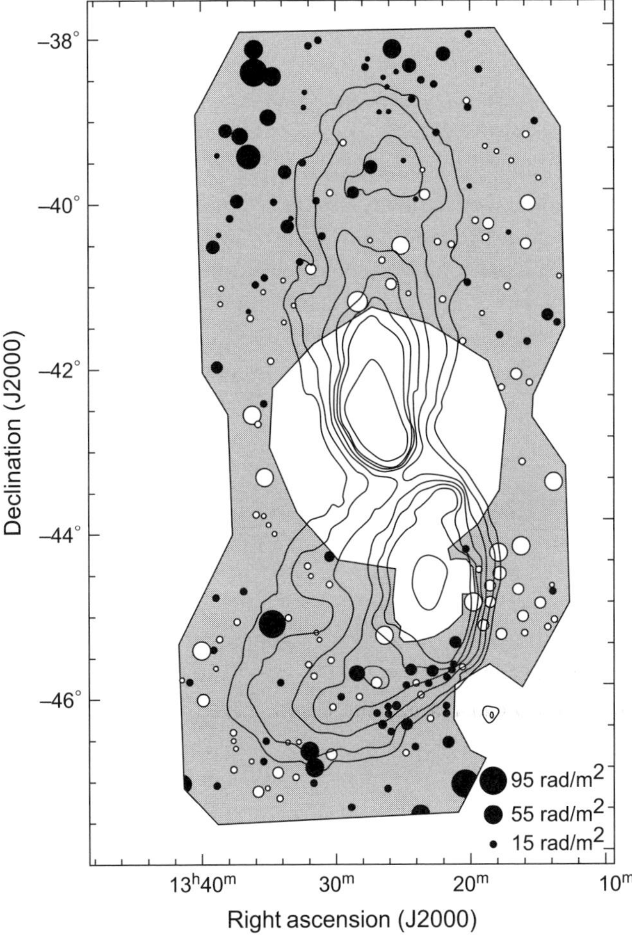

Figure 11.2. A radio contour image of Centaurus A, superimposed on the rotation measures of 281 background radio sources. The *RM* magnitudes scale as circle diameter, as shown in the legend. An overall average *RM* of -57 rad m^{-2} has been subtracted from the values shown. (Adapted from I. Feain *et al.* (2009).)

11.3   Energy losses of UHECR nucleons in intergalactic space

An intergalactic CR of energy E will lose energy due to the adiabatic expansion of the Universe; $E(t) = E_0/R(t)$, where R is the scale factor of the Universe and t is a function of redshift, z, through the adopted cosmological model. This is the first loss mechanism, and it is constant for a fixed distance to the UHECR source. As discussed above, the highest energy CRs are not visible over sufficiently large values of R because they are subject to distance-dependent losses, caused by interactions with the intergalactic photon field, most prominently the CMB photons and the EBL. A good discussion of the important UHECR proton–photon interactions can be found in Stanev *et al.* (2000). In the rest frame of a UHECR nucleus, CMB photons are blueshifted to high energy γ rays, and participate in the reactions listed below. The rate of each reaction is CR energy dependent in a different way.

A relatively "broadband" CR energy loss process is Bethe–Heitler (B–H) pair production loss, which occurs when a CR proton interacts with an intergalactic photon:

$$p + h\nu \rightarrow p + e^+ + e^-. \tag{11.5}$$

The fractional energy loss per B–H pair production interaction is of order m_e/m_p, and it is mildly energy dependent over 10^{19}–10^{21} eV (Stanev *et al.* 2000, Waxman & Miralda-Escudé 1996). It is nonetheless a significant UHECR loss process between $\sim 3 \times 10^{18}$ and $\sim 5 \times 10^{20}$ eV (Fig. 11.3), and has a maximum loss rate (minimum x_{loss}) at $\sim 3 \times 10^{19}$ eV, which decreases (x increases) as E exceeds $\sim 2 \times 10^{20}$ eV. The trajectories of the e^+ and e^- particles produced in each such interaction are subject to the Lorentz force of an intergalactic magnetic field, but the amount of each B_{IGM} deflection, and the corresponding time delays, will be different than for the primary UHECR proton.

Another important reaction chain occurs as follows, producing neutrons, pions, muons and neutrinos:

$$p + h\nu \rightarrow n + \pi^+; \ \pi^+ \rightarrow \mu^+ + \nu_\mu. \tag{11.6}$$

And finally, an increasingly effective CR energy loss at the highest energies comes from the reactions

$$p + h\nu \rightarrow p + \pi^0; \ \pi^0 \rightarrow \gamma + \gamma. \tag{11.7}$$

The loss length, x_{loss}, can be expressed in terms of a mean photon–particle interaction length, λ, and a corresponding elasticity, κ, where

$$x_{\text{loss}} = \frac{E}{dE/dx} = \frac{\lambda(E)}{\kappa(E)} \tag{11.8}$$

and

$$\kappa = \frac{\langle \Delta E \rangle}{E}. \tag{11.9}$$

A UHECR proton near $\approx 10^{20}$ eV loses about 20% of its energy ($-\Delta E$) for each p–$h\nu$ interaction and the average λ is only ≈ 15 Mpc, a relatively modest intergalactic distance. This interaction distance sharply decreases above $\sim 5 \times 10^{19}$ eV. These proton–photon interactions producing pions are commonly known as the GZK effect after the authors who first pointed it out (Greisen 1966, Zatsepin & Kuz'min 1966).

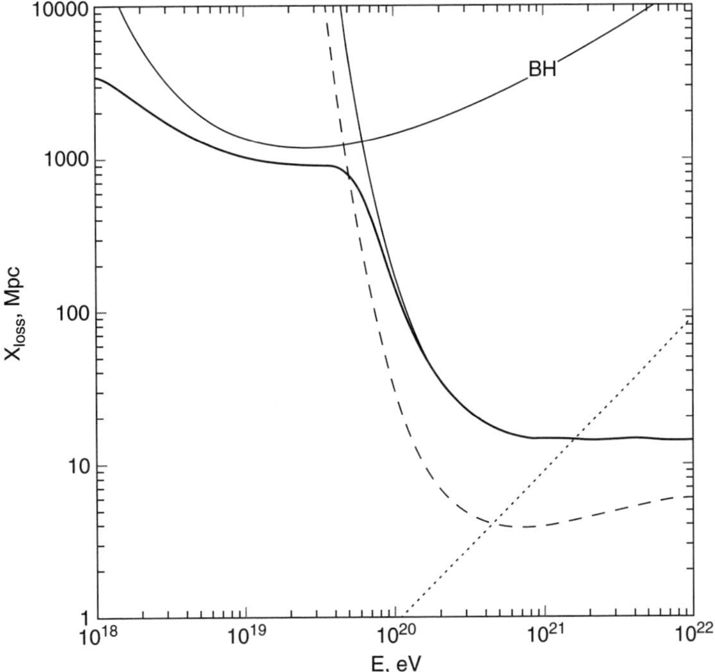

Figure 11.3. The comparative loss lengths, x, of UHECR protons and secondary neutrons as a function of the original UHECR energy (Stanev 2008). B–H indicates x_{loss} due to Bethe–Heitler pair production. The thin black curve is for hadron production, and the heavy back curve shows its combination with B–H losses. The neutron decay length (lower dotted line) *increases* with source nucleon energy. The hadron interaction length ($\lesssim 10$ Mpc for $E > 2 \times 10^{20}$ eV) is shown by the dashed line. Losses due to cosmological expansion (not shown) are independent of E, i.e. x would be a constant (horizontal line), and have a vertical location at $x \approx 4000$ Mpc for distances within a GZK horizon. $H_0 = 75$ km/s/Mpc is assumed (Stanev *et al.* 2000).

The GZK effect sets a "horizon" on the visibility of the production sites of UHECRs above $\approx 10^{20}$ eV which is ~100 Mpc, independent of B_{IG}. The kinematics of the pion production process generates a Poissonian distribution of interactions in the course of propagation, rather than a continuous energy loss approximation (see Hill & Schramm 1985).

The x_{loss} of neutrons produced in the $p + h\nu$ interaction (Equation 11.6) will increase with the original UHECR nucleon energy. The cross section for photoproduction of neutrons is of comparable order to that for protons, however the energy dependence of x_{loss} for neutrons is very different. Below $\approx 10^{20.5}$eV neutrons have lower x_{loss} than protons, but their x_{loss} increases above that energy (Engel et al. 2001). Figure 11.3 shows the relative importance of line-of-flight loss mechanisms for UHECR protons and neutrons up to ~$10^{21.5}$ eV. Adiabatic losses are energy independent, and give a component of $x_{loss} \approx 4000$ Mpc $(1 + z)^{-3/2}$ for an Einstein–deSitter Universe and $H_0 = 75$. In Fig. 11.3, this would be a horizontal line at $x_{loss} \approx 4000$ (not shown) for the cosmologically small distances that apply to the highest energy interactions. Incorporation of cosmological evolution in the CR spectrum causes some pile-up in energy space near the intersection of the pair production and photo-pion (GZK) loss curves at ~5 × 10^{19} eV (Berezinskii & Grigor'eva 1988). A lesser pile-up effect happens around the intersection of the pair production curve and the adiabatic loss line (not shown) at ~10^{18} eV (Stanev *et al.* 2000). The remarkable rapidity with which x_{loss} declines as we approach 10^{20} eV is fundamentally due to the steepness of the high frequency tail of the very blueshifted CMB Planck distribution (Greisen 1966). The calculations required to produce these curves are complex, and beyond the scope of discussion here. For calculations similar to the ones shown in Fig. 11.3, and related discussions, the interested reader could consult Berezinskii & Grigor'eva (1988), Rachen & Biermann (1993), Yoshida & Teshima (1993), and Protheroe & Johnson (1996a,b). Differences among published calculations of x_{loss} have typically ranged from 0–~40% and depend on the UHECR energy.

Finally, there could be components of EeV – level events due to photons and/or to neutrinos of cosmic origin, which future detectors should be able to clarify. These possibilities are in addition to remaining uncertainties on the mix of nucleon Z-values in the EeV range (Section 11.4), and the additional possibility that not-yet-specified *source-intrinsic* energy cutoffs may exist, and that they could partially or completely overwhelm the GZK photo-pion line-of-flight losses. For further reading, the reader could consult Letessier-Selvon & Stanev (2011) and Letessier-Selvon *et al.* (2013).

We next illustrate some model UHECR observations for simple cases involving only protons, to give an idea of the richness of information they contain, given a sufficiently powerful UHECR "telescope" and excellent event statistics over a wide CR energy range. The left panel of Fig. 11.4 shows the observed distribution of arrival energies for a source of mono-energetic protons of $E = 10^{21.5}$ eV, injected at various cosmological distances, and subject to the loss processes described above. The "spike" at the injection energy can be understood as due to the low interaction probability when sources are close by.

The right panel shows propagation delays relative to a light ray propagating along a geodesic. They apply to protons injected at the same chosen distances, for a 10^{-9} G random (by orientation) intergalactic magnetic field with a reversal scale $l_0 = 1$ Mpc.

Because of the Poissonian nature of the *p-hν* GZK interactions, many of the $10^{21.5}$ eV protons from an 8 Mpc distant source, and ~25% of those from 32 Mpc distance, will be detected at Earth with their undiminished emitted energy. Given sufficiently small intergalactic magnetic field strengths, our hypothetical UHECR "telescope" could directly register the source's (α, δ) location on the sky and original emitted energy. In Section 11.11 we discuss the results of UHECR detection statistics using some preliminary estimates of intergalactic magnetic field strengths. In a real situation the measured distribution of arrival energies in Fig. 11.4(left) are convolved both by the broader UHECR source energy function, $N_{emitted}(E)$, and a function representing the structure and strength of B_{IG} along the pathlength to us. A random B_{IG} will tend to increase the propagation path length. The relative numbers can be appropriately scaled depending upon the propagation environments (1)–(5) listed in Section 11.2.

Figure 11.5, also reproduced from Stanev *et al.* (2000), shows the hypothetical energy distribution of UHECR events as a function of time delay, again for the illustrative example of a monoenergetic source of $10^{21.5}$ eV protons propagating over 32 Mpc of IGM that contains a randomly oriented 10^{-9} G magnetic field. Considering what can actually be verified by measurement, this plot illustrates what we might hope to observe and, together

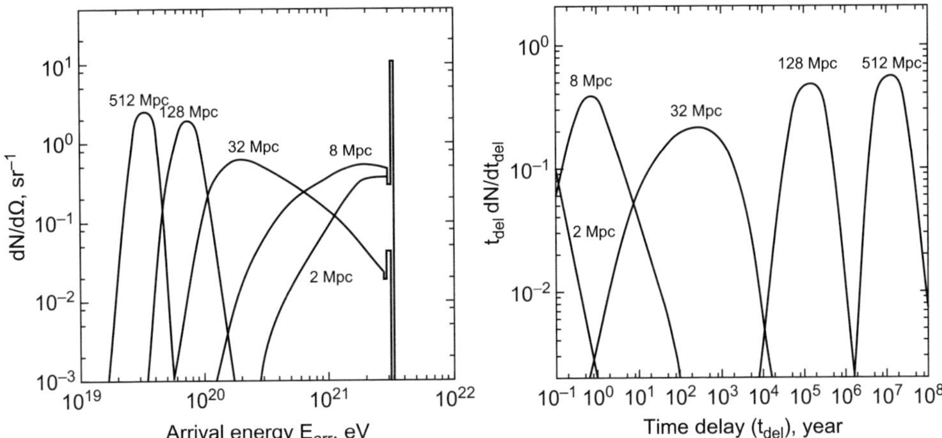

Figure 11.4. (Left) The modified CR energy distribution on arrival at Earth for a monoenergetically injected proton energy of $10^{21.5}$ eV for a randomly orientated $B_{IG} = 10^{-9}$ G at progressively larger distances up to 512 Mpc. The energy is reduced by the GZK effect (which has the most pronounced effect), B–H pair production losses, and adiabatic losses (Right). t_{del} is the relative time delay for protons injected at the same distance and propagated through a randomly oriented magnetic field of 10^{-9} G, where $l_0 = 1$ Mpc. (Reproduced from Stanev *et al.* 2000.)

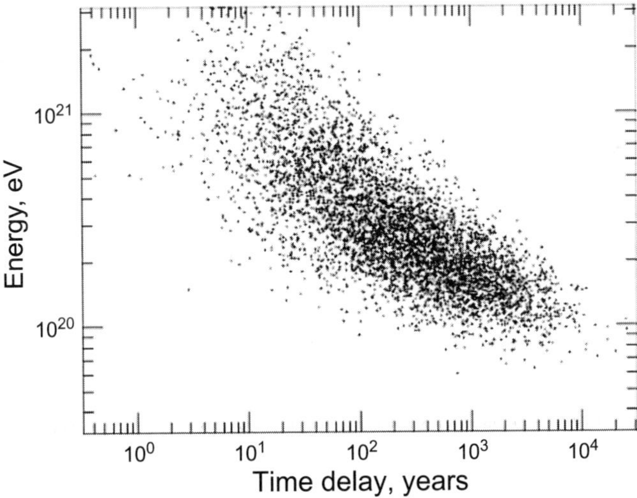

Figure 11.5. For a random magnetic field of 10^{-9} G with $l_0 = 1$ Mpc and a monoenergetic source of $10^{21.5}$ eV CRs at 32 Mpc distance, the observed distribution of relative time delays vs. *E*. (Source: Stanev *et al.* 2000.)

with Fig. 11.4(R), shows how time delays might be measured over months, years, or decades, depending on parameters of the propagation environment. Figure 11.6 below shows the distance-dependent distribution of arrival *directions*, relative to an undeviated light ray for $10^{21.5}$ eV protons in a 10^{-9} G field for the same standard distances as above.

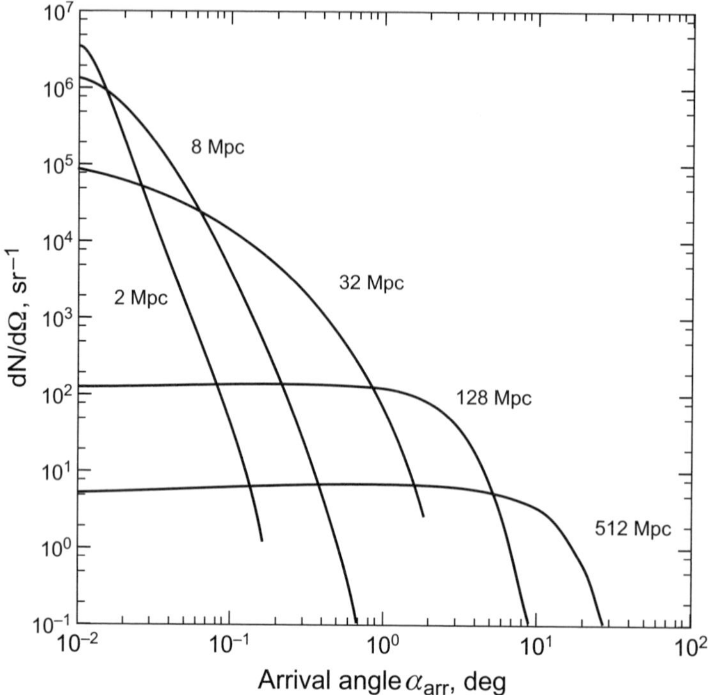

Figure 11.6. Distributions of arrival angles for monoenergetic injection of $10^{21.5}$ eV protons at the same standard distances as in Fig. 11.4, propagating through a 10^{-9} G intergalactic magnetic field. (Stanev 2008.)

These model detection results give some encouragement for our ultimate ability to analyse UHECR data and draw conclusions about B_{IG} within some parts of observational parameter space. It is especially encouraging that UHECR statistics have increasingly better resolution in energy, position, and composition. These advances combine, in turn, with an improved understanding of interactions within the Earth's atmosphere for nuclei of different atomic charge.

11.4   CR nuclei of helium and heavier nuclei

So far we have discussed only protons, but note that the Larmor radius of an atomic nucleus decreases as Z^{-1} (Equation 11.1), and that the relative propagation time delay increases as Z^2 (Equation 11.4). These facts may be pessimistic for identifying extragalactic sources that accelerate heavier CR nuclei, since r_g for a proton is already an inconveniently small extragalactic distance. It would exacerbate any difficulties in "pointing to" (identifying) the extragalactic UHECR source. Likewise, the interpretation of CR time-of-arrival delays for heavier nuclei to infer IGM magnetic field strengths within ~50 Mpc (Fig. 11.4 (R)) becomes more challenging. This is simply because these timescales can be inconveniently long relative to a human lifespan, and would be further stretched out by the factor of Z^2 just mentioned.

But the above is only part of the story, since it has ignored the x_{loss}, for heavier nuclei. In Section 11.1 we mentioned the "chemical switchover" that may occur near 10^{18} eV, in the sense that heavier element UHECRs may be more rare at $E \gtrsim 10^{18}$ eV. This question is still under investigation. The uncertainty is due in part to some degree of photodisintegration of nuclei against CMB photons. When 10^{19} eV is reached, the interaction distance for photo-disintegration of a $Z > 1$ nucleus is only 2×10^{22} cm, which is ~9 kpc, comparable to the radius of the Milky Way! However, for CR energies at ~10^{17} eV and below, the photo-disintegration cross sections are lower, and elements up the periodic table such as Fe, Ni, and Co are expected to be observed.

This leads to a brief mention of the connections between chemical abundances of CR nuclei at lower energies (1–10 GeV) and the Galactic magnetic field. A given CR species with (E, Z) has a corresponding r_g in the Galactic magnetic field. Depending on the strength and structure of the field, this will determine how many Milky Way disk crossings it has done over the Galaxy's lifetime. Combined with the column densities and cross-sections of target species in the Milky Way disk, one might link the Galactic magnetic field to nuclear spallation and to the age of the CR nucleus in question. This is because each crossing of the Galactic disk has a probability of altering a species or isotope due to interstellar collisions. Some CR spallation products such as Li, Be, and B have been thought to originate *only* in such interstellar reactions, because their relative abundances have appeared inconsistent with production by star-interior nucleosynthesis. However, some recent revisions to stellar nucleosynthesis calculations show that some of the above "interstellar" species might have been produced in stars after all. Consequent modifications to the arguments above require us to distinguish between species produced purely in interstellar CR collisions, and those that come from stellar winds – i.e. synthesised in a star.

Prominent instruments for measuring the chemical composition of UHECR nuclei at different energies are the KASCADE and KASCADE-Grande arrays, built and operated at the Kernforschungszentrum Karlsruhe, Germany (Kascade-Grande Collaboration 2008). For further reading on the basics of this extensive topic, the reader could begin with an important review article by Simpson (1983) and an update by Hörandel (2008).

11.5 Candidate production sites for UHECRs

The preceding discussion assumes that UHECRs are atomic nuclei and not some exotic species, such as X-particles associated with magnetic monopoles. Acceleration sites have been proposed to be: (1) close to the neutron star surface in Galactic pulsars, (2) in the bright, compact outer hotspots of some powerful radio galaxies such as Cygnus A (e.g. Biermann & Strittmatter 1987), (3) in extragalactic gamma ray bursters, and (4) in the jets and lobes of powerful radio sources. The latter contain two sub-possibilities: either (4a) within AU to parsec distances from the central Black Hole accretion disk – size régimes accessible to VLBI techniques; or, (4b) on much larger, kiloparsec scales along the colli-mated jet that carries the energy from the BH beyond the galaxy or quasar into the intergalactic medium (e.g. Lovelace & Kronberg 2013 and references therein). A further possibility is (5) the hottest, outer zones of large scale accretion shocks of infalling intergalactic gas into galaxy filaments (e.g. Ryu *et al.* 2007). A review of some of these possibilities can be found in Ostrowski (2002).

An insightful tool has been used to constrain possibilities for mechanisms that could accelerate CRs. Hillas (1984) introduced a plot of $|B|$ (G) vs. L (pc) illustrated in Fig. 11.7. In a "one-shot" direct acceleration model the EMF is ~LvB/c, where v is the velocity in some electrical conductor in a magnetic field, L is the scale of the system, and B the strength of the

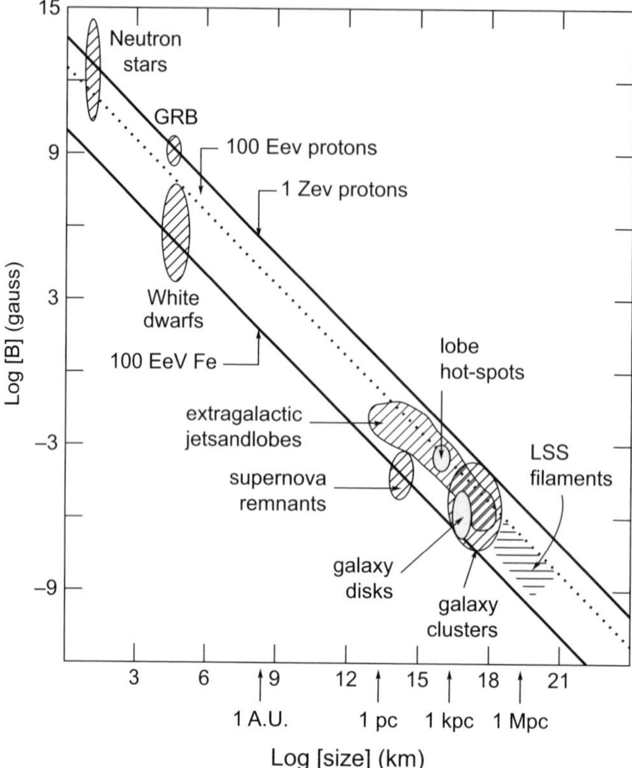

Figure 11.7. A plot for testing the *a priori* plausibility of different potential astrophysical host sites for UHECR production. Adapted from Hillas (1984) with some modifications.

magnetic field. In a rotating system, L could be the diameter of a pulsar envelope or a rotating neutron star. It could also be the scale of a diffusive shock zone where CRs are energised by magnetic reflection in some version of Fermi acceleration. These general considerations could apply to supernova shells, radio galaxy lobes, or the turbulent medium of a galaxy cluster. In general, the product $B \cdot L$ represents the maximum particle energy that the "system" could produce; Increasing the energy (for a given Z) advances the diagonal line toward the upper right in the "Hillas plot" in Fig. 11.7.

The above type of argument rules out, *prima facie*, magneto-plasma systems in sunspots, most supernova remnant shells, interplanetary space, magnetic stars, white dwarfs, and galactic halos as candidate acceleration sites for protons at 10^{20}–10^{21} eV.

Examples of objects that do qualify as "Hillas plot candidates" include some pulsars or magnetars, supermassive central black holes of galaxies, γ-ray bursters, extragalactic AGN-produced radio jets, the unusually intense hotspots of some luminous FRII radio galaxies such as Cygnus A, and some extended extragalactic radio source lobes.

Galactic sites such as rotating magnetars associated with millisecond pulsars have been suggested as "local" UHECR acceleration sites, along with some bright SNR filaments. Given this, as occasionally proposed, their sky distribution might be associated with some stellar population component of the Milky Way. On the other hand, UHECR observations do not yet show evidence for this. Also, any such star-associated system is likely to be buried deep in a dense, and highly magnetised environment, where any UHECRs produced would quickly give up their energy before they had a chance to propagate over kpc distances in the Galaxy. Further, UHECRs at energies around 10^{16-17} eV, the would-be "ashes" of such interactions, do not show a concentration to the centre of the Milky Way. Rather, they appear to be quite isotropic. So a significant Galactic component above ~10^{19} eV appears unlikely, and only the extragalactic candidates for the highest energy UHECRs remain, few as they are, and they must be well within a GZK horizon unless the high energy cutoff is due to some other effect, not yet specified.

As an alternative to the UHECR phenomena discussed above, it has been proposed that other, more "exotic", non-baryon particles are responsible. They too have potential links to cosmic magnetic fields but they would not be subject to GZK photopion reactions in intergalactic space (i.e. their x_{loss} could be $\gg$ 100 Mpc). Thus, their horizon for observation might enclose a large fraction of the observable universe. For a review of the possible roles of topological defects, cosmic strings, superconducting strings, and magnetic monopole-anti-monopole pairs, see Berezinsky & Vilenkin (1997).

11.6   *γ*-Rays, neutrons, and neutrinos at high energies

We now turn attention to products of the photo-pion reactions of a UHECR proton in Equations (11.6) and (11.7). Keeping the $10^{21.5}$eV reference energy for primary accelerated protons, the neutrino and gamma ray products have energies that extend all the way from ~10^{21} eV to $\lesssim$1 Tev, depending on the neutrino flavour. Electron neutrinos (v_e) have a particularly wide energy range, to 0.1 TeV. However, at the higher energies, above $10^{17.5}$ eV, they have lower fluences than muon neutrinos (v_μ). Once produced, most neutrinos are not subject to a limited x_{loss} horizon, and so can be "seen" to their source at great distances in the Universe. The observational question is whether we can detect UHECR neutrinos at a meaningful fluence level in energy ranges of interest. For completeness, we note in passing that Equations (11.6) and (11.7) have other branch reactions. The omitted ones usually have lower cross-sections, and for the purpose of this book we restrict this complex subject to reactions that appear to be more relevant to intergalactic magnetic fields.

The v_e, $\bar{v}_\mu$, v_μ $\bar{v}_\mu$ are produced out of photopion decays, including some in the course of propagation. This means that flux enhancements could occur for a particular species along the line of sight. This fact "frustrates" the inverse square law of fluences, but it can work to our advantage by allowing a deeper reach into the UHECR source population in a magnetised IGM. An example of relatively enhanced fluence occurs when some secondary particle or photon has a larger x_{loss} distance than the entity that produced it.

In processes that produce a secondary UHECR proton and an electron neutrino v_e (or antineutrino $\bar{v}_e$), the secondary UHECR proton inherits most of the energy, leaving correspondingly less for the $\bar{v}_e$. This means that a UHECR neutrino detector would see the observed v_e's with a relatively larger spread downward in energy. The energies of muon neutrinos (v_μ) produced in Equation (11.6) are relatively less diminished, so that they arrive with higher average energies. This represents a kind of "neutrino oscillation" in space. For further reading see Engel, Seckel, & Stanev (2001). A more in-depth discussion of β-decay, muon interactions and propagation, can be found in the book by Gaisser (1990) and references therein.

An important difference between UHECR protons on the one hand, and UHE neutrinos and γ-rays on the other is the different $x_{\mathrm{loss}} \lesssim 1$ Gpc above $\sim 10^{19}$ eV. Here, for protons $x_{\mathrm{loss}} \lesssim 0.1$ Gpc, corresponding to the "GZK horizon". In principle, given sufficiently sensitive neutrino detectors and γ-ray telescopes we can "see" high energy particle reactions to larger distances in the Universe. For example, γ-rays below ~ 1 TeV have very long interaction lengths. Combining them with ~ 100 TeV γ-rays, which have shorter x_{loss}, it is possible, as discussed in Chapter 10, to probe, or limit $|B_{\mathrm{IG}}|$ in very weak field régimes such as cosmic voids, where the magnetic fields are otherwise well below detection limits of either Faraday rotation, or faint synchrotron radiation.

In addition, diffuse magnetic fields, γ-rays, and neutrinos over various energies have some deep interconnections to the cosmogony of the early universe, as well as to some fundamental physics questions. For example, one might test for the existence of magnetic monopoles, and also for violation of Lorentz Invariance. Depending on whether Lorentz Invariance holds or not at the relevant energies, the γ–γ's produced in the main branch of π^0 decay (Equation (11.7)), would produce e^+e^- in very different quantities if Lorentz Invariance is broken – in fact more than 10 times fewer e^+e^- pairs, if Lorentz Invariance is broken. This test has not yet been conclusively done, partly because the detection threshold of existing particle physics experiments for the test is much higher than what the CR proton/nucleon flux from space is likely to achieve (Kampert 2008). A conclusive Lorenz invariance test from π^0 decay might very well first come from CR experiments.

We have seen that high energy neutrinos of different flavours, as high as $\sim 10^{20}$ eV, also accompany the nucleons, γ-rays, and leptonic pairs. Their detection would further extend the variety of diagnostics discussed in this chapter. Unfortunately, neutrino interaction cross sections are extremely low for most nuclei – however, not impossibly low, given that some neutrino flavours have resonant interactions with deuterons at higher energies (as in the SNO experiment), and with gallium at lower energies (as in the GALLEX experiment (Bahcall 1997)). A sufficient column density of water, as in the Superkamiokande experiment, or in the IceCube experiment near the South Pole (Abbasi *et al.* 2013), can improve the detectability of some interactions, making some experiments in neutrino astronomy and physics increasingly possible.

Atmospheric detection of UHE neutrinos can be identified from air showers of "grazing incidence" neutrinos that traverse the maximum amount of atmospheric mass. An overview of CR neutrino detectors and their relevance to high energy physics and astrophysics, including astrophysical magnetic fields, can be found in a comprehensive article by Becker (2008) and references therein. A briefer, likewise informative overview by Halzen & Klein (2008) is also recommended. Next we briefly describe some instrument developments that are providing further answers on intergalactic magnetic fields.

11.7   Detection instruments for UHECRs

The relatively small x_{loss} distances of extragalactic UHECR nucleons with $E \gtrsim 10^{19}$ eV are associated with a plethora of particle interactions. For $E \gtrsim 10^{20}$ eV the GZK interactions (Section 11.3) produce neutrinos of several flavours, mesons, and γ-rays, some with initial energies not far below those of the primary UHECR nucleon. All of this turns intergalactic space into a kind of particle physics "laboratory" at $E \gtrsim 10^{18}$ eV and investigation of the many related particle physics processes goes beyond the scope of this book. We have focused on some that can be related to the strength and structure of intergalactic magnetic fields.

Over the past 20 years, new instruments have been used, developed, and planned to detect UHECR nucleons, γ-rays, and neutrinos at the highest energy ranges. Pieces of information from each of these "messengers" of the UHE universe often complement each other. Just as the range of physics topics associated with Equations (11.5)–(11.9) goes beyond the scope of this book, so also does the range of instruments, planned and in use. A list in Fig. 11.1 of several UHECR nucleon detectors includes AGASA (The Akeno Giant Air Shower Array), KASKADE Grande, HiRes (The High Resolution Fly's Eye experiment), AUGER (The Pierre Auger Observatory), and the Telescope Array (TA).

An additional "messenger" of the high energy universe is low frequency radio (Čerenkov and curvature) radiation from atmospheric CR showers. Radio atmospheric Čerenkov pulses from CRs were first discovered by Jelley *et al.* (1965), and the early history of CR-radio detections is nicely reviewed by Weekes (2001). Fifty years after the earliest atmospheric Čerenkov pulse observations, radio techniques for UHECR analysis are undergoing a renaissance, with several new and planned experiments. For further reading on the LOPES radio detector, along with other current and planned low-frequency radio instruments, see Falcke (2007), and for descriptions of detection methods.

The number of recent and current detectors of these new messengers justifies a book in itself, and the interested reader can access information on them beginning with a web search on the names/acronyms mentioned in this book. The brief compilation in Fig. 11.1 gives only a starting list, especially for detectors of γ-rays and neutrinos. In summary, the expansion of experimental particle physics into space gives "access" to energy ranges and particle interactions that are sometimes well beyond what Earth-bound accelerators can produce. As mentioned at the outset, the detection techniques and instruments described here give relative emphasis to proton UHECRs at the highest observable energies because of their more direct connection to local-Universe IGM magnetic fields, e.g. through angular deflections.

11.8 Some additional comments on UHECR detection

CR protons can be directly detected with high altitude balloon and satellite experiments below $\sim 10^{14}$ eV, where their flux is relatively high (Fig. 11.1). Above these energies, as their fluence steadily decreases, large area air shower detectors are important. The column density of air is in the range ~ 500–800 gm cm^{-2} depending on the altitude of the detection device and the Zenith Angle of the CR primary's arrival direction. Especially above $\sim 10^{17}$ eV, a single UHECR initiates an intense cascade (shower) of elementary particles (photons, pions, muons, neutrinos, e^+e^-, etc.), which produce up to $\sim 10^{11}$ particles. These showers are what we detect from the ground, although the primaries could also be detected from near-space, given adequate collecting area. When the primary UHECR enters the atmosphere, the thickness of the resulting shower line distance, grows to a maximum at $X_{\max}$ (X in gm cm^{-2}), and then declines if X is sufficiently long. Detection and analysis of atmospheric showers is made either from the resulting atmospheric fluorescence, or by Čerenkov radiation. Fluorescence and Čerenkov radiation can also be observed in denser media such as water (e.g. the MILAGRO and HAWC γ-ray detectors) and ice (IceCube CR detector – The IceCube Collaboration, Achterberg *et al.* 2008). HiRes detects atmospheric fluorescence radiation (which is isotropically emitted) with angular resolution close to 1° in the 320–420 nm UV band, combined with a smaller fraction of atmospheric Čerenkov photons (which are directed).

The arrival direction, energy, and atomic species of UHECR primaries can be measured only *indirectly* (e.g. by detecting a nucleon in the shower and its Z, or γ-ray, photon, muon, or neutrino). To deduce the primary energy, species, and charge, etc. from the ground requires a complex modelling that incorporates the nuclear cross-sections of all particle species involved in the shower interactions. In addition, the geomagnetic field also influences the trajectories of the charged muons and lepton shower components. A proton can be distinguished from a heavy nucleus primary by the fact that proton-induced showers have a deeper X_{max} and smaller muon/electron ratio, and the muon shower develops relatively deeper into the atmosphere. Heavy nucleus primaries on the other hand have a smaller (higher up) X_{max}, a larger muon/electron ratio, and earlier development of the muon shower. An overview of X_{max} and $\sigma_{X\mathrm{max}}$ and their relation to nucleon composition estimates (still subject to changes) can be found in Letessier (2013).

Figure 11.8 summarises different types of atmospheric UHECR shower detection. Uncertainties in the modelling process and consequent varying results from different groups have been significant, particularly above 10^{19} eV, to the point where they have sometimes led to contradictory estimates of the primary UHECR energies.

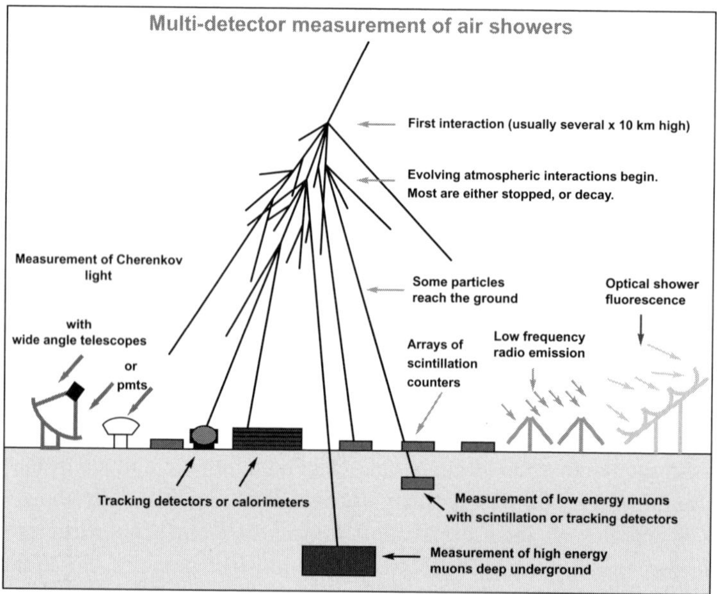

Figure 11.8. Illustration of the variety of detection methods for CR air showers. The left side shows photomultiplier (pmt) and optical telescope detectors of Čerenkov light. Other surface detectors (proceeding to the right) are: water tank surface scintillation counters, e.g. HAWC, tracking detectors and multi-layer (e.g. lead) calorimeters, antennas to detect lower frequency radio emission, and optical detectors and imagers of omni-directional shower fluorescence photons. Deep underground detectors can measure high energy neutrinos (e.g. the 1.5 km deep Ice Cube 3-D array at the South Pole (Abbasi et al 2008)), and high energy muons. (Figure adapted with permission from Andreas Haungs, Karlsruhe Institute of Technology.)

Given the energy-dependent arrival rates in events/unit area/year shown in Fig. 11.1, a major challenge is to collect and calibrate enough UHECR counts at the *largest* energies. Above $\sim 10^{19.5}$ eV the arrival rate is only ~ 1 event/km^2/year! This helps us to understand why, though the first 10^{20} eV CR was detected in 1962 by Linsley (1963 at the MIT Volcano Ranch Observatory), the CR energy spectrum at 10^{19}–10^{21} eV has become well defined only within the past 5 years. In order of their construction, some major UHECR nucleon observatories are the AGASA array in Japan (Yoshida *et al.* 1995), which closed in 2004, the HiRes detector array in Utah, USA (Boyer *et al.* 2002), the AUGER South array in Argentina (Cronin 1992, Abraham *et al.* 2004), and the Japan-USA Telescope Array (TA – Kawai *et al.* 2008).

The AGASA array consisted of scintillation counters on the ground to detect Čerenkov radiation from the shower. The newer HiRes-II uses two arrays of optical reflectors to detect and locate atmospheric fluorescence radiation from UHECR nucleons (and also neutrinos). The Japanese-American "Telescope Array" (TA) in Utah, has a ground array of 576 scintillation detectors combined with atmospheric fluorescence detectors. The AUGER detector (Fig. 11.9) combines 1600 ground water tanks containing Čerenkov radiation detectors covering 3000 km^2, with four atmospheric fluorescence detectors on the array's periphery. These "look" inward and upward at the same shower tracks that are simultaneously detected in the water-Čerenkov tanks. This combination, or "hybrid" provides important redundancy

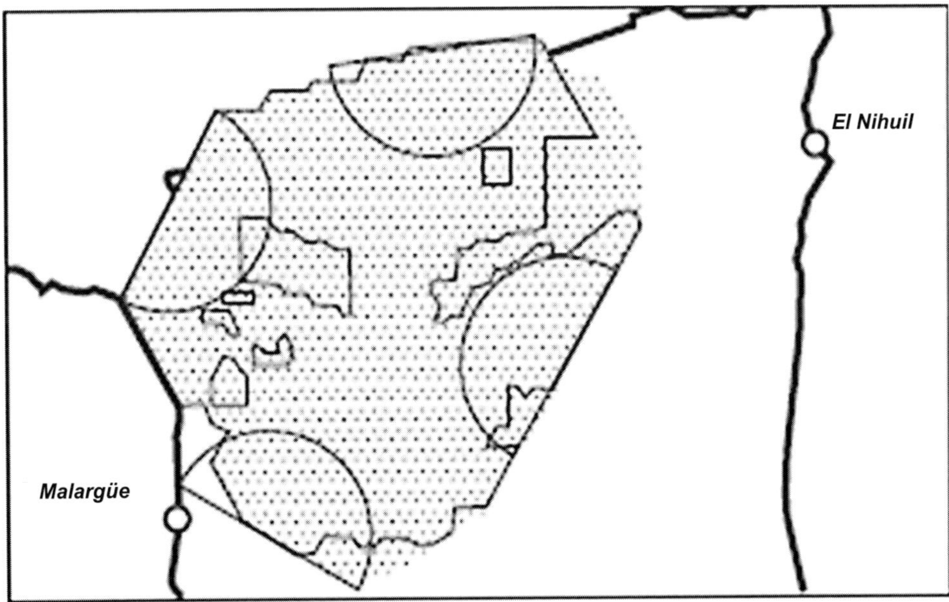

Figure 11.9. Sketch showing the layout of the AUGER cosmic ray detector array telescope near Malargüe, Argentina (The Pierre Auger Collaboration – Abraham *et al.* 2007). The fields of view of the four air fluorescence detectors on the periphery are illustrated by semicircles. Each dot represents a water Čerenkov detector. The hybrid nature (Čerenkov and fluorescence) of the instrument permits an angular resolution of $\sim 0.5°$. (Adapted from The Auger Collaboration – Abraham *et al.* 2007.)

and enhances the angular resolution down to ~30 arcminutes. It is described further in Sections 11.9 and 11.10 below.

Directional resolution is obviously important for measuring magnetic field deflection in the IGM and in the Galactic halo (Equations 11.1 and 11.2). As mentioned, we normally detect the secondary air shower particles from the ground, and these typically originate from a primary γ-ray photon, or a UHECR primary nucleon, whose energy, E, arrival direction (θ, φ), and nuclear charge, Z, must be derived from modelling of the observed shower particles and photons and their interactions.

If the UHECR detector were in space or on the moon, i.e. no atmospheric showers of secondaries, then E, (θ, φ), and Z of the primary CR particles could in principle be directly detected. But for a hypothetical balloon-borne detector or orbiting UHECR telescope, the required several thousand km^2 collecting area is an obvious problem. Space-based detectors that look down over an n,000 km^2 area of dark ocean are a promising alternative for atmospheric shower detection. Although such detectors have been proposed, they have not yet been implemented. A future large detector array on the moon could conceivably achieve the large collecting areas required to directly detect the primaries.

11.9   The Pierre Auger UHECR telescope

The large collecting area, dimensions, and the combination of arrival direction determinations – by simultaneous fluorescence and Čerenkov detectors – give AUGER an angular resolution potentially capable of analysing magnetic deflections of UHECR sources. The layout of AUGER (South) is shown in Fig. 11.9.

11.10   Recent astrophysical results from UHECR observations

If UHECRs are indeed associated with AGNs or extragalactic gamma ray bursters, then information of the kind displayed in e.g. Fig. 11.6 (and in Sections 11.10 and 11.11) leads to qualitative predictions of the arrival direction distributions for UHECR nucleons. Recall that at energies above ~$5 \cdot 10^{19}$ eV, the GZK interaction distance is less than ~100 Mpc. For reference, this is comparable to that of the Coma cluster of galaxies. Notably, the large radio galaxy Centaurus A (at 3.8 ± 0.1 Mpc (Harris, Rejkuba & Harris 2010)), falls well within the UHECR proton x_{loss} distance (Fig. 11.3). Thus, any extragalactic UHECR events at $E \gtrsim 5 \cdot 10^{19}$ eV can only likely come from a few relatively nearby "candidates". It also implies that measureable anisotropies near and above this energy range might be expected.

In the same context we expect an energy turnover, or cutoff, due to GZK losses above ~10^{19} eV. Because of the complexities of calibration indicated above, there have been contradictory claims since the early 1990s on whether this cutoff actually occurs. In fact, analysis of the AGASA array events had suggested an *increase* above ~$10^{19.5}$ eV – i.e. no GZK falloff. This alleged lack of a GZK falloff prompted further ideas that are consistent with no GZK cutoff, while still constraining the intergalactic magnetic field structure, but still within the framework of baryonic UHECRs. The "no GZK cutoff" claim had the effect of stimulating several alternative explanations for UHECRs, such as exotic heavy particles, magnetic monopoles, superconducting cosmic strings, hybrids of monopoles and strings (e.g. Farrar & Piran 2000). The ideas proposed cover some interesting themes in particle physics and cosmology. A good, brief summary of these theories, an introduction to some relevant literature, including a discussion of "cosmic necklaces" are given by Berezinsky &

Vilenkin (1997). Another "no GZK" possibility is consistent with a *source-intrinsic*, sharp fall-off above $\approx 10^{19.7}$ eV (Letessier-Salvon 2013), and for that reason could mimic photopion (GZK) interactions along the propagation path to us.

If extragalactic jets are the main UHECR accelerators, then their high energy cutoff would be determined by jet energetics (e.g. Lovelace & Kronberg 2013). All of this illustrates that a few alternative explanations may exist to explain the very highest energy observed count rates in Fig. 11.1.

The HiRes air fluorescence UHECR array (The High Resolution Fly's Eye Collaboration – Abbasi *et al.* 2008a), and the Auger array (The Auger Collaboration – Abraham *et al.* 2007) have more recently confirmed the evidence for a GZK-like cutoff above $10^{19.6}$ eV (Fig. 11.1). The turnover energy is slightly lower than straightforward theoretical expectation, which may be partly or entirely due to the intrinsic cutoff effect mentioned above. It is therefore not yet clear how this affects our physical understanding of p-$h\nu$ interaction physics at these energies. It may or may not remove the possibility of new, exotic particles and an entirely different framework of physical interpretation. More event statistics and composition determinations out to $\approx 10^{20.5}$ eV will do much to settle the uncertainties mentioned above. Around this energy range, the very slow event rates of $< \sim 1/\text{km}^2/\text{year}$ (Fig. 11.1) require more time and/or bigger collecting areas.

The UHECR trajectories could be bent within a segment $l < x_{\text{loss}}$ by a local intergalactic magnetic field. In any case, the relatively small number of expected events ought to be non-randomly distributed on the sky if, for example, magnetic fields only modestly change the direction of arrival from cosmologically nearby sources. Referring again to Fig. 11.3, we see that UHECRs in the range $10^{18} \lesssim E \lesssim 10^{19.5}$ eV have an x_{loss} of ~1 Gpc or more. In addition, within this much larger volume, we will see many sources with trajectories magnetically deflected by varying amounts. On this expectation, the UHECR nucleon arrivals in this energy range should "asymptote" to observed isotropy as the source distances increase beyond $\simeq 100$ Mpc and toward ~1 Gpc. A detailed experimental test of such statistical predictions should be possible in future. Richer statistics and better model and instrument resolution will permit better resolution in energy and atomic species composition of the primary events that are shown in Fig. 11.10 below.

Figure 11.10 shows the 69 AUGER UHECR events with energies above 6×10^{19} eV, collected up to 31st December 2009, and plotted on Galactic (l,b) coordinates. Virtually no events come from the direction of the Virgo cluster (20 Mpc distant), but there is an interesting apparent event clustering near the Centaurus A (NGC5128) radio galaxy at ~3.8 Mpc, which is the closest large AGN/radio galaxy to us. Another apparent "feature" in the AUGER 2009 data in Fig. 11.10 is a possibly significant deficit of UHECRs near $l \sim 260°$, $b \sim -40°$. More recent statistics to 2012 so far have reduced the statistical significance of these initially apparent effects. We also note that any angular coincidence with Cen A would be greatly degraded if the Z's of the UHECR nuclei were all $\gg 1$ (Equation 11.1). This implies a straightforward conclusion, namely that, for any high-energy anisotropy to exist, a significant fraction of protons must be present in the primary composition spectrum, relative to heavy nuclei such as He and Fe.

UHECRs from Centaurus A in this highest energy event range are relatively "pure" in the sense that they have not likely undergone significant B–H or GZK losses. Over time there have been persistent suggestions and also denials, of anisotropies associated with positions in the Veron-Cetty (2006) catalog of galaxies. This question was most recently revived in a

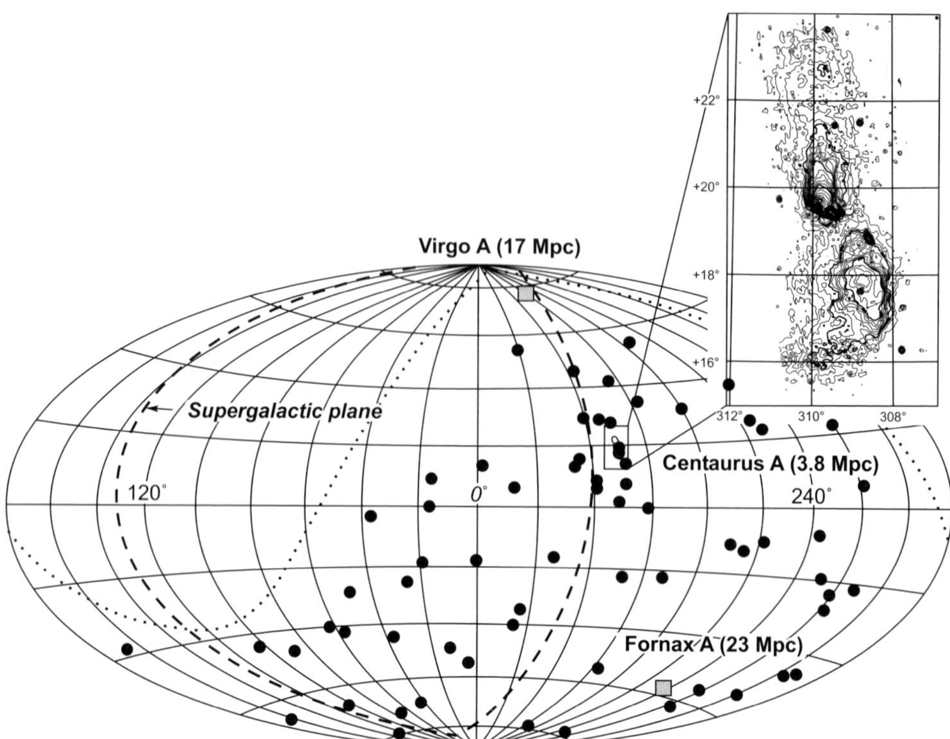

Figure 11.10. The arrival directions in Galactic coordinates of the UHECR events having $E \gtrsim 6 \times 10^{19}$ eV, detected by the Pierre Auger telescope in its first phase of operation up to 31 December 2009 (Abreu *et al.* 2010). Also shown are the locations and distances of three other relatively nearby AGN radio sources, the Supergalactic Plane (dashed line), and the instrument "horizon" of the AUGER array (dotted line). At these highest energies, there has been no clear association with the Supergalactic Plane, nor with Virgo A or Fornax A., although some recent re-analyses have revised the possibility of some correlation (see below). Centaurus A, the closest, large AGN-fed radio source, with its 0.6 Mpc size radio lobe structure does show a tentative association with a grouping of UHECR arrival directions (see Fig. 11.10). The inset 4.7 GHz image of Centaurus A (Junkes *et al.* 1993) is also shown in (l,b) coordinates.

new discussion of Pierre Auger highlights by Letessier-Selvon (2013). A typical coincidence acceptance criterion is $\Delta\theta \lesssim 3°$ given the current angular event density on the sky and instrumental resolution.

It is interesting that these first AUGER South AGN-UHECR statistics were not corroborated by a similar investigation in the Northern Hemisphere by the HiRes Collaboration (Abbasi *et al.* 2008b). Specifically, the latter find no correlation within their (similar) energy range. This discrepancy might be simply explained by the fact that the Northern (HiRes) sky has no candidate AGNs as close as Centaurus A in the Southern sky. One might argue that this is the only candidate source whose loss distance for GZK + BH pair production is $\ll x_{\mathrm{loss}}$ for the highest observed energy ranges (Fig. 11.3).

If we were to remove the UHECR events near Cen A (which are also in the Supergalactic Plane), visual inspection of Fig. 11.9 suggests that the AUGER correlation would be much weaker, a conclusion that would probably agree with the HiRes statistics. If the "close but not quite" coincidences with Centaurus A in the first published AUGER results in Fig. 11.10 are borne out with more count statistics, it would put us within range of estimating or constraining magnetic field strengths and structure in the nearby universe and the Galactic halo. This possibility is elaborated further in Section 11.11 – see also Fig. 11.11. The earlier angular scatter of UHECR events near Centaurus A would be consistent with modest deflections of their trajectories by up to $\pm\sim18°$ due to magnetic fields along the way. The fields could plausibly be a combination of a Milky Way-associated field and a weaker, more extended intergalactic field over the 3.8 Mpc between us and Centaurus A.

We have also discussed non-charged "messengers" such as γ-rays, neutral pions, neutrons, and neutrinos, whose x_{loss} distance can be much greater than for the protons above 6×10^{19} eV. In this sense, possibilities for directly identifying sources by pointing have not been exhausted. By extension, the prospect of future extragalactic magnetic deflection measurements seems promising. These aspects and the discussion in the previous paragraph are subject to a renewed inspection when we incorporate a different method of VHECR/UHECR analysis, described in Section 11.11 below.

The above overview of UHECR results has emphasised the AUGER and HiRes results because they are new and because they have already eliminated some previous ambiguities in previous, more limited data. For example, they have confirmed the "Ankle", and a cutoff at $\sim6 \times 10^{19.6}$ eV – see Fig. 11.1 and Letessier (2013).

It will be clear that the UHECR data at this time of writing still do not appear watertight on important questions of (1) anisotropies over the sky, (2) the true composition mix of the nucleons – e.g., what, if any, is the heavy nucleon component?, and (3) the lingering question of whether *Galactic* CR acceleration sites might, after all, be responsible for much of the UHECR sky. As mentioned, the arrival rate of only ~1 nucleon/km^2/century above 10^{19} eV has limited the rate of data gathering in this important energy range. An alternative conceptual framework for multi-parameter analyses of UHECR data is introduced in Section 11.11 below.

To conclude, the initial Auger and HiRes results, along with other VHE/UHE CR, neutrino and γ-ray detectors, have set the stage for a new era in astrophysics, which incorporates new "tools" for exploring the magnetic universe. A plethora of additional UHECR, γ-ray, neutrino and radio detectors at these highest energies are extending the exploration of the high energy Universe.

11.11   An intergalactic magnetic field probe based solely on UHECR measurements

A new analysis framework and graphical display has been introduced by Yüksel *et al.* (2012) for the interpretation of very high energy CR nucleon detections. A new type of VHECR/UHECR "all-sky" sphere is introduced to help clarify the multi-parameter aspects of high energy CR analysis: For the purpose of illustration, we assume that there is a single UHECR source which emits isotropically, and that this emitter is the dominant source of the observed event distributions such as in Fig. 11.10. The coordinate frame origin of this computed sky is *a priori* unknown. The VHECR/UHECR events can be modeled either as a single species such as hydrogen nuclei or some combination of H, He, Fe, etc. The sphere's radius can be assigned – for our purposes 3.8 Mpc, CenA's distance from the Earth. In this case we are located at some unknown position *on* this sphere – a 4π all-sky sphere (Fig. 11.11a (upper)) as it would be seen by an observer placed at Cen A.

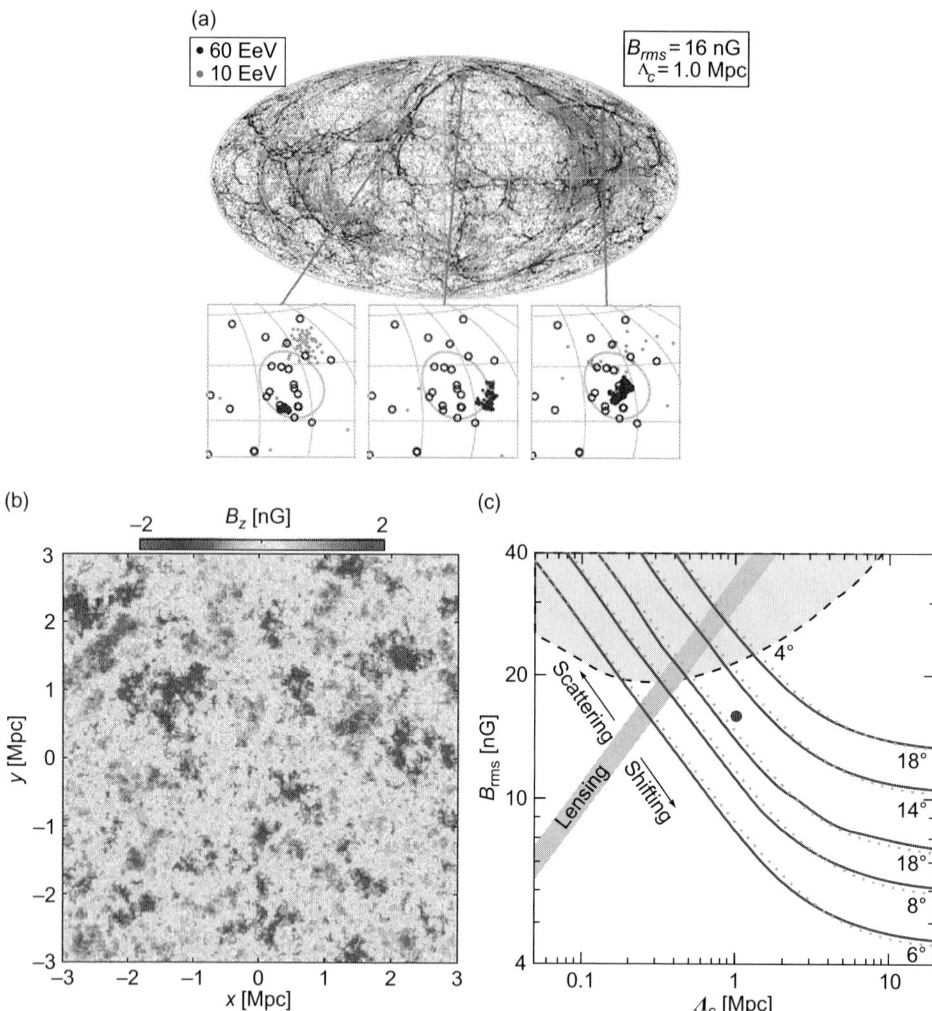

Figure 11.11. (a) *Upper*: Modelled UHECR arrivals on a "detector" sphere of radius 3.8 Mpc, centred on the Centaurus A radio galaxy, and which have propagated through a modelled magnetic field between Cen A and us. The UHECR are shown for isotropic emission at CenA at energies of 10 EeV (light grey dots) and 60 EeV (dark grey dots) for proton nuclei (Yüksel *et al.* 2012). (a) *Lower*: Computed (for 10 EeV and 60 EeV), and observed, CR arrival events for an observer placed at 3 shown upper image locations on the meta-sphere, now in (*l,b*) Galactic coordinates. Each shows a 13° circle that encloses the Cen A radio galaxy. Observed events (from AUGER, black circles) are compared with the computed events, selecting those at 10 EeV (light grey) and 60 EeV (dark grey). (b) A slice of the modelled 3-D magnetic field configuration (see text), projected onto the *x–y* plane. (c) For the adopted model parameters, a 2-D plot of computed magnetic field fluctuations in the Λ_B–Λ_C plane, and compared with UHECR data from the AUGER UHECR detector at epoch AD2010. It is described in the text how the relative predominance of "shifting" and "scattering" are inferred from the arrival event data, and how CR lensing (evident from the parameter combination chosen in (a)) can break the degeneracy of the oblique lensing zone in Fig. 11.11c to directly specify the strength and structure of the IGMF. This assumes that a concentration of events originating in one object is confirmed in future. (Figure components (a), (b), and (c) are reproduced and, in part, adapted from Yüksel *et al.* (2012).)

One, but not the sole purpose of this general framework is, for whatever number of emitters and their distances, to deduce the strength and structure of the intergalactic magnetic field. Fig. 11.11(a) shows the result of a UHECR propagation model within this 3.8 Mpc radius sphere. Here, a 3-D Kolmogorov spectrum was adopted to parameterise the IGM magnetic field fluctuation structure. To further simplify the number of parameters, two UHECR energy ranges were chosen in the propagation model, 1×10^{19} and 6×10^{19} eV, and these are denoted at the top left of Fig. 11.11(a). It is immediately evident that the CR arrival patterns on the sphere are very sensitive to CR energy, and the parameters of the IGM magnetic field.

Assuming further that the primary source of UHECRs is protons at these two chosen energies, we can imagine that most of the observed arrival events in Fig. 11.10 have been bent from their original propagation direction due to a fluctuating intergalactic magnetic field, described below, over the ~3.8 Mpc intervening propagation path (refer to Yüksel *et al.* 2012 for details).

If these assumptions are approximately correct, including the proton-only assumption, the above facts can be combined with a modeled IGM magnetic field strength and structure within 3.8 Mpc of the Sun in our Galaxy. The magnetic field strength was normalised to a quadratic mean value of 1 nG, and the (assumed Kolmogorov) spectrum of B fluctuations was truncated at maximum and minimum l-scales of 0.04 Mpc and 2 Mpc, respectively. A 2-D slice of this magnetic field configuration is shown in Fig. 11.11(b). For computations that produced the upper 4π pseudo-all sky shown in Fig. 11.11a (upper) , the B-coherence length, $\Lambda_{\rm c}$, was chosen to be 1 Mpc, and $B_{\rm rms}$ to be 16 nG, as indicated in the upper right corner. $\Lambda_{\rm c}$ and $B_{\rm rms}$ are model variables, and plots for three other combinations of $(\Lambda_{\rm c}, B_{\rm rms})$ can be found in Yüksel *et al.* (2012).

Two striking outcomes of this model are: (1) the very different arrival patterns on the sphere for the two chosen energy ranges –10 and 60 EeV, and (2) perhaps more significantly, the fact that the UHECR sky that we observe (Fig. 11.10) is not necessarily representative of the arrival direction statistics – which depend on $E_{\rm CR}$, $B_{\rm IGM}$, and Λc. This latter point may have been widely overlooked up to now.

Illustrating point (2) above we show, in Fig 11.11a (lower), segments of the *observed* (l,b) sky for three chosen locations of the Earth on the upper meta-sphere. Black circles show observed AUGER events relative to an $18°$ circle around Cen A's true location on the (l,b) sky, for each of three chosen locations on the (upper) meta-sphere. This illustrates how the *observed* arrival directions of the UHECR events can alter dramatically, depending on *our* location (upper plot) on the meta-sphere around Cen A. Also, as mentioned they illustrate how the event distributions are very energy sensitive over this 10–60 EeV range. To get around this problem, more resolution in energy space is required, especially at the high-E end, where magnetic deflections are minimized. For further details and discussion, the interested reader could consult Yüksel *et al.* (2012).

Furthermore, with more UHECR event data, an additional important clue to $B_{\rm IG}$ can be adduced: The prospect that the magnetic properties of the intergalactic (and Galactic halo) medium can be further specified using VHECR/UHECR data is illustrated in Fig. 11.11(c). This plots $B_{\rm rms}$ (in nanogauss) vs. $\Lambda_{\rm c}$ (in Mpc) and applies, in this case, within the 3.8 Mpc-radius meta-sphere in Fig. 11.11(a). Information contained in Fig. 11.11(c) can be complemented by the dark grey "caustic" zones in the 4π plot of arrival directions in Fig. 11.11a (upper centre). These represent relative concentrations of (in this case) of 60 Eev events due to magnetic focussing. Let us now imagine that an event pair is separated by only a few degrees. If the arrival energies of each event pair are the same, an estimate of B_{IGM} could be made from the implied r_{L} (or r_{g} – Equation 11.1). In fact, possible lensing effects might be inferred from the relative clustering in position of some event pairs seen in current AUGER UHECR data.

These have an average pair separation of ~4°(dashed line in Fig. 11.11(c)). Subject to other assumptions, this implies an approximate lower limit to the r.m.s. intergalactic magnetic field strength, $B_{\text{IGM}} \gtrsim 20$ nG between us and Centaurus A (Yüksel *et al.* 2012), in that it limits the relative predominance of event location "shifting", as opposed to "scattering". Finally, if magnetic focusing caustics, suggested for the model parameter combination in Fig. 11.11a (upper) can be verified, this could break the degeneracy in the "lensing" zone stripe of Fig. 11.11(c) and give an unambiguous estimate of the rms B_{IGM}. In the sense of what was already said, the above analysis procedures collectively provide a "starting template" for future fruitful analyses with many more VHECR/UHECR arrival events than what exist at the present time of writing.

In summary, the relationship between magnetic fields and UHECR propagation has yet to be fully understood. In the case of UHECRs above ~10^{19} eV, important answers appear to be gradually emerging. This highest energy range is where the accumulation of event counts is slowest! It is clear that several additional high energy messengers mentioned here and in Chapter 10 have yet to be better characterised, and they will eventually resolve some important uncertainties. In the long term we can be very optimistic, despite the "thickening plot" impression with the current VHECR/UHECR data that some readers might have gained at this point.

11.12　Concluding comments on the high energy universe of Chapters 10 and 11

It is helpful to recall that both GZK and Bethe–Heitler interactions have $x_{\text{loss}} \simeq 1$ Gpc, for UHECR protons between ~$10^{18.3}$ and ~$10^{19.6}$ eV (Fig. 11.3). Thus, if r_{L} for these protons is also large (low $|\boldsymbol{B}_{\text{IG}}|$, as in the large cosmic void regions), a variety of combined investigations of UHECR nucleons, and gamma rays (and neutrinos) could be made. Some detailed scenarios of IGM high energy cascades involving γ-rays, leptonic pairs and UHECR nucleons in multi-phase IGM environments have been outlined by e.g. Lee *et al.* (1995), Waxman & Coppi (1996), Waxman & Miralda-Escudé (1996), Elyiv, Neronov, & Semikoz (2009), and Neronov & Vovk (2010).

Instrumental detection of extragalactic γ-rays, X-rays and neutrinos is also a question of sensitivity, in particular the requirement for sufficient collecting area, which is usually expensive. It is hoped that the foregoing discussion might stimulate readers' to imagine further ways to probe intergalactic magnetic fields with large detectors using UHECR nucleons, γ-rays, and neutrinos. In some cases, observations can be done in conjunction with radio, IR, optical, and X-ray instruments. It underlines the importance that CR and high energy astrophysics are assuming in general, in a new era of exploration of intergalactic magnetic fields. Other recent discoveries arising from high energy particle and photon interactions were described in Chapter 10.

References

Abbasi, R. U. *et al.* (The High Resolution Fly's Eye Collaboration). 2008a, First Observation of the Greisen-Zatsepin-Kuzmin Suppression, *Phys. Rev. Lett.*, 100, 101101

Abbasi, R. U. *et al.* (The High Resolution Fly's Eye Collaboration). 2008b, *Search for Correlations between HiRes Stereo Events and Active Galactic Nuclei, arXiv*, 0804.0382

Abbasi, R. *et al.* (The IceCube Collaboration). 2013, All-Particle Cosmic Ray Energy Spectrum Measured with 26 IceTop Stations. *Astroparticle Phys.,* 44, 40

Abraham, J. *et al.* (The Pierre Auger Collaboration). 2004, Properties and Performance of the Prototype Instrument for the Pierre Auger Observatory, *Nucl. Inst. Meth. A,* 523, 50

Abraham, J. *et al.* (The Pierre Auger Collaboration). 2007, Correlation of the Highest-Energy Cosmic Rays with Nearby Extragalactic Objects, *Science,* 318, 938

Abreu, P. *et al.* (The Pierre Auger Collaboration). 2010, Update on the Correlation of the Highest Energy Cosmic Rays with Nearby Extragalactic Matter, *Astroparticle Phys.,* 34, 314

Abreu, P. *et al.* (The Pierre Auger Collaboration). 2011, The Pierre Auger Observatory I : The Cosmic Ray Energy Spectrum and Related Measurements. 32nd International Cosmic Ray Conference, Beijing, China. http://arxiv.org/abs/1107.4809

Achterberg, A. *et al.* (The IceCube Collaboration). 2008, IceCube Contributions to the XIV International Symposium on Very High Energy Cosmic Ray Interactions (ISVHECRI 2006), *Nucl. Phys. Suppl.* 175, 407

Antoni, T. *et al.* 2004, The Cosmic Ray Experiment KASCADE, *Nucl. Inst. Meth. A,* 513, 490

Apel, W. D. *et al.* (The KASCADE-Grande Collaboration). 2011, Kneelike Structure in the Spectrum of the Heavy Component of Cosmic Rays Observed with KASCADE-Grande, *Phys. Rev. Lett.,* 107, 171104

Bahcall, J. N. 1997, Gallium Solar Neutrino Experiments: Absorption Cross Sections, Neutrino Spectra, and Predicted Event Rates, *Phys. Rev. C,* 56, 3391

Becker, J. K. 2008, High Energy Neutrinos in the Context of Multimessenger Astrophysics, *Phys. Rep.,* 458, 173B

Becker Tjus, J. K. 2013, *Private communication*

Berezinsky, V. & Grigor'eva, S. I. 1988, A Bump in the Ultra-High Energy Cosmic Ray Spectrum, *Astron. Astrophys.,* 199, 1

Berezinsky, V. & Vilenkin, A. 1997, Cosmic Necklaces and Ultrahigh Energy Cosmic Rays, *Phys. Rev. Lett.,* 79, 5202

Biermann, P. L. & Strittmatter, P. A. 1987, Synchrotron Emission from Shock Waves in Active Galactic Nuclei, *Astrophys. J.,* 322, 643

Bird, D. J., *et al.* (The High Resolution Fly's Eye Collaboration). 1994, The Cosmic-Ray Energy Spectrum Observed by the Fly's Eye, *Astrophys. J.,* 424, 491

Boyer, J. H. *et al.* (The High Resolution Fly's Eye Collaboration). 2002, FADC-Based DAQ for HiRes Fly's Eye, *Nucl. Inst. Meth.,* A 482, 457

Cronin, J. W. 1992, Summary of the Workshop, *Nucl. Phys. B Proc. Suppl.,* 28, 213

Elyiv, A., Neronov, A., & Semikoz, D. V. 2009, Gamma-Ray Induced Cascades and Magnetic Fields in the Intergalactic Medium, *Phys. Rev. D.,* 80, 023010

Engel, R., Seckel, D., & Stanev, T. 2001, Neutrinos from Propagation of Ultrahigh Energy Photons, *Phys. Rev. D,* 64, 093010

Falcke, H. 2009, Radio Detection of Ultra-High Energy Cosmic Rays, in *Proceedings of the 30th International Cosmic Ray Conference*, ed. R. Caballero, J. C. D'Olivo, G. Medina-Tanco, L. Nellen, F. A. Sánchez, & J. Valdés-Galicia (Mexico City: Universidad Nacional Autonóma de México), 6, 79

Farrar, G. R. & Piran, T. 2000, Violation of the Greisen-Zatsepin-Kuzmin Cutoff: A Tempest in a (Magnetic) Teapot? Why Cosmic Ray Energies about 10^{20} eV May Not Require New Physics, *Phys. Rev. Lett.* 84, 3527

Feain, I., Ekers, R. D., Murphy, T., Gaensler, B. M., Macquart, J.-P., Norris, R. P., Cornwell, T. J., Johnston-Hollitt, M., Ott, J., & Middelberg, E. 2009, Faraday Rotation Structure on Kiloparsec Scales in the Radio Lobes of Centaurus A, *Astrophys. J.,* 707, 114

Gaisser, T. K. 1990, *Cosmic Rays and Particle Physics*, (Cambridge: Cambridge University Press)

Greisen, K. 1966, End to the Cosmic Ray Spectrum?, *Phys. Rev. Lett.,* 16, 748

Halzen, F. & Klein, S. R. 2008, *Phys. Today,* 61, no. 5, 29

Harris G. L. H., Rejkuba, M., & Harris, W. E. 2010, The Distance to NGC 5128 (Centaurus A) *PASA,* 27, 457

Hill, C. T. & Schramm, D. N. 1985, Ultrahigh-Energy Cosmic-Ray Spectrum, *Phys. Rev. D,* 31, 564

Hillas, A. M. 1984, The Origin of Ultra-High-Energy Cosmic Rays, *Ann. Rev. Astron. Astrophys.,* 22, 425

Hörandel, J. R. 2008, Cosmic-Ray Composition and Its Relation to Shock Acceleration by Supernova Remnants, *Adv. Space Res.,* 41, 442

Icetop – Abbasi, R. *et al.* (The IceCube Collaboration). 2013, All-Particle Cosmic Ray Energy Spectrum Measured with 26 IceTop Stations. *Astroparticle Phys* 44, 40

Jelley, J. V., Fruin, J. H., Porter, N. A., Weekes, T. C., Smith, F. G., & Porter, R. A. 1965, Radio Pulses from Extensive Cosmic-Ray Air Showers, *Nature,* 205, 327

Junkes, N., Haynes, R. F., Harnett, J. I., & Jauncey, D. L. 1993, Radio Polarization Surveys of Centaurus A (NGC 5128). I - The Complete Radio Source at 6.3 CM, *Astron. Astrophys.,* 269, 29

KASCADE-GRANDE Collaboration, The, Apel, W.D. *et al.* 2013, "KASCADE-Grande measurements of energy spectra for elemental groups of cosmic rays" arXiv:1306.6283 (Phys. Rev. D87, 081101 ?)

Kawai, H. *et al.* (The TA Collaboration). 2008, Telescope Array Experiment, *Nucl. Phys. Proc. Suppl.*, 175, 221

Lee, S., Olinto, A. V., & Sigl, G. 1995, Extragalactic Magnetic Field and the Highest Energy Cosmic Rays, *Astrophys. J.*, 455, L21

Letessier-Selvon, A., & Stanev, T. 2011, Ultra High Energy Cosmic Rays , *Rev. Mod. Phys*, 83, 907

Letessier-Selvon, A., For the pierre auger collaboration 2013, Highlights from the Pierre Auger Observatory, *Proc. 33rd ICRC*, Rio de Janeiro (arXiv1310.4620)

Linsley, J. 1963, Evidence for a Primary Cosmic-Ray Particle with Energy 10^{20} eV, *Phys. Rev. Lett.*, 10, 146

Lovelace, R. V. E., & Kronberg, P. P. 2013, Transmission Line Analogy for Relativistic Poynting Flux Jets, *MNRAS*, 430, 2828

Nagano, M., Hara, T., Hatano, Y., Hayashida, N., Kawaguchi, S., Kamata, K., Kifune, T., & Mizumoto, Y. 1984, Energy Spectrum of Primary Cosmic Rays Between $10^{14.5}$ and 10^{18} eV, *J. Phys. G*, 10, 1295

Navarra, G. *et al.* 2004, KASCADE-Grande: A Large Acceptance, High-Resolution Cosmic-Ray Detector up to 1018 eV, Nucl. *Instr. Meth. Phys. Res. A*, 518, 207

Neronov, A. & Vovk, I. 2010, Evidence for Strong Extragalactic Magnetic Fields from Fermi Observations of TeV Blazars, *Science*, 5974, 73

Ostrowski, M. 2002, Mechanisms and Sites of Ultra High Energy Cosmic Ray Origin, *Astroparticle Phys.*, 18, 229

Protheroe, R. J. & Johnson, P. A. 1996a, Propagation of Ultra High Energy Protons and Gamma Rays Over Cosmological Distances and Implications for Topological Defect Models, *Astroparticle Phys.*, 4, 253

Protheroe, R. J. & Johnson, P. A. 1996b, Propagation of Ultra High Energy Protons and Gamma Rays Over Cosmological Distances and Implications for Topological Defect Models (Astroparticle Physics 4 (1996) 253) [Erratum], *Astroparticle Phys.*, 5, 215

Rachen, J. & Biermann, P. L. 1993, Extragalactic Ultra-High Energy Cosmic-Rays - Part One - Contribution from Hot Spots in Fr-II Radio Galaxies, *Astron. Astrophys.*, 272, 161

Ryu, D., Kang, H., Hallman, E., & Jones, T. W. 2003, Cosmological Shock Waves and Their Role in the Large-Scale Structure of the Universe, *Astrophys. J.*, 593, 599

Sigl, G., Miniati, F., & Enßlin, T. E. 2003, Ultrahigh Energy Cosmic Rays in a Structured and Magnetized Universe, *Phys. Rev. D*, 68, 043002

Simpson, J. A. 1983, Elemental and Isotopic Composition of the Galactic Cosmic Rays, *Ann. Rev. Nuclear Particle Phys.*, 33, 323

Stanev, T. 2008, *private communication*

Stanev, T., Engel R., Mücke, A., Protheroe, R. J., Rachen, J. 2000, Propagation of Ultrahigh Energy Protons in the Nearby Universe, *Phys. Rev. D*, 62, 093005

Waxman, E. & Coppi, P. Delayed GeV–TeV Photons from Gamma-Ray Bursts Producing High-Energy Cosmic Rays, *Astrophys. J.*, 464, L75

Waxman, E. & Miralda-Escudé, J. 1996, Images of Bursting Sources of High-Energy Cosmic Rays: Effects of Magnetic Fields, *Astrophys. J.*., 472, L89

Weekes, T. C. 2001, Radio Detection of Cosmic Ray Extensive Air Showers, in *Radio Detection of High Energy Particles, Proceedings of First Int. Workshop RADHeP 2000, Los Angeles, California, 16–18 November 2000*, ed. D. Saltzberg & P. Gorham, AIP Conf. Ser. (New York: American Institute of Physics), 579, 3

Yoshida, S. & Teshima, M. 1993, Energy Spectrum of Ultra-High Energy Cosmic Rays with Extra-Galactic Origin, *Prog. Theoret. Phys.*, 89, 833

Yoshida, S. *et al.* 1995, The Cosmic Ray Energy Spectrum Above 3×10^{18} eV Measured by the Akeno Giant Air Shower Array, *Astroparticle Phys.*, 3, 105

Yüksel, H., Stanev, T., Kistler, M. D., & Kronberg, P. P. 2012, The Centaurus a Ultrahigh-Energy Cosmic-Ray Excess and the Local Extragalactic Magnetic Field, *Astrophys. J.*, 758, 16

Zatsepin, G. T. & Kuz'min, V. A. 1966, Upper Limit of the Spectrum of Cosmic Rays, *JETP Lett.*, 4, 78

12

Magnetic fields in cosmologically distant galaxy systems

12.1 Magnetic fields associated with absorption line systems in quasars

12.1.1 Introduction and some background

The Galaxy-corrected Faraday rotation measure of a distant quasar can be broadly divided into five or more unrelated components in Equation (1.3). These are (1) RM that is intrinsic to the source (IRM); (2) a component added by some *intervening* galaxy system (e.g. galaxy disc or halo, galaxy group, or cluster of galaxies); (3) intergalactic overdense zones such as an IGM filament of large scale structure (LSS); (4) that due to an all-pervading, cosmologically scaled magneto-ionic medium; and (5) various processes accompanying both the formation and evolution of the first stars and galaxies have been proposed to either seed, or amplify the first magnetic fields in the post-Recombination Universe. This could happen during the infall collapse to the first stars via associated shearing and turbulent amplification. An additional source could be in the fireballs of the first supernovae or hypernovae and/or GRB sources. The shock regions of these first stellar explosions might, too, amplify or seed a magnetic field. In studying these processes relativistic two-stream instabilities can be modelled, and have been described. More than one of these indicate candidates for early, post-Recombination amplification or seeding of IGM magnetic fields.

Given a combination of estimated n_e, with an estimated or measured spectrum of field reversals, the RM associated with any of these five components could give an estimate of, or limit to the associated magnetic field. The RM from a *widespread* IGM (4) appears below current levels of detectability by conventional radio methods, though not by γ-ray techniques, discussed in Chapter 10. We focus here on what can be learned about magneto-ionic gas in *discrete* intervening systems at intermediate-to-large redshifts – those galaxy systems that can produce absorption lines at z_a between us and z_e, that of the background emitter, e.g. a distant quasar or galaxy. In some cases the intervening material can be near z_e, and associated with the background quasar.

In the 1960s accurate radio interferometry (of 3C48) and lunar occultation techniques (on 3C273) revealed the precise coincidence of a strong radio source (quasi-stellar radio source, or "quasar") with a luminous star-like object, whose optical spectrum showed it to be extragalactic (Hazard *et al.* 1963, Schmidt 1963). The terms "QSO" and "quasar" were thereby introduced into the literature. In this book we use both terms, sometimes interchangeably.

Absorption lines at intermediate redshifts, z_a, in the optical, infrared, and at 21 cm, reveal quasar lines of sight passing through intervening galaxy systems. When this happens, the column densities, N_e, and excitation conditions can be inferred from the absorption species and their line equivalent widths, W, at z_a from suitably chosen absorption line transitions. In a combined analysis of available quasar RMs and absorption line data for the same sources, clear correlations have been found between high column depth absorption at z_a ($\ll z_e$) and the tendency for the background quasar at z_e to have an excess RM (Kronberg & Perry 1982, Welter, Perry, & Kronberg 1984, Watson & Perry 1991, Bernet *et al.* 2008), thereby confirming the existence of magnetic fields in galaxy systems at large redshifts. Where estimates can be made of the column density of *free electrons*, N_e, one could make the first crude estimates of magnetic field *strengths* in intervening systems. The intervenor magnetic field strengths are scaled by N_e, as shown in Equation (12.1), and they vary from a few μG to nearly a milligauss.

To relate B more closely to observable quantities, it is convenient to re-arrange the Faraday rotation expression in Equation (2.7), replacing $\int n_e dl$, the electron density n_e and physical depth, l, by the electron *column* density, N_e, in an intervening cloud at z_a. That is, the observable RM and the observationally derivable N_e, based on the measured absorption line spectrum, can be combined and related to the cloud's line-of-sight weighted magnetic field as follows:

$$<B_\parallel> = 2.6 \times 10^{-13} RM \, N_e (1 + z_a)^2 \quad \text{G},\tag{12.1}$$

where N_e is in cm^{-2}. $<B_\parallel>$ is defined by

$$< B_\parallel > = \frac{\int n_e B_\parallel dl}{\int n_e dl},\tag{12.2}$$

where n_e is in cm^{-3} and l in pc.

At most a small number of large galaxies will intersect a typical line of sight out to $z_e \sim 3$. The present-epoch density of galaxies having halos of $r \sim 45\ kpc$ is $0.017 h_{75}$ Mpc^{-3} (e.g. Burbidge *et al.* 1977), and a few quasar lines of sight can be expected to have a intersections with large, high column density galaxies. A detectable rotation measure normally requires an intersection column density of at least $N_e \sim 10^{20}$ cm^{-2}, if the galaxies contain μG-level fields.

Equation (12.1) can be taken a step further to estimate the most likely actual magnetic field strength in a small number of clouds, having running number (i), and assumed to have similar N_e^i and RM^i, which are intersected by the sightline to a background quasar. We assume that the intervenor redshift range, Δz_a^i, is not cosmologically large. The RM^is add algebraically, so that the average magnetic field in a single, "i th" cloud can be approximately expressed as

$$<|B_c|> \sim \frac{8.4 \times 10^{12}(1 + <z_a>)^2 |RM|}{i^{1/2} N_e} \quad \text{G}\tag{12.3}$$

(Kronberg & Perry 1982). As above, the N_e^i's are assumed to be similar, and the field directions of $\boldsymbol{B}_i$ have an expectation value of 0.5 for $\cos \theta$ over i randomly oriented cloud magnetic field directions. The RM and the column density are assumed not to be averages over very *different* sightlines to an extended background quasar. This would cause a misleading estimate of $<|B|>$ in Equation (12.1). As an example, the *RM* in one or more of the sightlines in an extended image might be significantly larger than in the integrated *RM*.

12.1.2   *Cross section, optical depth, and Faraday depth of intervening clouds to quasars*

Here we discuss the column depth, and optical depth, of lines of sight to, and/or through a quasar, which could be an extended radio source.

It is possible, even likely, that some radio structures extend well beyond the scale represented by the emission line column at z_e, and thus be only partially relevant to the sightlines where the *RM* is generated. Another possibility is that the Faraday rotating magnetoplasma contributing to N_e is very hot, i.e. $\gtrsim 10^{6.5}$ K, characteristic of a hot galactic halo, or the intracluster gas of a galaxy cluster or group. This would make typical *optical* absorption transitions very weak, effectively "hiding" the gas. The result is that $<|B|>$ in the intervening cloud would be overestimated. On this point, we note that an intervening cloud in this temperature range would be detected in thermal X-rays in the 0.1–1 keV range – which is how the hot ICM in galaxy clusters is normally detected. Note that clusters are relatively rare compared to galaxy discs and halos (i.e. they have a cosmologically low n_{c0} in Equation (12.5) below).

In conclusion, sightlines at z_a of *RM* and optical absorption may differ in different ways, and this needs to be kept in mind in statistical studies that compare integrated *RMs* and absorption lines. The number of sources with detailed $RM(\alpha,\delta)$ images that are directly resolved within the angular size of the intervenor at z_a is relatively small, since sub-arcsecond-level resolution is typically needed for this. Some examples that qualify on these criteria are illustrated in Section 12.3. Most of the following discussion deals with interpretations of the *integrated RMs*, which are available in relatively large numbers. In a statistical sense, the projected size of many polarised background radio sources at z_e may be fairly representative of the bundles of ray paths through a modestly extended intervenor at z_a.

More precisely said, there is a probability, $P(z_a)$, that a cloud at z_a (e.g. a galaxy disc or halo) will intersect the path between the observer and a background quasar at z_e. The intersection depth to the background object at z_e (as distinct from a Faraday depth) can be straightforwardly represented as an *optical depth*:

$$\tau(z_e) = \int P(z_a)\mathrm{d}z. \tag{12.4}$$

In a simple model, the space density of clouds at z_a is

$$n_c(z_a) = n_{c0}(1 + z_a)^3. \tag{12.5}$$

We can also define a dimensionless cross-section at $z = 0$,

$$A_{0a} = \frac{cn_{0a}\Sigma_{0a}}{H_0}, \tag{12.6}$$

and parameterise its cosmological evolution as

$$\Sigma_a(z) = \Sigma_{0a}(1 + z)^{\eta_a}. \tag{12.7}$$

Σ_{0a} is the intervening clouds' average physical cross section at $z = 0$, having space density n_{0a}. The probability that a quasar line of sight will intersect an absorbing cloud is (adopting, in this case, a Friedmann universe model)

$$P_a(z)\mathrm{d}z = \frac{A_{0a}(1 + z)^{(1+\eta_a)}\mathrm{d}z}{(1 + 2q_0)^{1/2}} \tag{12.8}$$

where a, as above, denotes an absorbing (not Faraday rotating) intervenor, and η_a allows for any cosmological evolution of the intervenor cross-section (see Khare-Joshi & Perry 1982, Welter *et al.* 1984), and q_o is the cosmological deceleration parameter.

12.2   Probes of cosmological magnetic field evolution using quasar Faraday rotation statistics

We now extend the basic concepts of *absorption depth* associated with Equations (12.4)–(12.8) to Faraday depths, *Faraday depth* cross-sections, and magnetic fields that we wish to probe over a wide range of z_a.

Magnetic field "*RM* signals" from a widespread, all-pervading intergalactic plasma or cosmic void above $\sim 10^{-9}$ G (co-moving) have been searched for but not yet detected in very distant source *RM*s. A possible exception is in a relatively dense nearby filament if the Perseus-Pisces supercluster, where $<|B|>$ may be 10^{-7}–10^{-8} G. Another exception has been the detection of diffuse intergalactic synchrotron radiation in a ~2 Mpc region of galaxy overdensity in the Coma cluster region of the "Great Wall", where the implied B_{IG} is in a comparable range (see Fig. 10.1).

However, we showed in Chapter 10 how intergalactic γ-ray observations can probe B_{IG} at levels of $\ll 10^{-12}$ G, and even as weak as $\sim 10^{-18}$ G! This is far below what *RM*-based probes can achieve, and sets an important context here, distinguishing (for example) widespread IGM and void fields from those associated with collapsed systems such as galaxies and galaxy clusters which were discussed in earlier chapters.

The following sections focus on *discrete* "magnetised cloud systems", associated with galaxy systems that have collapsed out of the IGM at some past cosmological epoch. Similar to the Milky Way and galaxy clusters, these typically possess μG-level magnetic fields. Population models can be constructed, or simulated, for galaxies and tested against *RM* observations out to significant z_a lookback times in the Universe.

12.2.1   *Models of Faraday intervenors using only the radio and redshift data*

Here we extend the concept of the summed *RM* from a small number of intervenors in Equation (12.3), to describe the collective *RM* statistics for populations of Faraday rotated radio sources over the largest observable redshift range. There are two reasons why the more familiar optical depth concepts of Equations (12.4)–(12.8) are not sufficient to describe probes of Faraday depth. First, contributions to Faraday depth, unlike optical depth, can be negative as well as positive (Chapter 2). Secondly, a Faraday rotation generated at $z > 0$ is reduced in the observer's frame by a significant factor of $(1 + z)^{-2}$. To illustrate this, a Faraday rotation generated at $z = 2$ is reduced for the observer at $z = 0$ by a (dramatic) factor of 9! The "–2" exponent straightforwardly arises from the Faraday rotation law (Equation 2.7). Thus, a significant *RM* detection in an extragalactic system can imply a substantial increase in the magnetic field strength estimate at any significant z_a (Equations 12.1 and 12.3).

In the following, we initially consider the case of an *intervening* population of Faraday rotating clouds, and ignore the "intrinsic" Faraday rotating clouds close to z_e. One assumes that the former, at z_a^i, are not correlated with Faraday rotation at z_e, whereas the latter might be correlated with the background host quasar environment properties.

By analogy with P_a in Equation (12.8) for absorption line clouds, we can define $P_r(z)dz$ for Faraday rotation r (a short notation for Galaxy-corrected *RM*, or *RRM*) as the probability that the radiation from a quasar at z_e will intersect a Faraday active cloud and undergo an incremental *RM* (+ or −) while passing from z to z-dz. Following Welter *et al.* (1984), we define $R_C(r, z)dr$ as the *fraction* of clouds at some z ($<z_e$) whose *RM* lies between r and $r + dr$, as measured *at* $z = 0$. It includes the $(1 + z)^{-2}$ dilution factor. The form of $R_C(r, z)$, which is model parameter free, is determined by the magneto-ionic and local geometric structures of all members of an ensemble of intervening clouds. The subscript C denotes cloud.

The probability density function ($R_e(r, z_e)$) for the total *RM* due to the intervenors between us and z_e must include the sum over all combinations of RM_i (r_i). These, in turn, must be summed over all combinations of z_a. For example, two clouds of comparable RM_i at two large z_a's (geometrically more likely to occur) will generate a smaller combined *RM* than the same two clouds at low z_a (less probable). This applies also for no cosmological evolution of $<RM_i>$. $R_e(r, z_e)$ is the observed distribution of the total rotation measure for radiation emitted by sources at z_e. It is a key tool for analysing the origin and evolution of discrete Faraday rotating clouds in intervenors (mostly galaxy-scale). These are illuminated by a background polarised source population over a range of z_e.

From the observed distribution $R_e(r, z_e)$ the goal is to deduce the *probability density function* for sources at z_e, for n intervenor intersections. It can be defined as $R_{en}(r, z_e)$, again following Welter *et al.* (1984). It is related to $R_C(r, z)dr$, above, as follows:

$$R_{en}(r, z_e) = \tau_e^{-n} C_n \int_0^{z_e} R_C(r, z) P_r(z) dz, \tag{12.9}$$

where the *Faraday* intersection depth is

$$\tau_e = \int_0^{z_e} P_r(z) dz. \tag{12.10}$$

This is the analogue of the optical depth in Equation (12.4). C_n [F(r)] represents the convolution on r of any function with itself ($n - 1$) times. In this case it is the integral in Equation (12.9).

The derivation of the observed $R_e(r, z_e)$ can be found in the appendix of Welter *et al.* (1984). The result is

$$R_e(r, z_e) = f_0(z_e)\delta(r) + \sum_{n=1}^{\infty} f_n(z_e) R_{en}(r, z_e) \tag{12.11}$$

where

$$f_n(z_e) = \frac{\tau_e^n e^{-\tau_e}}{n!}. \tag{12.12}$$

We note that in this general analysis described above, there are no *a priori* assumptions on the form of $R_C(r, z)$.

Finally, in analysing real data it can be both convenient and instructive to use the *cumulative* distribution function of *RM*, I_e:

$$I_e(r, z_e) = \int_{-r}^{r} R_e(r', z_e)\mathrm{d}r'. \tag{12.13}$$

As long as the number of intersections (i.e. τ_e – Equations 12.10 and 12.12) is not too large, there will be a detectable fraction, f_0, of zero intersections. The f_0 term in the summation in Equation (12.11) represents specifically the component of $R_e(r, z)$ that experiences no intersections with an *RM* intervenor. This "no intersection" subset can place important constraints on the interpretation of *RM* probes at cosmologically significant distances. The f_0 subset can be identified in the actual data, as illustrated below with examples of actual observed $R_e(r, z)$ data in Fig. 12.2.

It follows that if a significant *quasar-local RM* exists at $z \approx z_e$ for *all* quasars then, in principle, we might not detect any "zero intersections", and this is testable, as just described. But we are dealing in this section with *RRM* (or "*r*") intervenors at z_a^i, not at z_e. Of course for modest *RMs* at z_e, say $r \lesssim 30$ rad m^{-2}, the strong dilution factor of $(1 + z_e)^{-2}$ at very large z_e will tend to reduce them to the eventual point of undetectability in the observed $R_e(r, z)$ data. Furthermore, the above analysis has been laid out for the noise-free case. A fully realistic $R_e(r, z)$ dataset would need to be deconvolved by a function representing measurement uncertainties and other "noise" such as the astrophysical source-to-source scatter in r.

Figure 12.1 shows sample model plots of the components of R_e (Equation 12.9–12.11) for the error-free case, again from Welter *et al.* (1984), and then of the cumulative probability distribution I_e (Equation 12.13).

The next figure shows a 3-D plot of actual observed $R_e(r, z)$ data, to illustrate a typical *RM*–redshift probe using a recent data sample.

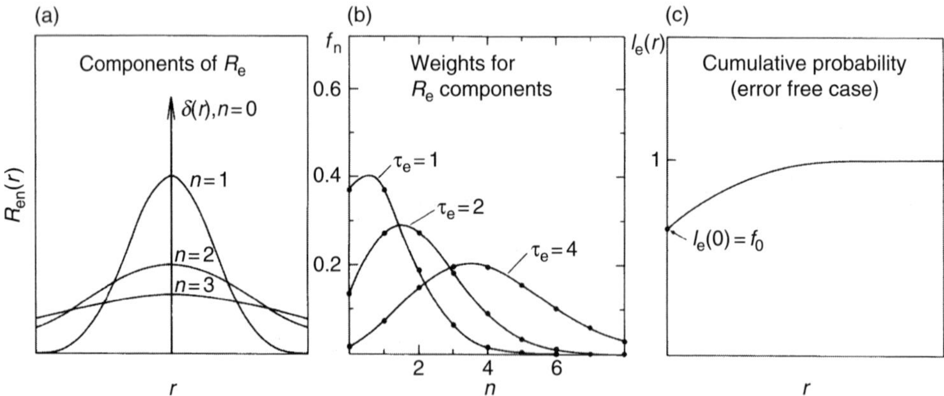

Figure 12.1. Left: A sample summation of the components of R_{en} (Equation 12.9), summed for 1, 2, and 3 cloud intersections in R_e (Equation 12.11). Centre: The weights, f_n, for 3 values of τ_e (Equations 12.11 and 12.12). Right: The cumulative probability plot $I_e (r)$ as a function of rotation measure (r), for the error-free case (see Equation 12.13). (Source: Welter, Perry, & Kronberg (1984).)

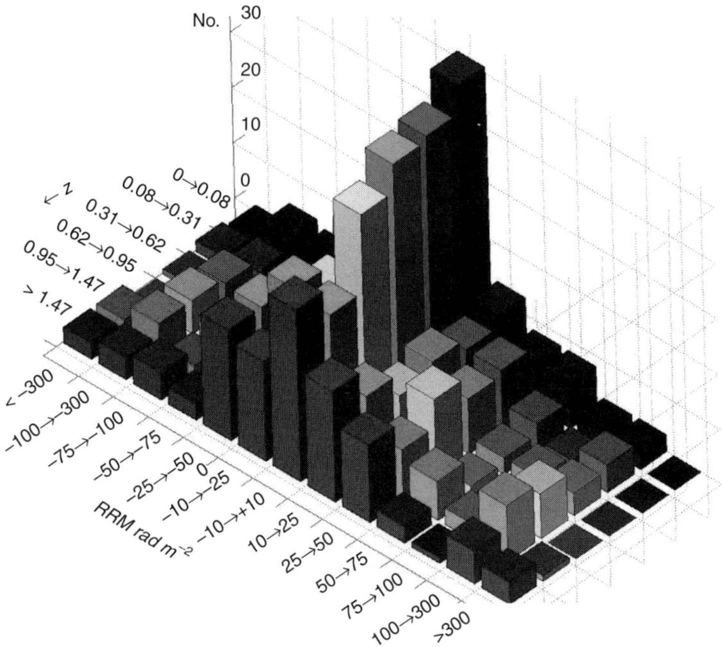

Figure 12.2. Histograms of actually observed, Galaxy-corrected Faraday rotation distributions (*RRM*, or *r*) for extragalactic radio sources out to $z \simeq 3$, representing an experimental approximation to R_e (r, z), in which the typical error of best straight line fit on the individual *RM*'s is $\mathcal{O}$ 2 rad m^{-2}. They are drawn from 2200 accurately determined, integrated emission *RM*s (r) (see text). The distributions are coded by redshift range.

The experimental representation of R_e (r, z) in Fig. 12.2 reveals some features that may be emerging for the first time with more accurately determined *RM*s. They illustrate aspects of Equations (12.9)–(12.11) and some additional phenomena, discussed below. The f_0 component near $z = 0$ is clearly apparent. It is broadened by an amount close to a typical, independently estimated source-to-source scatter of 10–15 rad m^{-2}. Another notable feature is the *form* of the $R_e(r)$ distributions: At $z \simeq 0$: they are clearly non-Gaussian, having very broad wings and other discernible outlier components. Thus, fits of e.g. Gaussian variances or Lorenzian widths to describe the z-evolution of R_e (r, z) have limited usefulness – though in the past they have been applied, in the absence of more definitive $R_e(r, z)$ data.

With future analyses in mind, all of the above underlines the importance of using non-parametric mathematical frameworks for interpreting complex *RM* behaviour over a large z-range. A basic reason for this is the complex mix of several influences and their redshift dependences, each of which can alter the *form* of $R_e(r, z)$. We now summarise some of them, referring to Figs. 12.1 and 12.2.

First, evolution can occur in the Faraday depth cross-section $\Sigma_r(z)$ by analogy with Equation (12.7), as well as in the strength and ordering of B within their evolutionary z-dependent cross-sections. This means, that the distributions, and the Faraday depths of increasing numbers of individual intervenors with z, will affect $R_e(r, z)$ for astrophysical reasons alone. This includes the possible evolution of the quasar-intrinsic clouds that we have left aside this point. To these, for high z we must add the effects of the cosmological geometry and the $(1 + z)^{-2}$ effect of wavelength transformation. As mentioned, the latter becomes very significant as $z \gg 1$. The astrophysical components of $R_e(r, z)$ evolution are not *a priori* predictable, and require additional measurement-based information, such as optical/IR spectra and imaging in the radio (including *RM*), and in IR, millimetre, optical, and X-ray bands.

Returning to Fig. 12.2, the relative amplitude of the "f_0 peak" appears to decrease beyond $z \approx 0.6$, at the expense of larger, but not extremely large, r values in the wings. This could be interpreted as genuine epoch evolution in Faraday depth and/or cross section of the intervenor population, or quasar-local "intrinsic" clouds. In the $z = 0.6$–1.5 range in Fig. 12.2 there appear to be fewer "zero r" intersections; Rather, a noticeable broadening occurs (i.e. relatively higher *RM*s in the range |20|–|60| rad m^{-2}). This could indicate higher co-moving cross sections and/or magnetic field strength evolution as cosmic look-back time increases within these ranges.

Proceeding to the $z \gtrsim 1.5$ range in Fig. 12.2, the f_0 peak appears to reverse itself; that is, to rise again at the higher z's. The *RM* "wings" discussed above cease to grow with z, and show a possible tendency to shrink. However, extreme *RM* outliers, though reduced by wavelength transformation, are still in evidence. A case in point is 3C191 at $z = 1.95$, shown in Fig. 12.4. Confirmation of all these trends will require more *RM* measurements both in this range and to $z \sim 4$.

If the above trends are confirmed in future *RM* data, they could be straightforwardly understood in the following way, in terms of the R_e models discussed above: As z further increases beyond $z \approx 1.5$, any co-moving cross section and/or Faraday depth increase of intervening clouds are further reduced by the strong $(1 + z)^{-2}$ factor – a reduction of 12.3 times at $z = 2.5$, which is close to the z limit reached in Fig. 12.2. That means that as z increases, the effect of the $(1 + z)^{-2}$ *RM* dilution alone will significantly "squeeze" successive $R_e(r, z)$ distributions toward $r = 0$, i.e. *"restoring" the low-z f_0 peak.*

Unless there is very strong positive evolution in the clouds' intrinsic Faraday rotation at the highest redshifts, the f_0 peak, which represents no intersections near $z = 0$, and which is quite discernable at low z in Fig. 12.2, may blend with the strongly reduced *RM*s that are represented by the second term in Equation (12.11) as z increases. Correspondingly, note that the source-to-source scatter in r is also reduced by $(1 + z)^{-2}$.

The essence of the above is that the evolving distributions illustrated in Fig. 12.2 represent sums of somewhat complicated and multivariate functions. Notwithstanding their complexity, they can constitute powerful diagnostics of the evolution of cosmic magnetism in discrete intervenor galaxy systems at substantial cosmological look-back times. That said, 2-D imaging as in Figs. 12.4 and 12.5 and/or spectroscopic information are needed for individual systems – in different radio bands and/or in IR, optical, UV, and X-ray bands. Much of this lies in the future. In Section 12.2.2, we describe some first attempts to combine optical spectroscopic observations with radio information for *integrated RMs*. In Section 12.3, we describe some sample *RM* images that transversely probe the magnetic fields in substantially redshifted systems at sub-arcsec resolution.

12.2.2   *Combination of RM with optical spectra to further clarify the properties of the magnetic intervenors and estimate <B>*

Optical and 21 cm absorption lines of distant quasars and radio galaxies have been analysed to reveal the species and excitation state of absorbing gas in intervenors of the types discussed above. Where the electron column density (N_e) is relatively large, it shows a striking correlation with excess Faraday rotation (sources in the wings in Fig. 12.3(a) and (b). This was discovered by Kronberg & Perry (1982), and confirmed in Fig. 12.3(b) for a more extended sample of quasars having both measured optical spectra and *RM* (Welter *et al.* 1984). The results are shown graphically in Fig. 12.3(a) and (b). Analysis of the optical spectra used a variety of line species, including those of heavier elements such as MgII, CIV,

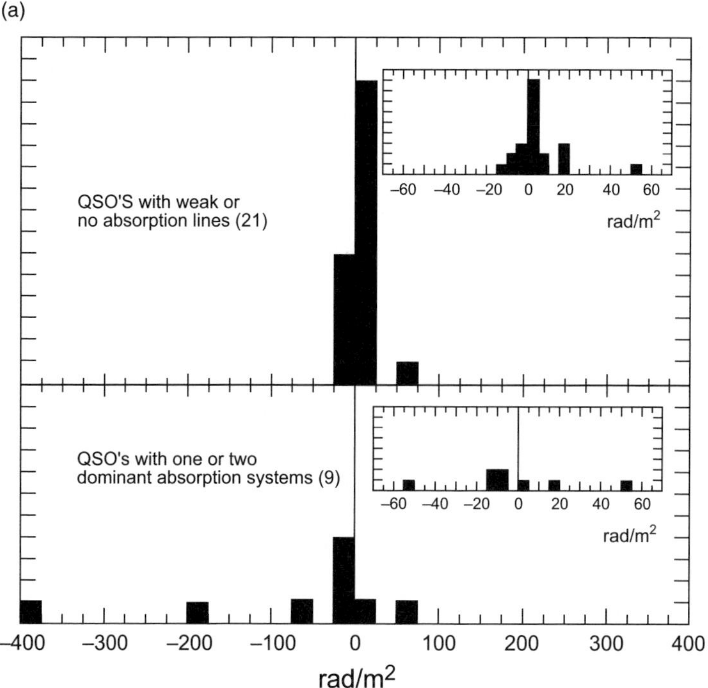

Figure 12.3. Three different investigations combining quasar Faraday *RM*s with spectral absorption strength for quasars having both *RM* and optical absorption line data:
(a) Distributions of RRM (= *r*) for quasars with (i) one or two dominant absorption systems, (*upper*), and (ii) weak or no absorption lines (*lower*) – after Kronberg & Perry (1982).
(b) The observed distribution of *r* in a sample of 119 quasars, split by their optical absorption properties. For those with rich absorption lines (*upper two distributions*), *r* is assumed to originate either (i) entirely at z_e, (*upper* – the most conservative assumption), or (ii) at the average of the observed z_a's (*middle*). The *lower* panel in (b) shows the distribution of *r* for those quasars with much weaker absorption lines. The z' notation indicates that the RRM is assumed generated at z_a, or at the average of an ensemble of z_a's ($<z_a>$). The actual *r*'s, including noise scatter, are smaller than shown, since they are plotted after correcting by $(1 + z_e)^2$ – i.e. on the *most conservative* assumption that *all* of the *RRM* occurs at z_e (Welter *et al.* 1984). The strong and weak absorption line sets are seen to have strikingly different *r* distributions. (c) Cumulative distributions of |*RM*| for a quasar sample whose sources had measured MgII absorption. The dependence of |*RM*| on the MgII absorption redshifts is clear, especially (*dashed line*) when two strong absorption systems are seen at different z_a's in the same quasar spectrum (source: Bernet *et al.* 2008).

(b)

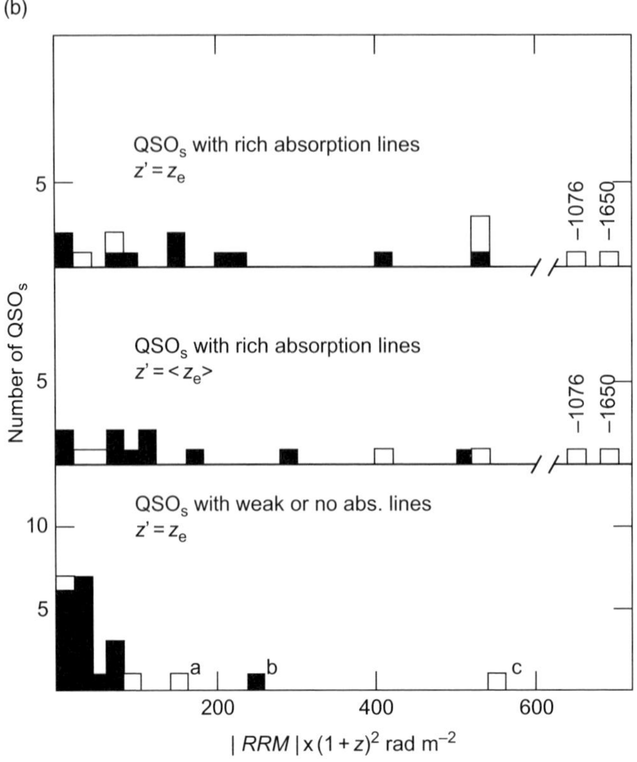

(c)

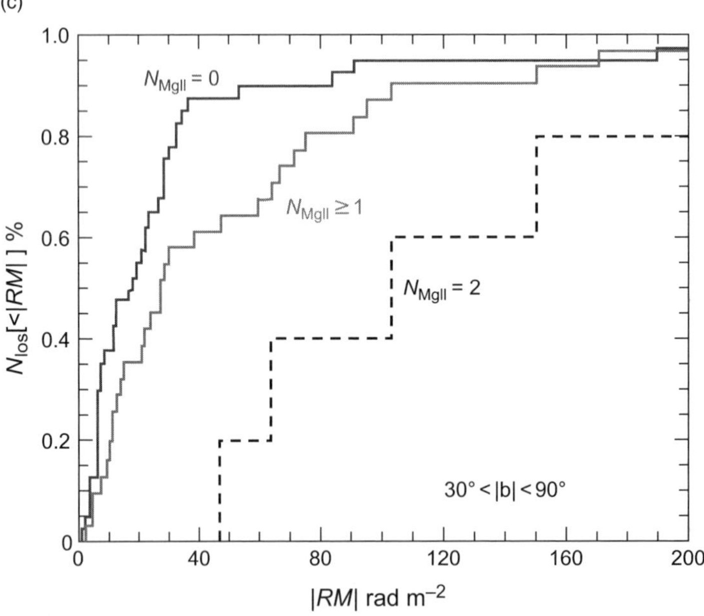

Figure 12.3. (*cont.*)

NII, SiIV, OVI, etc. Commonly used lines are neutral or near neutral hydrogen, and an MgII doublet whose excitation temperature is not much different from ionised hydrogen. They are transitions typically seen in intervening galaxy discs and sometimes in halos. For all higher excitation transitions, e.g. CIV, NII, SiIV, OVI, etc., calculation of N_e requires an estimate of the element's abundance relative to hydrogen, via the excitation temperature and an ionisation model – in order to convert to *electron* column densities, N_e, which are involved in the Faraday rotation.

More recently, with better spectrograph resolution and sensitivity (e.g. with the UVES spectrograph on the ESO 8-m telescope) it has been possible to compare *RM*s with larger and homogeneous quasar spectral line samples of a single species, MgII. Figure 12.3(c) shows *cumulative* RRM-z plots (e.g. in the sense of Fig. 12.1(right)), for quasars that have no, one, and two MgII absorption systems, respectively. For the latter, the *RM* comes from a combination of intervenors at two z_a's.

Analogous comparisons with a sub-class of damped Ly-α absorption lines have been made by Oren & Wolfe (1995) and Wolfe et al (1995). These also show a correlation with *RRM* at smaller redshifts, with uncertain results at the larger redshifts, possibly due to the $(1 + z)^{-2}$ reduction of the *RRM*. In consequence, and absent a positive co-moving $RRM(z_a)$ evolution, the *RRM* of a Milky Way-like galaxy disc becomes sharply reduced as z increases beyond 1 – refer to Section 12.1 and Fig. 12.2. This suggests that more damped Ly-*RRM* lines might be observed at $z \lesssim 1$ where space-based spectroscopy can measure the Ly-α transition in the *UV* and where $RRM(z_a)$ at lower z_a is not overly attenuated by the $(1 + z)^{-2}$ factor.

12.3 Magnetic field estimates from *transverse* probes of quasar intervenors

Here we investigate the potential for probing the state of extended, and magnetised gaseous systems at a substantial cosmological lookback times that can be "illuminated" by a similarly extended and polarised background radio source. In Section 12.4, we also discuss how multiply lensed images of a background galaxy/quasar can be exploited to explore the magnetisation state of a lensing galaxy, which may in this case be an elliptical galaxy (more spatially numerous), rather than a spiral galaxy.

(a) Transverse probe of 3C191 at z = 1.95 and magnetic fields in elliptical galaxies

3C191 at $z = 1.95$, was one of the first quasars to show an unusually rich optical absorption spectrum *near* z_e (Williams *et al.* 1975). Unlike the typical z_a cosmological intervenors described in Section 12.1, the absorbing material for this source is located close to the quasar. 3C191 has *RRM* fluctuations up to $\mathcal{O}$ 150 rad m^{-2} *in the observer's frame* within its ~4$''$ long jet. These correspond to *RM* fluctuations up to $\mathcal{O}$ 1300 rad m^{-2} in the emitted frame. The angle rotations are clearly seen in Fig. 12.4 below in a comparison of the 15 GHz and 4.5 GHz images at the same 0.39$''$ resolution. The estimated N_e column depths of 10^{21}–10^{22} cm^{-2}, combined with the observed *RRM* fluctuations gave magnetic field estimates in the range ≈ 0.4–4 μG via Equation (12.1). These magnetic field strengths, at $z \simeq 2$, though subject to further scaling adjustments of N_e, are comparable to what we measure in the Milky Way disc. But the similarity to the Milky Way may in this case be a coincidence. Nonetheless, the 3C191 system documents a significant $<|B|>$ at a cosmological lookback time that is only $\approx 15\%$ of the current age of the Universe (refer to the $\mathbf{T}(z)$ curve in Fig. 6.7).

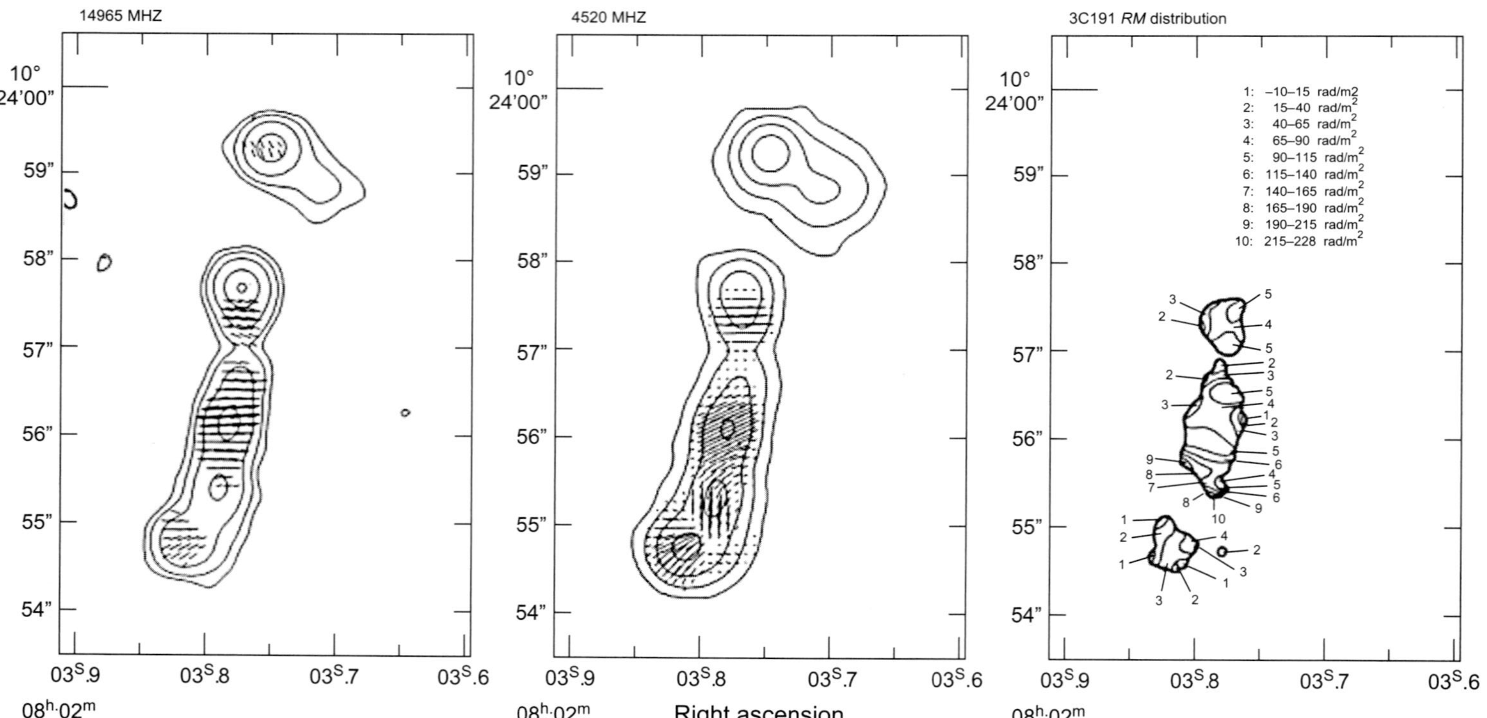

Figure 12.4. (Left 2 panels) VLA images of Stokes parameters *I, Q*, and *U* for 3C191 at the lowest common resolution of 0.39″, at the two frequencies shown. The 15GHz 0.12″ resolution VLA image not shown here reveals a narrow, polarised, and laterally unresolved jet "spine" with modest changes in direction along its 4″ length. (Reproduced from Kronberg, Perry, & Zukowski (1990).)

A physical interpretation of this interesting system has been constructed in two companion papers, Kronberg, Perry, & Zukowski (1990) and Perry & Dyson (1990). These provide a detailed analysis of both the radio properties and optical spectrum. 3C191 is surely not unique, and the narrow HeII emission lines in its spectrum have also been observed in the optical spectra of other QSOs (Foltz *et al.* 1988). However, 3C191 is unusual among known "associated" absorption line systems in that it is seen with a strong extended radio jet. This is a consequence of the fact that, globally, only a small fraction of optically luminous QSOs have both strong *and* extended radio emission. The strong bending and distortion of many radio jet morphologies at redshifts $z > 1.5$ may generally be caused by relatively large associated gas column densities, $N_e \sim 10^{21}$ cm^{-2}, near to the quasar. An alternative cause of the distortion could, in a few cases, be gravitational lensing by an intervening galaxy at $z \lesssim 1.5$ (see Section 12.4).

One possible conclusion is that we detecting a powerful wind-driven shell of radius $\mathcal{O}$ 16 kpc in the host quasar of 3C191. If so, the outflowing shell would be very thin, $\Delta R/R \sim 10^{-5}$, and the shell-internal gas density approaches $\sim 10^3$ cm^{-3}, with a bulk outflow velocity of ~ 800 km s^{-1} and a ratio of gas to magnetic pressure of $\approx 2 \times 10^{-3}$. These numbers are energetically compatible with the quasar's total luminosity of 10^{45}erg s^{-1}, a fraction of which is dissipated in the quasar wind-driven shell. The interested reader can consult Perry & Dyson (1990) for details and for a good bibliography of related papers. In this model, the outflowing shell in 3C191 has a substantial covering factor, but it is less than 100%.

The host of the 3C191 quasar at $z \approx 2$ appears to be an *elliptical* galaxy at a cosmologically large lookback time (Perry & Dyson 1990). This association is not surprising in a global context, since powerful extragalactic radio sources are generally associated with relatively massive elliptical galaxies. But in elliptical galaxies, unlike spirals and other late-type galaxies, there have been few opportunities to verify magnetic fields. When other systems similar to 3C191 are found, absorption spectroscopy and transverse *RM* probes may confirm the level of magnetic fields in elliptical galaxies in general, and especially at large cosmological lookback times. Further, similar sub-arcsecond, multi-frequency polarised radio images should become available with more sensitive radio interferometers. Of comparable importance are large optical telescopes with sensitive high resolution spectrographs, which also can extend into the UV and IR.

(b) A QSO intervening, Faraday rotating spiral galaxy at z = 0.38

A second example of transverse *RM* probes is the quasar PKS1229-02 at $z = 1.038$. This quasar has an extended radio source that happens to align closely with a foreground intervening galaxy at $z = 0.395$. As with 3C191, the absorption spectrum of this optically bright ($V \sim 16.7$) quasar has been analysed in detail – by Briggs *et al.* (1985). It includes a damped Ly-α line as well as 21 cm absorption at the same intervenor redshift. Over its 6″ inner radio jet, the *RRM* oscillates with an amplitude of ~ 35 rad m^{-2} (~ 145 rad m^{-2} at $z_l = 0.395$) and a angular periodicity that corresponds roughly to the inter-arm separation of a spiral galaxy at the intervenor's redshift (Kronberg, Perry, & Zukowski 1992). The juxtaposition is illustrated in Fig. 12.5(a) where the jet's radio structure is shown in grey, along with the sign of extrema of *RM* oscillation along the jet "spine". Further details not shown here are given in Kronberg *et al.* (1992), including the very extended radio structure of the background QSO. Despite the projected line-of-sight closeness at $z_a = 0.395$ of the background quasar nucleus, there is no sign of multiple imaging of the background jet – a topic discussed in the next section. However the jet morphology in PKS 1229-021 is likely to be

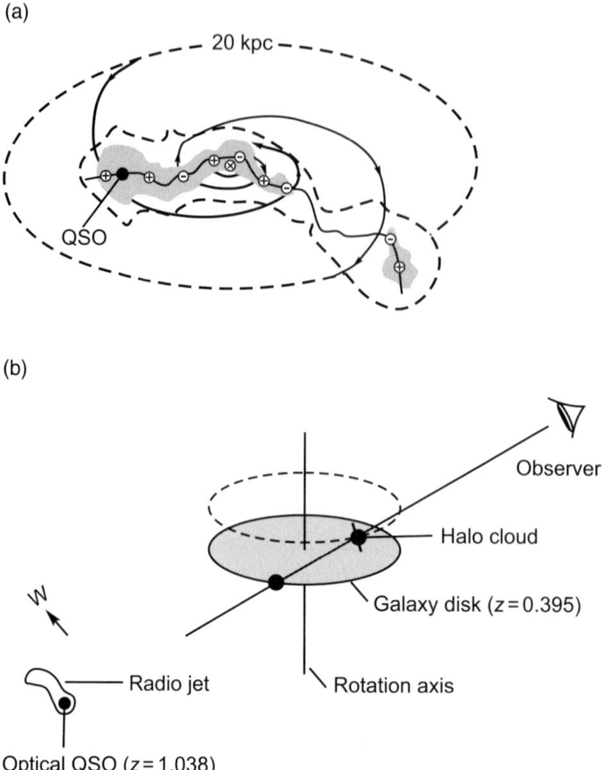

Figure 12.5. (a) Sketch showing the background quasar PKS 1229-021, its ~6″ long inner jet, and the disc and halo intersection points of the intervening galaxy at $z = 0.395$. The outer, fainter radio quasar structure at $z = 1.038$ (not shown) extends over arcminutes. (b) Cartoon illustrating the relative positions of the background quasar and foreground galaxy. (Source: Kronberg, Perry, & Zukowski (1992).)

also gravitationally affected by the intervening lens. The closeness of the quasar-galaxy juxtaposition indicates that future optical and radio images of higher sensitivity and resolution will also reveal some lensing effects (see Section 12.4) below.

Details of the optical absorption spectrum reveal distinct halo and disc components in the intervening galaxy consistent with a highly inclined ($i \gtrsim 60°$) disc (Briggs *et al.* 1985). Figures 12.5(a) and (b), reproduced from Kronberg *et al.* (1992), show a sketch (a) and a cartoon (b), respectively, of the inferred layout of both background quasar and intervening galaxy.

The *RM* variations are consistent with a highly inclined spiral galaxy – that is, at the *RM* extrema we are looking nearly parallel to the arms, and the *RRM* oscillations about zero along the jet suggest a possible bisymmetric magnetic spiral structure, akin to the nearby, well-studied spiral galaxy M81, which also happens to have a similar inclination angle. The dashed lines sketched in (a) show an M81-like galaxy transposed to $z = 0.395$, and tilted by its inclination of 60°.

The radio structure, *RRM*, and optical spectrum analysis above led to an estimate of 1–4 μG, similar to the Milky Way, for the mean disc magnetic field in the P1229-021 intervenor galaxy. The *difference* in cosmic time between the absorber-galaxy's formation time, T_{form}, and that at its observed redshift, $(T_{form} - T_{z=0.395})$ for the PKS 1229-021 intervenor can be compared to the dynamo amplification time required to amplify the field from, say, the hypothesised primordial seed field. Its lookback time at $z = 0.395$, approximately 40% of the Hubble time, and comes close to constraining the cosmological evolution of the galaxy disc's magnetic field as would be predicted by a large scale galactic α–Ω dynamo. It suggests that, if the putative $z = 0.395$ spiral galaxy in Fig. 12.5 is typical, then $<|B|>$ amplification has not occurred over the most recent ~40% of the age of the Universe. This situation is different from the 3C191 system in (a) at 85% of the lookback time: The P1229-021 intervenor contrasts with the apparent shell intervenor in 3C191 near z_e, which so far has no well-documented analogue in the nearby Universe.

The implied galactic magnetic field evolution of the PKS 1229-021 intervenor also relates to Chapter 5, and especially to articles by Elstner, Meinl, & Beck (1992), Elstner *et al.* (2000); and Braun, Heald, & Beck (2010). The latter includes our reproduced images in Figs. 5.10 and 5.11 (citations all in Chapter 5). The discussion above illustrates how radio-optical probes such as these "prototype" studies of 3C191 and P1229-021 can potentially probe galactic magnetic field strengths and structures in considerable detail, and over large lookback times.

Applications of the analyses described in this chapter are limited by the precision of *RRM* and optical data available at the time of the observations. For the reader wishing to apply future *RRM* data to these kinds of study, useful mathematical formalisms can be found in Welter, Perry, & Kronberg (1984) and its appendix. These formalisms are analytically derived, and they can be used as a basis for large and many-parameter numerical simulations, and at large redshifts. Future investigations will expand to multi-excitation absorption line data going, as mentioned above, beyond visible λ's to the UV and IR, to deeper images, and larger numbers of *RRM* and radio images over a wide z-range. To assist in the use of this book for these and similar purposes, we largely use the terminology introduced by Khare-Joshi & Perry (1982), Kronberg & Perry (1982), and Welter *et al.* (1984). Extensive numerical simulations, in combination with *RM* and absorption line data over a wide z-range, are now more available than when the above papers were written.

12.4 Transverse *B* probes and gravitational lensing

Readers interested in the wider subject of gravitational lenses could consult an excellent review by Blandford & Narayan (1992), containing a comprehensive reference list. Also, Zwicky (1937), Refsdal (1964), and Dyer & Roeder (1977) have written important foundation papers on this general subject.

In this section, we consider the relatively rare situation of a background quasar that happens to align very closely with the nucleus of an intervening galaxy with a relatively compact mass distribution. The latter characteristic applies to some elliptical galaxies, and makes them, as a class, more likely to produce *multiple* images. In the case of an extended or multi-component background source, lensing applies to each (polarised) pixel of the source. Favourable circumstances for multiple lensing include, importantly, a suitable combination of z_e and z_a. This provides separate ray paths through the intervenor, and thus separate *Faraday rotation paths* from a common polarised background emitter.

A multiply-lensed object will always produce an odd number of images, though with different relative amplitudes. The case of one image is trivial for the present purpose, since we are focussing here on multiple *RM* paths. The case of three images is potentially very interesting in the context discussed here, and higher lensing multiplicities (<3) probably occur too rarely to warrant further consideration.

The first multiply-lensed radio system, discovered by Walsh *et al.* (1979), was 0957+561, in which $z_{\text{lens}} = 0.36$. In another, remarkable example, MG1131+0456, $z_{\text{lens}} = 0.844$ the very close alignment transformed a polarised background jet into a ring-shaped lensed image (Chen & Hewitt 1993). This situation makes ellipticals of particular interest when more than one *RM* line of sight can be probed through them, given that the ISM of ellipticals is more difficult to probe than for spiral galaxies.

In general, the optimum "gravitational leverage" of a lens mass (L) on a background image (S) occurs when the lens' cosmological angular size distance, $D_<^{\text{L}}$, is $\mathcal{O}$ 1/2 of $D_<^{\text{S}}$, the angular size distance of the projected background source. In practice this is approximately satisfied when the lens redshift, z_{L}, is in the range ~0.2–0.8, and $z_{\text{S}} \gtrsim 1.3$.

An important property of an extended and *polarised* background object, e.g. an extended polarised quasar jet, is that its intrinsic polarisation angles are not rotated by an intervening lens (Kronberg *et al.* 1991), even when the jet's morphological shape is distorted by the lensing galaxy (see also Dyer & Shaver 1992, Kronberg *et al.* 1996). This might seem at first counter-intuitive, given that the lens *will* change the local orientation of the affected jet segments while the intrinsic polarisation angles within these same lensed jet segments are not rotated. In the case of different multiply lensed ray paths, they will have different *RM*s corresponding to their separate ray paths through the lensed galaxy. Whether the jet is multiply, or singly imaged (more common), such measurements can provide a kind of transverse magnetic field probe of a galaxy, provided the n_{e} distributions along the varying or separate ray paths can be measured separately or modelled.

Before leaving this section, some additional comments are appropriate. The first is that path-specific *RM*s are not easy to unambiguously determine, and this technique is still at an early phase of implementation. Better surface brightness sensitivity over a full range of angular scales in (α, δ) would allow better specification of the (projected) lensed image. This includes images of $Q(\alpha, \delta, \lambda_i)$ and $U(\alpha, \delta, \lambda_i)$ (Chapter 3), which will specify the linear polarisation at several (i) frequency channels for each sky position. Given this full set of multi-frequency image information, Faraday depth synthesis techniques (Chapter 3) could model $n_{\text{e}}B_{\parallel}$ along each sightline. The multi frequency imaging capability of e.g. the "Jansky VLA" (EVLA), especially with enhanced resolution, is capable of such magnetic field probes – for lensed or unlensed images. Finally, important complementary data will be the optical and millimetre wave spectra at comparable, down to sub-arcsecond resolution. The ultimate aim is to measure or model, through lensing, an average magnetic field strength distributed within the lensing galaxy at different galactocentric radii, in 3-D.

Another related point, though beyond the theme of this book, is that gravitational lensing, multiple or weak, can be combined with Faraday rotation to isolate purely gravitational phenomena. Deep optical imaging and spectra can, in principle, be combined to measure the global mass, mass distribution and the magnetic structure of a galaxy, using the essential information described above (see e.g. Kronberg, Dyer, & Röser 1996 for an early attempt at measuring the first two of these, including the weak lensing detection of more than one gravitational intervenor).

12.5 Rotation measure searches for a widespread, cosmologically co-expanding intergalactic magnetic field

Previous sections have discussed probes of magnetic fields in discrete extragalactic systems, including intergalactic clouds, galaxies, their halos, and clusters of galaxies. An obvious extension of these analyses, models, and observations might prompt the reader to ask: Can we identify an *RRM* signature from an all-pervading widespread, coexpanding intergalactic medium?

The first availability of large samples of extragalactic source rotation measures in the 1970s inspired tests for the existence of a widespread B_{IGM} (Rees & Reinhardt 1972, Nelson 1973, Kronberg & Simard-Normandin 1976). We can most simply parameterise the density of intergalactic ionised gas $n_{\mathrm{IG}}(z)$ by scaling it to the average baryonic matter (BAO) density, which increases with cosmological epoch as $(1 + z)^3$. A measured *RRM* (z) out to some maximum z_{max} can be related to a widespread magnetic field using Equations (12.14) and (12.15) below.

$$RRM(z_{\mathrm{max}}) = 8.1 \times 10^5 \int_0^{z_{\mathrm{max}}} n_0(1 + z)^3 B_{\|0}(1 + z)^2(1 + z)^{-2}\mathrm{d}l(z) \ \mathrm{rad \ m}^{-2} \qquad (12.14)$$

where, in a $\Lambda = 0$ Friedmann cosmology,

$$\mathrm{d}l(z) = 10^{-6}\frac{c}{H_0}\frac{1}{(1 + z)(1 + \Omega z)^{+0.5}}\mathrm{d}z \ \mathrm{Mpc} \qquad (12.15)$$

and the $(1 + z)^2$ factor describes the adiabatically scaled magnetic field evolution into the past. Equation (12.14) makes the further simplifying assumption that the magnetic field has no sign reversals.

More realistically, $RRM(z)$ will represent the summation of independent, and contiguous, cells of randomly changing $\boldsymbol{B}$. Thus, by incorporating a random walk in *RRM* through cells of co-moving size l_0, = 1 Mpc we can calculate a variance V_{RRM} (z), as shown in Fig. 12.6. In Fig. 12.6, the Universe has a closed geometry, a consequence of the assumption of $\Omega_{\mathrm{baryons}} \simeq 1$ (no dark matter). We note the contrast between these assumptions and with the current, more realistic ΛCDM cosmology. This incorporates dark matter and dark energy ($\Lambda \neq 0$) ($\Omega_{\mathrm{baryons}} \simeq 0.03$ and $\Omega_{\mathrm{M}} \simeq 0.3$). Describing earlier attempts within the Friedmann cosmology paradigm serves to illustrate the difficulty of performing the above probe in a ΛCDM cosmology, and also to review the history and motivation for searches for a widespread B_{IGM}. Figure 12.6 and Equation 12.14 factor out the already discussed discrete *RRM* intervenors at $z_{\mathrm{a}}{}^i$ and the distant quasars' own *RRM* contribution at $z_{\mathrm{e}}{}^i$. In this model, the Faraday rotation law is simply embedded into a widespread, coexpanding IGM using polarised background sources out to some $z_{\mathrm{e}}{}^{\mathrm{max}}$.

Under these assumptions the evolutionary and cosmological powers of $(1 + z)$ would overwhelm the $(1 + z)^{-2}$ *RRM* reduction due to the λ^2 transformation. Thus, the contribution to $<RRM>(z)$ and its variance would increase monotonically with z becoming, on these simple scalings, detectable at some observable large redshift (Fig. 12.6, solid curves).

A second model (dashed curve) shows the effect of a "discrete" intervenor population that ignores a widespread magnetised IGM. Here, the intervenors (galaxy systems, etc.) have a typical (halo) size and magneto-ionic parameters, and a global space density such that their

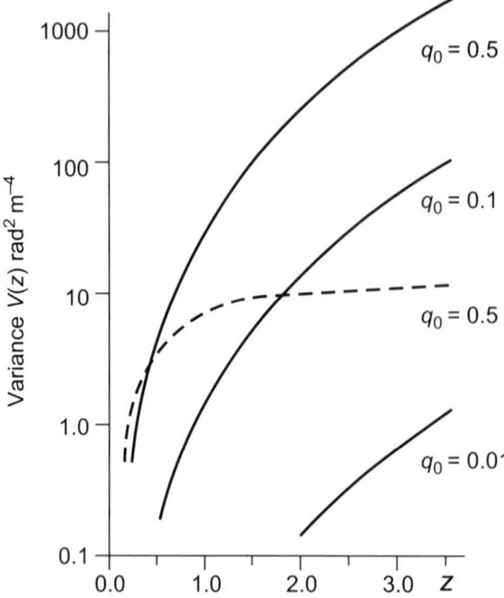

Figure 12.6. Solid curves show the variance of $V_{RRM}(z)$ for samples of radio sources due solely to a hypothetical widespread, coexpanding IGM at successively larger redshifts in a Friedmann $\Lambda = 0$ cosmology for three different values of Ω_m (=$2q_0$). They assume $B_0 = 1.8 \times 10^{-8}$ G, $H_0 = 75$ kms^{-1}kpc^{-1}, that 100% of Ω_m is in intergalactic gas, and a single magnetic field reversal scale, $l_0 = 1$ Mpc. The dashed curve is described in the text.

filling factor is less than unity at lower redshifts. In the simplest non-evolving model, the combination of the increasing filling factor ($f(z)$) and the $(1 + z)^{-2}$ watering down of the observed *RRM* causes a levelling off of ($V(z)$), shown by the dashed curve in Fig. 12.6. More realistic variations of the second class of model will obviously perturb the form of the dashed curve shown, but its general character will remain distinct from the first model. This dashed curve shows $V_{RRM}(z)$ for a sample "discrete intervenor" model. Here, $f = f_0(1 + z)^3$, and $f_0 = 1/64$ (Kronberg, Reinhardt, & Simard-Normandin 1977).

The curves in Fig. 12.6 indicate that a B_{IG} might have been detectable in a $\Omega_m \gtrsim 0.2$ Friedmann cosmology under some reasonable assumptions for $z_{max} \gtrsim 3$. However, these levels of B_{IG} detection in Fig. 12.6 are no longer expected to occur, since Ω_m is less than 5% in current cosmologies. Consequently, the expected V_{RRM} for $z \simeq 3$ is drastically reduced in proportion – to a level of ~ 1(rad m^{-2})^2! This is currently undetectable in the presence of noise, intervenors at $z_a^{\,i}$, and source-intrinsic effects at $z_e^{\,i}$. Note also that the model curves in Fig. 12.6 do not allow for sightlines to z_{max} that include a mix of LSS filament and void zones.

To summarise, this brief return to the original *RM* searches for a smooth co-expanding magnetised IGM demonstrates (1) why the $V(z)$ test in Fig. 12.6 is outdated in a ΛCDM cosmology, and (2) what *RM* datasets out to large z's might, in future, reveal about a widespread B_{IG}. For the latter, an order of magnitude estimate corresponds to $q_0 \approx 0.02$ in Fig. 12.6, where the *RRM* contributions to V_{RRM} might be only $\mathcal{O}\,2 - 4$ rad^2 m^{-4} at $z \simeq 3$. Future detectability might be possible beyond $z \approx 3$ with better data than what is available at this time of writing. This would include sources at $z_e > 3$, and which have an anomalously large "local", or intrinsic *RRM*.

12.6   Importance of the global star formation rate (SFR) over all galaxy redshifts

Another relevant conceptual framework is provided by modelling the input of stellar and SN-driven outflows into the IGM over cosmic time. Earlier results in this complex subject, calculated the rate of star formation to peak near $z = 2$. Now, better input data, supplemented by comprehensive simulations, such as those of Springel & Hernqvist (2003), show a global z-dependent SFR that peaks at $z \sim 5$–6 (Fig. 12.7). At $z < 4$, the specific SFR increases at the greatest rate up to $z \sim 3$ (Fig. 12.9). At earlier epochs ($z \sim$ 10–20) simulations indicate a monotonic increase with cosmic time (decreasing z). Estimates of the initial rate of rise in the SFR up to $z \sim 2$ have varied widely, and have converged to a global SFR $\propto (1+z)^x$, where $x \approx 1.3$.

The global analytical SFR fit in Fig. 12.7 was derived from the following double exponential equation:

$$\frac{d\rho_*}{dt}(z) = \frac{d\rho_m}{dt} \frac{\beta e^{[\alpha(z-z_m)]}}{\beta - \alpha + \alpha e^{[\beta(z-z_m)]}}, \tag{12.16}$$

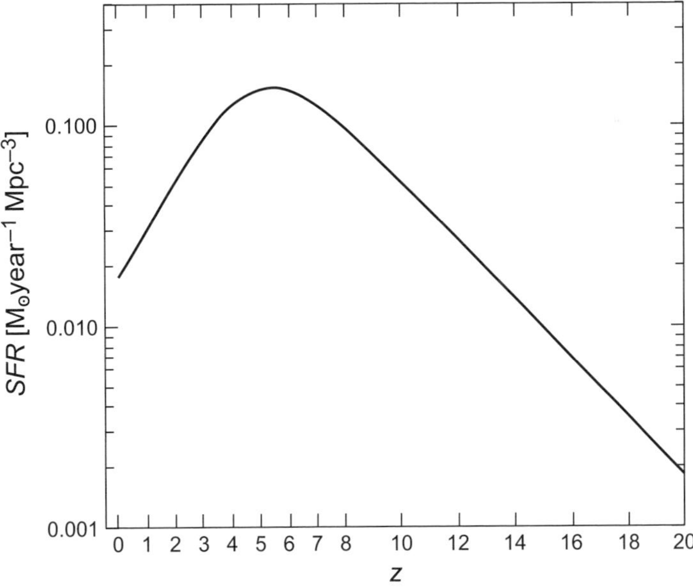

Figure 12.7. Computation of the global, specific SFR vs. redshift over the range $0 < z <$ 20 for a ΛCDM cosmology. It shows the rapid increase up to $z \sim 4$, a broad peak at $z \sim 5$–6, and a steady, steep decline from $z \approx 8$ to 20. It appears in good overall agreement with observational data as far as they are available, but they are sparse and uncertain at higher z's. (Adapted from Springel & Hernqvist (2003).)

where $\alpha = 3/5$, $\beta = 14/15$, $d\rho_{\rm m}/dt = 0.15 {\rm M}_\odot$ year^{-1} Mpc^{-3} sets the overall normalisation, and $z_{\rm m}$ denotes a break redshift (Springel & Hernqvist 2003).

It is instructive at this point to consider the magnetic energy associated with the early universe starburst scenario described above. It appears to be essentially stellar and thermonuclear, and hence the relevance of the SFR curve in Fig. 12.7. Additional energy must be added to the field if its present-epoch $B_{\rm IGM}$ builds up to 10^{-8}–10^{-7} G in galaxy zones of the IGM at the present epoch. This would occur if most of the subsequent energy source comes ultimately from gravity – from large scale infalls in the development of LSS, and/or the accompanying shearing and turbulence. In addition, and importantly, supermassive black holes also play an important, possibly even greater role, as discussed earlier, in supplying the IGM with magnetic energy. For a relevant discussion of early black hole and star formation, see Whelan *et al.* (2013).

12.7   Post-amplification of existing IGM fields

12.7.1   *Setting context*

To understand evolution of the structure and strength of magnetic fields that are "fed" into the IGM, we require observational data that characterise the gas temperature and turbulence in the IGM, and also the magnetic field strength and structure. These are more readily available for galaxy clusters (see Chapter 9). For the nearby Coma cluster of galaxies at 100 Mpc, the observed gas pressure fluctuations suggest that the energy density of gas turbulence, $\varepsilon_{\rm T}$, is $\gtrsim 0.1\ \varepsilon_{\rm th}$, and that it seems approximately Kolmogorov, at least at higher k-scales – see also Fig. 9.5 for the case of the Hydra cluster. The temperature ranges from 10^3K in or near the cluster core to $\sim 10^8$K near the periphery

Contrasting with ICM densities of $10^{-1} \lesssim \rho_{\rm ICM} \lesssim 10^{-3}$ cm^{-3} (Chapter 9), the much lower baryonic density in the wider IGM, $10^{-3} \lesssim \rho_{\rm IGM} \lesssim 10^{-6}$ cm^{-3} generally requires more sensitive measurements using different methods to detect it, and these are not all yet available. They are improving, and await future generations of instrumental improvements. Nonetheless, much can still be learned from the plasma parameters of well-studied clusters. Up to this time much of our insight into the cosmic evolution of $<|B|>$ in the IGM relies on theoretical modelling based on measured and inferred plasma parameters (Section 6.3). An early effort in this direction is described in Ryu, Kang, & Biermann (1998).

The IGM outside of clusters has different phases and temperature régimes. X-rays in the keV range reveal hot gas at $T \gtrsim 10^7$K, as is observed in galaxy clusters. Another temperature phase is the range 10^5–10^7K, commonly called the "warm-hot intergalactic medium" (WHIM), and most prominently visible in the extreme ultra-violet (EUV) and low energy X-rays. For further reading see e.g. Cen & Ostriker (1999) and Kang *et al.* (2005). The EUV has been more difficult to access observationally, and it is affected by Milky Way foregrounds. The WHIM is an important tracer of the IGM in large scale filaments of galaxies. Below 10^5 K, closer to the ionisation potential of hydrogen and at sufficiently high density, recombination lines and bremsstrahlung are detectable. This temperature range also occurs at the denser and cooler cores of some galaxy clusters.

In cosmic voids the very low thermal gas density and also low $|B_{\rm VOID}|$, make these difficult to detect in most of the wavebands mentioned here, especially Faraday rotation. However, in Chapter 11 it was shown how γ-ray cascades involving (charged) e^+e^- can be exploited to probe very low magnetic fields down to $|B_{\rm VOID}| \sim 10^{-18}$G.

Different modelling efforts have been undertaken to explore how initially weak intergalactic magnetic fields can be amplified by a combination of intergalactic shocks, turbulence, diffusion, and shearing flows. All of these phenomena may occur widely in the IGM, even though often difficult to observe with present plasma-probing techniques and instrumentation.

12.7.2 Some specifics of intergalactic |B| evolution models

Turbulent magnetic field amplification is expected to be connected to the number of eddy turnovers in a turbulent flow structure of the IGM, in which the vorticity, $\vec{\omega}$, is the curl of the local $\vec{v}$,

$$\vec{\omega} = \nabla \times \vec{v}. \tag{12.17}$$

Vorticity is most prominent in and around intergalactic shocks occurring, for example, in the course of gravitational infall in LSS. In simulations of LSS evolution in a dark matter dominated universe, $\vec{\omega}$ is computed to be most prominent in curved infall shocks, and progressively less prominent in LSS filaments and voids, respectively (see e.g. Ryu *et al.* 2008).

Along with B amplification on different eddy scales, cosmic rays are believed to be accelerated in the IGM by the process of diffusive shock acceleration. Mechanisms of diffusive shock acceleration have been worked out for the Solar System, supernova remnants, galactic, galaxy cluster, and intergalactic medium contexts. In the intergalactic context discussed here, energetic charged particles can be scattered off self-generated Alfvén waves in the vicinity of a shock front. The magnetic scattering process also serves to confine the particles to the region of the shock. This confinement permits multiple scatterings, enabling the particles to gain energy by first-order Fermi acceleration. It also prevents them from streaming far upstream of the shock (see e.g. Bell 1978). For further reading on diffusive shock acceleration using particle-in-cell simulations the interested reader could also consult Dieckmann *et al.* (2008) and its bibliography.

Computational models of the cosmic evolution of turbulent eddy-generated magnetic fields, and cosmic ray acceleration, have produced a result that most of these eddies occur near large scale shocks, where the density and temperature are highest. These begin at a $\mathbf{T}(z)$ corresponding to $z \sim 6$, approximately the epoch of re-ionisation, and the cumulative number of eddy turnovers is calculated as a function of cosmic time (Ryu *et al.* 2008). The results of a typical simulation run are shown for $z = 0$ in Fig. 12.8.

In the models shown, the average magnetic field strength and the average vorticity for a given temperature range increase slightly as $\mathbf{T}$(year) proceeds (z decreases). In a given z range the vorticity increases with temperature and density, so that it roughly traces the LSS filaments. The growth of B_{IGM}, and its structure (k-scale distribution) uses a separate turbulent dynamo model simulation. The larger scale structures (lower k-scales) are proposed to develop through an inverse cascade process along the lines of that described by Cho & Vishniac (2000). Dependence on the inverse cascade process is important, since a turbulent dynamo by itself seems unlikely to generate large scale B_{IGM} components. The reasons for this are essentially those presented by Kulsrud & Anderson (1992), who point out that a turbulent galactic dynamo cannot explain the large scale components of the Milky Way disc field (Chapter 3).

In the simulation results in Fig. 12.8 most of the IGM volume develops a magnetic field strength in the range $10^{-9} \rightarrow 10^{-6}$ G whereby a small fraction reaches microgauss

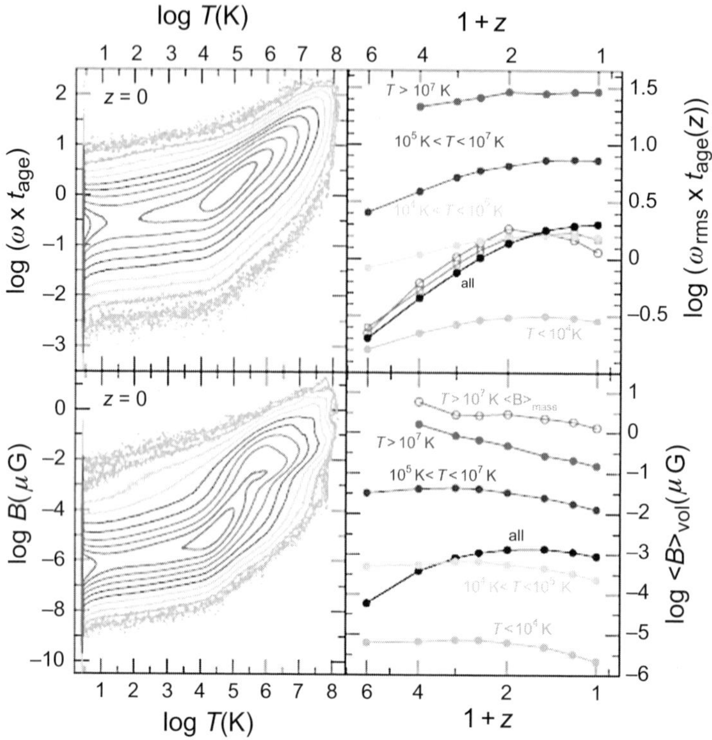

Figure 12.8. The predicted volume fraction of IGM vorticity, $\omega \times t_{\rm eddy}$ (upper plots), and magnetic field strength (lower plots), shown separately as a function of relative gas density and gas temperature $\log(T)$. Source: Ryu *et al.* (2008). "$t_{\rm age}$" denotes time, normalised to cosmological time, **T**. (Kindly provided by D. Ryu).

levels – close to those of clusters and galaxies. However, for reasons discussed earlier, clusters have their own complex dynamics and interactions *within* their ICM (Chapter 9). This means that the cluster-internal $B_{\rm ICM}$ evolution is not, in all aspects, a straightforward extension of LSS evolution in the IGM. A more complex extension of cluster-internal simulations is required. They could possibly bring simulated ICM field strengths (and structure) to the observed levels of $\gtrsim 10\ \mu$G, which do occur near the "cool cores" of some clusters (Chapter 9).

We note that the evolution of $B_{\rm IGM}$ that has been described and modelled so far does not include the important component of $B_{\rm IGM}$ (z) that is introduced by supermassive BH jets and lobes (Chapter 7). A *total* $B_{\rm IGM}$ and its k-scale structure has not yet been calculated, but the required supermassive BH–generated component would obviously add to $B_{\rm IGM}$ (z) arising from processes described in this section – that is, from star/SN-driven galactic outflows and other large scale IGM-generated fields.

12.8   Seed fields produced during and following the appearance of the first stars and galaxies: some general, basic calculations

The vorticity described above could have been generated either at large scale, curved IGM shocks, or by pressure variations resulting from entropy variations induced at the shocks. Such conditions could produce a seed field by a Biermann battery mechanism similar to that illustrated in Fig. 3.1 Recalling that a Biermann battery arises fundamentally where a local gradient in gas pressure differs from that of the local velocity field, $\nabla \vec{p} \neq \nabla \vec{v}$, one can imagine different scenarios in the IGM (and elsewhere) in which such a non-coincidence can arise: For example, the shock front in Fig. 3.1 could encounter a stationary cloud with some central density concentration. In this example the direction of $\nabla \vec{p}$ would be unrelated to that of $\nabla \vec{v}$. But the situation $\nabla \vec{p} \neq \nabla \vec{v}$ would similarly cause a current and hence a magnetic field, even if none had previously existed. These straightforward examples serve to illustrate that this type of battery process is quite universal, going far beyond its original stellar interior application. This universality was also recognised and predicted by L. Biermann (1950). For some simulation details, and relevant physical assumptions the reader could consult Gnedin *et al.* (2000) – see also Chapter 3 – and Ryu *et al.* (2012).

Aside from these applications of the Biermann battery, there are extragalactic circumstances in which the Weibel instability, possibly combined with Langmuir instabilities, is independently able to seed and grow magnetic fields (e.g. Weibel 1959, Medvedev & Loeb 1999, Gruzinov 2001, Schlickeiser & Shukla 2003). Given the right conditions these instabilities can occur in association with intergalactic shocks. Their importance for both seeding and growing a B_{IGM} depends importantly on their growth times. These, in turn, depend on the plasma parameters and specific aspects of the plasma dynamics of the situation in question.

Favourable conditions for the Weibel instability include a collisionless shock environment, where normally the maximum shock density ratio is ~4. Additionally, a high Mach number is required for a non-radiating, cold beam of density n_b and bulk streaming velocity u, when it collisionlessly interpenetrates some other plasma mass. At this point very high velocity gradients can result. These can produce a longitudinal two-stream instability, and an anisotropic bi-Maxwellian electron-ion distribution. Such systems are unstable against the production of growing electromagnetic perturbations, even in an originally neutral plasma, as argued by e.g. Medvedev & Loeb (1999).

Following Schlickeiser & Shukla (2003), we can write an expression for the *maximum* growth *rate*, $\Gamma_4^{\mathbf{max}}$, of the "secular Weibel instability", where 4 denotes the maximum shock compression ratio. It is given by

$$\Gamma_4^{\mathrm{max}} = \frac{u\omega_{\mathrm{pe}}}{c} = 0.54 n_{-4}^{0.5} M T_7^{0.5} \ \mathrm{s}^{-1} \tag{12.18}$$

(Schlickeiser & Shukla 2003), where T is in units of 10^{-7} K, n is in units of 10^{-4} cm^{-3}, M is the Mach number ω_{pe} is the plasma frequency:

$$\omega_{\mathrm{pe}} = \sqrt{\frac{4\pi e^2}{m_{\mathrm{e}}}} = 5.64\sqrt{n} \ \mathrm{s}^{-1}. \tag{12.19}$$

This leads to a corresponding minimum growth time, τ:

$$\tau \approx \frac{1}{\Gamma_4^{\mathrm{max}}} = \frac{2}{M T_7^{0.5} n_{-4}^{0.5}} \ \mathrm{s}. \tag{12.20}$$

This time is short compared with any cosmological time scale, though it is long compared with the maximum Langmuir growth time, $\Gamma_{\mathrm{L}}^{\mathrm{max}}$:

$$\Gamma_{\mathrm{L}}^{\mathrm{max}} \simeq \frac{\sqrt{3}}{2^{4/3}} \alpha^{1/3} \omega_{\mathrm{p,e}}, \qquad (12.21)$$

where α is the density ratio of the two shearing plasma streams (Medvedev & Loeb 1999). We note that M may not be much greater than ~45, based on simulations by Miniati (2002).

However, in extragalactic structure formation shocks the range of (high) Mach numbers is likely large. The B_{IG} level at which Weibel instability-seeded fields saturate has been discussed by Medvedev & Loeb (1999); the instability will grow as long as free-streaming particles are able to cross magnetic field lines. It is suppressed when the scale of the particles' gyroradii in the freshly excited magnetic fields approximates the scale length of unstable modes, $k^{-1}_{\mathrm{s,max}}$. This sets

$$B \simeq \frac{u^2}{v_{\mathrm{th}}} \sqrt{4\pi n_{\mathrm{e}} m_{\mathrm{e}}} \qquad (12.22)$$

or, in energy-related terms,

$$\frac{B^2/8\pi}{n_{\mathrm{e}} m_{\mathrm{e}} u^2} = \eta \frac{u^2}{2 v_{\mathrm{th}}^2}, \qquad (12.23)$$

where η (~0.01–0.1) corresponds to slightly sub-equipartition magnetic fields. The resulting Weibel instability-produced fields are

$$B \simeq 1.3 \times 10^{-7} T_7^{0.5} \left(\frac{M}{43}\right)^2 n_{-4}^{0.5} \ \mathrm{G}. \qquad (12.24)$$

This result incorporates simulations by Wallace & Epperlein (1991), Yang, Arons, & Langdon (1994) and others – see Schlickeiser & Shukla (2003). It is interesting to note that the strength of these Weibel instability seeded magnetic fields is not much different from those tentatively established from the observational analysis of Xu *et al.* (2006). They analysed Faraday rotation in a low-z intergalactic filament. It is also of comparable order to the B_{IG} simulations of Ryu *et al.* (2008), shown in Fig. 12.8.

12.9   Indirect observational indicators of extragalactic magnetic fields at large cosmological distances

12.9.1   *The radio-far infrared correlation for galaxies in the nearby universe*

Harwit & Pacini (1975) predicted a correlation between the far infrared (FIR) emission from dust in starburst galaxies and the coextensive synchrotron emission, which involves the interstellar magnetic field. This was subsequently corroborated by others (e.g. Dickey & Salpeter 1984, Kronberg, Biermann, & Schwab 1985, Helou, Soifer & Rowan-Robinson 1985). The correlation was surprisingly robust (Völk 1989, Helou & Bicay 1990, 1993), and suggests a close relationship between the interstellar dust and the magnetic field. Empirically it implies that, globally, $B \propto n^\beta$ over all galaxies (Helou & Bicay 1993) ($1/3 \leq \beta \leq 2/3$), where n is the interstellar gas density associated with the far infrared emission).

Because the *spectral density* of FIR emission over 40–100 μm in starburst-like galaxies is very high, it is more easily visible to larger redshifts, compared with the cm-wave synchrotron radiation. Consequently, millimetre and submillimetre telescopes and detectors can detect FIR emission at $z \approx 5\text{--}10$, where primeval starburst galaxies, relevant to Section 12.8 may be found, since this strong emission is redshifted to observationally accessible *mm* wavelengths. The apparently universal empirical relationship to magnetic fields and cosmic rays permits an indirect estimate of, or "proxy" for, the magnetic field strengths in primeval galaxies out to these large redshifts during, and perhaps before the quasar epoch.

12.9.2 Side-to-side RM differences of double radio sources over a cosmologically significant z range

Another, somewhat more direct way to explore intergalactic magnetic fields on the scales of double radio sources, and over a range of epochs, was explored by Garrington & Conway (1991). It exploits a general fact that the average depolarisation ratio between opposite lobes of extended double radio sources is intrinsically small. Applying the differential depolarisation results to galaxy clusters at $z > 1$, they were able to infer μG-level fields on scales up to ≈ 100 kpc.

Analogously, we can search for side-to side differences in *RM* on scales ~ 10 kpc (galaxy halo) to ~ 1 Mpc, the physical size range of the largest extended radio sources. An example of side-to-side *RM* differences can be seen in Mpc-scale giant radio sources 3C326 (Willis & Strom 1978) and 0634-20 (Chapter 10, Fig. 10.2). Here the two sides are separated by a significant fraction of a Mpc. An example of very small intrinsic *RM*s on much smaller physical scales, only ~ 0.05 Mpc, is seen in 3C303, outside of a galaxy cluster (Chapter 8). At resolutions and sensitivities now available at cm to metre λ's, techniques of "Faraday rotation synthesis" (Chapter 2) at low radio frequencies can also be applied to extend these methods for probing the magnetism of the IGM, as mentioned earlier.

12.10 Seed fields between Recombination and the early galaxy formation epoch

12.10.1 Magnetic fields in the "dark ages" before the first stars and galaxies

During this period, which is between Recombination and $z \approx 20$, the universe was dominated by near-neutral, primordial gas composed of about 75% hydrogen, 25% helium and traces of lithium and beryllium (e.g. Kolb & Turner 1990). The residual low level of ionisation permits a high enough conductivity to prevent existing magnetic fields from dissipating by Ohmic or ambipolar diffusion. This means that any widespread magnetisation that might have been "inherited" from the time of recombination or before (Chapter 13) could, at least indirectly, influence the processes of gravitational collapse that set in during this period. The characteristic temperature is ~ 3000 K, and estimates of the densities prior to and during the first stages of gravitational collapse cover a wide range of $\sim 100\text{--}10^{12}$ cm^{-3}.

The principal means of probing this epoch containing near-neutral plasma is via the redshifted 21 cm line of neutral hydrogen at radio frequencies below 1.4 GHz. For fiducial redshifts of 200 and 50, the 21 cm line observing frequencies are 7.067 and 27.85 MHz, respectively. Ground-based low frequency radio telescopes coming into operation at this time of writing will be capable of frequencies around 20–40 MHz, applicable to the lower of these two redshift ranges. For the higher dark age redshifts, future radio arrays on the far side of the Moon will be able to reach frequencies down to $\mathcal{O}$ [1MHz]. At more modest redshifts, the ubiquitous, redshifted 21 cm line can also be used to explore baryonic "acoustic" oscillations BAO after Recombination, and during the primeval galaxy period.

12.10.2 General comments

In recent years much effort has been focussed on the possibility that magnetic fields in the present day galaxies and LSS may have had their origin *at* or *before* the epoch of Recombination. In the latter case, they might have originated during the plasma epoch and/ or during the period of inflation when the Universe age, **T**, was $\mathcal{O}$ [sub-microseconds to seconds] (refer to Chapter 13, and the reviews of e.g. Kronberg 1994, Grasso & Rubenstein 2001, Widrow 2012.) Indeed, *B*-seeding scenarios at or before Recombination have been sometimes proposed as necessary to produce the observed post-Recombination galactic magnetic fields, though this remains to be clarified.

Given the combination of better recent theoretical understanding, simulations, and observational analysis described in this chapter, it might alternatively be concluded that magnetic field seeding in the primordial, pre-Recombination epoch has limited importance for seeding magnetic fields in present-day astronomical systems. As above, this is not entirely clarified, and does not diminish the likely role of magnetic fields for understanding the physics of the primordial universe from the plasma epoch to the Planck scale. Some ideas on this question are briefly discussed in Chapter 13.

References

Bell, A. R. 1978, The Acceleration of Cosmic Rays in Shock Fronts. *I, MNRAS*, 182, 147

Bernet, M. L., Miniati, F., Lilly, S. J., Kronberg, P. P., & Dessauges-Zavadsky, M. 2008, Strong Magnetic Fields in Normal Galaxies at High Redshift, *Nature*, 454, 302

Biermann, L. 1950, Über den Ursprung der Magnetfelder auf Sternen und im interstellaren Raum (mit einem Anhang von A. Schlüter), *Zeitschrift für Astrophysik*, 5a, 65

Blandford, R. D. & Narayan, R. 1992, Cosmological Applications of Gravitational Lensing, *Ann. Rev. Ast. Astrophys.*, 30, 311

Braun, R., Heald, G., & Beck, R. 2010, The Westerbork SINGS Survey. III. Global Magnetic Field Topology, *Astron. Astrophys.*, 514, 42

Burbidge, G. R., O'Dell, S. L., Roberts, D. H., & Smith, H. E. 1977, On the Origin of the Absorption Spectra of Quasistellar and BL Lacertae Objects, *Astrophys. J.*, 218, 33

Cen, R. & Ostriker, J. P. 1999, Where Are the Baryons?, *Astrophys. J.*, 514, 1

Chen, G. H. & Hewitt, J. N. 1993, Multifrequency Radio Images of the Einstein Ring Gravitational Lens MG 1131 +0456, *Astron. J.*, 106, 1719

Cho, J. & Vishniac, E. T. 2000, The Generation of Magnetic Fields through Driven Turbulence, *Astrophys J.*, 538, 217

Dickey, J. M. & Salpeter, E. E. 1984, 1.4 GHz Continuum Sources in the Hercules Cluster, *Astron. J.*, 284, 461

Dieckmann, M. E., Shukla, P. K., & Drury, L. O. C. 2008, The Formation of a Relativistic Partially Electromagnetic Planar Plasma Shock, *Astrophys J.*, 675, 586

Dyer, C. C. & Roeder, R. C. 1977, Optical Scalars and the Spherical Gravitational Lens, *MNRAS.*, 180, 231

Dyer, C. C. & Shaver, E. G. 1992, On the Rotation of Polarisation by a Gravitational Lens, *Astrophys. J. Lett.*, 390, L5

Elstner, D., Meinel, R., & Beck, R. 1992, Galactic Dynamos and Their Radio Signatures, *Astron. Astrophys. Suppl. Ser.*, 94, 587

Elstner, D., Otmianowska-Mazur, K., von Linden, S., & Urbanik, M. 2000, Galactic Magnetic Fields and Spiral Arms. 3D Dynamo Simulations, *Astron. Astrophys.*, 357, 129

Foltz, C. B., Chaffee, F. H., Weymann, R. J., & Anderson, S. F. 1988, QSO Absorption Systems with Z(ABS) = about Z(EM), in *QSO Absorption Lines: Probing the Universe; Proceedings of the QSO Absorption Line Meeting, Baltimore, MD, May 19–21, 1987* (Cambridge; New York: Cambridge University Press), 53

Garrington, S. T. & Conway, R. G. 1991, The Interpretation of Asymmetric Depolarization in Extragalactic Radio Sources, *MNRAS*, 250, 198

Gnedin, Y., Ferrara, A., & Zweibel, E. N. 2000, Generation of the Primordial Magnetic Fields during Cosmological Reionization, *Astrophys. J.*, 539, 505

Grasso, D. & Rubenstein, H. R. 2001, Magnetic Fields in the Early Universe, *Phys. Rep.*, 348, 163

Gruzinov, A. 2001, Gamma-Ray Burst Phenomenology, Shock Dynamo, and the First Magnetic Fields, *Astrophys. J.*, 563, L15

Harwit, M. & Pacini, F. 1975, *Astrophys. J.*, 200, L127

Hazard, C., MacKay, M. B., & Shimmins, A. J. 1963, Investigation of the Radio Source 3C 273 by the Method of Lunar Occultations, *Nature*, 197, 1037

Helou, G., Soifer, B. T. & Rowan-Robinson, M. 1985, Thermal Infrared and Nonthermal Radio – Remarkable Correlation in Disks of Galaxies, *Astrophys. J.*, 298, L7

Kang, H., *et al.* 2005, Shock-Heated Gas in the Large-Scale Structure of the Universe, *Astrophys. J.*, 620, 21

Khare-Joshi, P. & Perry, J. J. 1982, Cosmological Implications of the Redshift Distribution of QSO Absorption Systems, *MNRAS*, 199, 785

Kolb, E. W. & Turner, M. S. 1990, *The Early Universe, Frontiers in Physics 69* (Boulder, CO: Westview Press)

Kronberg, P. P. 1994, Extragalactic Magnetic Fields, *Rep. Prog. Phys.*, 57, 325

Kronberg, P. P., Biermann, P. L., & Schwab, F. R. 1985, The Nucleus of M82 at Radio and X-Ray Bands – Discovery of a New Radio Population of Supernova Candidates, *Astrophys. J.*, 291, 693

Kronberg, P. P., Dyer, C. C., Burbidge, E. M., & Junkkarinen, V. T. A. 1991, Technique for Using Radio Jets as Extended Gravitational Lensing Probes, *Astrophys. J.*, 367, L1

Kronberg, P. P., Dyer, C. C., & Röser, H.-J. 1996, Estimates of the Global Masses of Two Distant Galaxies Using a New Type of Astrophysical Mass "Laboratory", *Astrophys. J.*, 472, 115

Kronberg, P. P. & Perry, J. J. 1982, Absorption Lines, Faraday Rotation, and Magnetic Field Estimates for QSO Absorption-Line Clouds, *Astrophys. J.*, 263, 518

Kronberg, P. P., Perry, J. J., & Zukowski, E. L. H. 1990, The 'Jet' Rotation Measure Distribution and the Optical Absorption System Near the Z = 1.953 Quasar 3C 191, *Astrophys. J.* 355, L31

Kronberg, P. P., Perry, J. J., & Zukowski, E. L. H. 1992, Discovery of Extended Faraday Rotation Compatible with Spiral Structure in an Intervening Galaxy at Z = 0.395 – New Observations of PKS 1229 – 021, *Astrophys. J.*, 387, 528

Kronberg, P. P., Reinhardt, M., & Simard-Normandin, M. 1977, On the Intergalactic Contribution to the Rotation Measures of QSO's, *Astron. Astrophys.*, 61, 771

Kronberg, P. P. & Simard-Normandin, M. 1976, New Evidence on the Origin of Rotation Measures in Extragalactic Radio Sources, *Nature*, 263, 653

Kronberg, P. P. Unpublished.

Kulsrud, R. M. & Anderson, S. W. 1992, The Spectrum of Random Magnetic Fields in the Mean Field Dynamo Theory of the Galactic Magnetic Field, *Astrophys. J.*, 396, 606

Lavery, R. J., Seitzer, P., Suntzeff, N. B., Walker, A. R., & Da Costa, G. S. 1996, Distant Ring Galaxies as Evidence for a Steeply Increasing Galaxy Interaction Rate with Redshift *Astrophys. J.*, 467, L1

Medvedev, M. V. & Loeb, A. 1999, Generation of Magnetic Fields in the Relativistic Shock of Gamma-Ray Burst Sources, *Astrophys. J.*, 526, 697

Miniati, F. 2002, Intergalactic Shock Acceleration and the Cosmic Gamma-Ray Background, *MNRAS*, 337,199

Nelson, A. H. 1973, On the Interpretation of Faraday Rotation Data, *Pub. Ast. Soc. Japan*, 25, 489

Oren, A. L. & Wolfe, A. M. 1995, A Faraday Rotation Search for Magnetic Fields in Quasar Damped Ly Alpha Absorption Systems, *Astrophys. J.*, 445, 624

Perry, J. J. & Dyson, J. E. 1990, 3 C 191 Revisited – Circumquasar Shells and Radio Jets, *Astrophys. J.*, 361, 362

Rees, M. J. & Reinhardt, M. 1972, *Astron. Astrophys.*, 19, 104

Refsdal, S. 1964, The Gravitational Lens Effect, *MNRAS*, 128, 295

Ryu, D., Kang, H., & Biermann, P. L. 1998, Cosmic Magnetic Fields in Large Scale Filaments and Sheets, *Astrophys. J.* 335, 19

Ryu, D., Kang, H., Cho, J., & Das, S. 2008, Turbulence and Magnetic Fields in the Large-Scale Structure of the Universe, *Science*, 320, 909

Ryu, D., Schleicher, D. R. G., Treumann, R. A., Tsagas, C. G., & Widrow, L. M. 2012, Magnetic Fields in the Large-Scale Structure of the Universe, *Space Sci. Rev.*, 166, 1

Schlickeiser, R. & Shukla, P. K. 2003, Cosmological Magnetic Field Generation by the Weibel Instability, *Astrophys. J.*, 599, L57

Schmidt, M. 1963, 3C 273: A Star-Like Object with Large Red-Shift, *Nature*, 197, 1040

Springel, V. & Hernqvist, L.2003, The History of Star Formation in a LCDM Universe, *MNRAS*, 339, 312

Völk, H. J. 1989, The Correlation Between Radio and Far-Infrared Emission for Disk Galaxies – A Calorimeter Theory, *Astron. Astrophys.*, 218, 67

Wallace, J. M. & Epperlein, E. M. 1991, Weibel Instability with Constant Driving Source, *Phys. Fluids B Plasma Phys.*, 3, 1579

Walsh, D., Carswell, R. F., & Weymann, R. J. 1979, 0957+561 A, B: Twin Quasistellar Objects or Gravitational Lens?, *Nature*, 279, 381

Watson, A. M. & Perry, J. J. 1991, QSO Absorption Lines and Rotation Measures, *MNRAS*, 248, 58

Weibel, E. 1959, Spontaneously Growing Transverse Waves in a Plasma Due to an Anisotropic Velocity Distribution, *Phys. Rev. Lett.*, 2, 83

Welter, G. L., Perry, J. J., & Kronberg, P. P. 1984, The Rotation Measure Distribution of QSOs and of Intervening Clouds - Magnetic Fields and Column Densities, *Astrophys. J.*, 279, 19

Whelan, Daniel J., Johnson, Jarrett L., Meiksin, Avery, Heger, Alexander, Even, Wesley, Fryer, Chris L. 2013, The Supernova that Destroyed a Protogalaxy: Prompt Chemical Enrichment and Supermassive Black Hole Growth, *Astrophys. J.*, 774, 64

Widrow, L. M. 2012, The First Magnetic Fields, Space Sci. *Rev.* 166, 37

Williams, R. E., Strittmatter, P. A., Carswell, R. F., & Craine, E. R. 1975, Splitting of Absorption Lines in 3C 191 *Astrophys. J.*, 202, 296

Willis, A. G. & Strom, R. G. 1978, Multifrequency Observations of Very Large Radio Galaxies. I - 3C 326, *Astron. Astrophys.*, 62, 375

Wolfe, A. M., Lanzetta, K. M., Foltz, C. B., & Chaffee, F. H. 1995, The Large Bright QSO Survey for Damped Ly Alpha Absorption Systems, *Astrophys. J.*, 454, 698

Xu, Y., Kronberg, P. P., Habib, S., & Dufton, Q. W. 2006, A Faraday Rotation Search for Magnetic Fields in Large-scale Structure, *Astrophys. J.*, 637, 19

Yang, T.-Y. B., Arons, J., & Langdon, A. B. Evolution of the Weibel Instability in Relativistically Hot Electron-Positron Plasmas, *Phys. Plas.*, 1, 3059

Zwicky, F. 1937, Nebulae as Gravitational Lenses, *Phys. Rev.*, 51, 290

13

Cosmic magnetic fields and our ultimate origins

13.1 Introduction: Why pre-Recombination magnetic fields are of interest

An important question to ask is whether magnetic fields produced around the time of inflation and in the early quantum Universe are really relevant to what we measure in the post-Recombination Universe. More generally, can we identify a "link", through observations and experiments, between interstellar and intergalactic magnetic fields and particle physics processes in the early universe? It has been supposed that the extreme weakness of some estimated primordial magnetic fields make them irrelevant for the post-Recombination Universe that we can directly observe. But even if there were no direct magnetic link to galactic magnetic fields, primordial magnetogenesis appears inextricably linked to the physics of the early universe, including the appearance of the first baryons and leptons. Ideas and results summarised in the following sections may convince the reader that magnetic fields and the physics of post-Recombination, and primordial epochs are ultimately not separable. A wider review of inflation era physics, can be found in an article by Binétruy *et al.* (2012), a review article more specifically on primordial magnetism by Widrow (2012), and cited in the following sections.

The Biermann battery was discussed as a widely-invoked field-seeding process in Section 3.3, and elsewhere in previous chapters. Battery-like processes can be effective in providing seed magnetic fields in many contexts. These include the solar tachocline and elsewhere in the Sun (Biermann 1950), reconnection layers, shearing and shock structures on various scales, and the Weibel instability. The seed field strengths, especially if $\gtrsim 10^{-12}$ G, can plausibly grow, often exponentially, in plasma motions up to saturation levels at μG level. These are familiar for the interstellar medium and in intracluster gas.

Given the creation of some seed field, it has been proposed that, in the primordial universe dynamo activity could have acted in a tangled magnetic field which has an original scale l, $\mathcal{O}$ cm, at a first-order QCD phase transition where $kT \sim 200$ Mev. In this scenario, the B fluctuations might have been larger than those of adiabatic or thermal energy density fluctuations.

The possibility that a fluctuating magnetic field, $B(l)$, is divergence-free would lead to a magnetic scaling law,

$$B(L, T_{\text{rec}}) = B(L, T) \left(\frac{T_{\text{rec}}}{T}\right)^2 \left(\frac{l}{L}\right)^{\beta}, \tag{13.1}$$

where $\beta = 5/2$ (Durrer & Caprini 2003). T is the cosmic temperature (T_{rec} at Recombination), and L is a larger system scale.

It is increasingly apparent that inherent couplings exist at a basic level between nucleogenesis, magnetic fields, and gravitational waves. Indeed, the earliest emergence of magnetic fields appears to have been closely associated with the origin of atomic nuclei (baryogenesis), other primordial processes, and gravity. This would make primordial fields of great interest for, and relevant to, the "mature" universe. In another context, it might answer the question of whether gravity was quantised in this early phase of "origins".

The "last scattering surface" of the CMB implies that the primordial fluid or "soup" is impenetrable to direct electromagnetic detection, and therefore not *directly* observable at still larger redshifts. However, gravity waves *can* penetrate the Universe at these primordial times (and redshifts). For this reason primordial gravity wave detection takes on great importance, and instruments such as eLISA and successors will be indispensable in this context.

Much of this brief chapter centres on magnetic field generation around the epochs of inflation in the time range of the electroweak and QCD phase transitions, and the appearance of the first hadrons at $\mathcal{O}\,1\ \mu$s. The relative time ranges of different phase transitions (Fig. 13.1) are not firm numbers as the reader will appreciate. The following few sections can, at best, offer some "pointers of understanding" in unravelling the early stages of magnetic field origins. It should be read in this context.

13.2   CMB fluctuations

It is convenient to denote angular fluctuations in multipole scales (l), in inverse radians. The strongest quadrupole components of the brightness temperature fluctuations at the last scattering surface come from Doppler shifts associated with the velocity field in the plasma. These in turn define the dynamics at the time of decoupling. The components have zero curl, and produce E-modes of the CMB polarisation.

Gravity waves, in contrast, can produce a polarised component of CMB fluctuations, and these have a component with non-zero curl (Kamionkowski *et al.* 1997, Seljak & Zaldarriaga 1997). They are contained in B-modes of the polarised component of the CMB radiation. However, gravity waves are not the only possible source of B-modes. A gravitationally induced "foreground" perturbation of the B-mode signal can also be introduced in intervening post-Recombination structures (e.g. Dyer & Shaver 1992, Hu & Okamato 2002, Knox & Song 2002).

Faraday rotation in a polarised component observed at the last scattering surface could reveal pre-Recombination magnetic structures and an associated $\nabla \times \boldsymbol{B}$ component. The illuminating polarised "surface" at this epoch arises when the radiation field has a local quadrupole component, having been Thompson-scattered by electrons (Rees 1968). Predictions for the amount of CMB polarisation have been made by Bond & Efstathiou (1984), Polnarev (1985), and Hu & White (1997). In particular, knowledge of the CMB polarisation structure would extend CMB tests, giving better constrained cosmological parameters. As just mentioned, CMB polarisation fluctuations can also be modified by post-Recombination galaxy systems of various types, and these need to be identified in order to be separated. This includes foreground gravitational lensing due to massive proto-clusters and galaxies, which modify the space-time metric at some level between the last scattering surface and us.

The angular power spectrum of the CMB and its *E*- and *B*-mode components are closely bound up with the primordial Universe, and it includes fluctuation statistics for Stokes parameters *I*, *Q*, and *U* over different angular scales and frequency ranges (Hu & White 1997). Clear signatures from the Cosmic Background Explorer (COBE) and more recent Wilkinson Microwave Anisotropy Probe (WMAP) satellites have been successful in confirming the principal features of standard inflationary cosmology.

Higher multipole modes ($l \gtrsim 1000$), corresponding to CMB fluctuations on arcminute-level scales, were measured with the ground-based Cosmic Background Interferometer (Mason *et al.* 2003). It is at these higher multipole scales, up to $l \sim 3500$, where some additional features of the Standard Model can be tested. *B*-modes, in particular, will test primordial magnetic field structure at the last scattering surface as well as gravitational lensing effects. These should be detectable at the level of $\sim 10^{-6}$ of the scalar CMB fluctuations.

A *polarised* component of the CMB was measured by the Degree-Angular Scale Interferometer (DASI) (Kovac *et al.* 2002), WMAP, and the PLANCK microwave background explorer. The first verification of vector *B*-modes was achieved by the South Pole Telescope, SPTpol (Hanson *et al.* (2013). Fluctuations of vector *E* and *S* modes are small, and this is where interesting tests of the Standard Model can be made, including traces of magnetic fields. As discussed below, CMB fluctuations have so far placed approximate upper limits on a primordial cosmic magnetic field, which are typically at μG levels on co-moving scales of ~ 1 Mpc. They are subject to ever more precise verification of the foreground features, discussed in Section 13.3. These numbers appear consistent with predictions of the standard model of Big Bang Nucleosynthesis (BBN). The concordance with BBN can be used as an upper limit, in the sense that it can constrain *B* generation models in the primordial Universe which are otherwise difficult to verify by direct experiment.

13.3   The problem of foreground CMB fluctuations

Fluctuations in the observed background sky radiation can be caused e.g. by polarising dust grains. These emit in the infrared, and are a ubiquitous component of the galactic interstellar medium. Dust emission can also originate in metagalactic regions of the Milky Way, in other galaxies, and possibly in the intergalactic filaments of LSS discussed earlier in the book. A further contributor is dust emission in the primeval outflows from the first stars and galaxies. CMB background variations of *E*- and *B*-mode radiation can be difficult to isolate from foreground thermal dust emission. Cosmic dust emits in the infrared and microwave bands, similar to the frequency range of the (cosmological) CMB radiation, redshifted from the last scattering surface. A complete separation of foreground effects from the true CMB remains as a challenge, which has not been surmounted as this book goes to press.

13.4   Faraday rotation at the last scattering surface

One specific probe of the magnetic field at the last scattering surface can come from Faraday rotation of the polarised component of the CMB. In turn, and with sufficient added information, this could be connected to primordial magnetic structures.

The λ^2 dependence in the Faraday rotation law (Equation 2.7) requires that we choose an adequately large "baseline" in (frequency)^2, $v_1^2 - v_2^2$, $v_2 > v_1$, where v_1 is as low as possible for the purpose of detecting rotation in the polarisation angle, $\Delta\chi(x, y)$. $\Delta\chi(x, y)$ is a *B*-mode-like distribution, where (x, y) are angular coordinates. But the lower frequency, v_1, must not be so low that unwanted foregrounds of (e.g. synchrotron) radiation, including galactic and extra-galactic Faraday rotation can confuse or dominate the desired *RM* signal from the CMB.

A convenient expression was given by Kosowsky & Loeb (1996) for the RMS fluctuation of Faraday rotation of polarised CMB radiation between observed frequencies v_2 and v_1:

$$\left\langle \Delta\chi_{v_1,v_2}{}^2 \right\rangle^{1/2} = 1.1° \left(1 - \frac{v_1{}^2}{v_2{}^2}\right) \left(\frac{\boldsymbol{B}_0}{10^{-9}G}\right) \left(\frac{30 \text{ GHz}}{v_1}\right). \tag{13.2}$$

Here $\boldsymbol{B}_0$ is assumed to be coherent over a few Mpc (co-moving) at the last scattering surface. Conversion of a measured $\Delta\chi$ to a $\boldsymbol{B}_0$ requires knowledge of the free electron density in the associated region (see Chapter 2). In this case the electron density has been incorporated into Equation (13.1) using a numerical code that incorporates the Recombination physics scenario described by Hu *et al.* (1995). By setting $v_1 \lesssim 30$ GHz, we choose approximately the lowest frequency that is considered free of the foreground "pollution" described above. The $\Delta\chi$ fluctuations due to Faraday rotation require measurement of CMB polarisation at (at least) two frequencies, and at each frequency preferably on different scales – for multipole scales, l, up to $\approx$3000.

Equation (13.2) demonstrates that a field strength of 10^{-9} G (referred to the current epoch) would produce a barely-detectable Faraday rotation fluctuation, $\mathcal{O}\,1°$, in the CMB polarised signal between v_2 and v_1 around 30 GHz. This gives an idea of how $\boldsymbol{B}_0$ might be detected in future such experiments down to $\approx 10^{-10}$ G, if ordering is on co-moving Mpc scales. Definitive measurements have not yet been made. However as mentioned above, CMB polarisation fluctuations have been detected in a broad band near 30 GHz on l scales of $\sim$140 to $\sim$900 using an interferometer at the South Pole (Kovac *et al.* 2002). This is an important early step toward defining CMB Faraday *RM* fluctuations at the last scattering surface, hence magnetic fields.

13.5   Ideas and theories on seed fields originating around the QCD and/or electroweak phase transitions in the primordial fluid

13.5.1   *Introduction and background*

Given that weak intergalactic fields exist on a scale up to a few Mpc, it was argued by Turner & Widrow (1988) that inflation is a good candidate process for producing them. A key idea here is that, during inflation, very large scale phenomena can arise from microphysical processes that operate on a sub-horizon scale. The low conductivity during the inflation epoch precedes the highly conducting plasma epoch. In the latter epoch, *E*-fields would be suppressed, but not the *B*-fields. A context of cosmic time scales is given in Fig. 13.1, below.

We next summarise some ideas which propose how seed fields could have been generated near the time of the first-order QCD phase transition. That is thought to occur at a cosmic time $T \approx 10^{-36}$ s (Fig. 13.1) at which point the temperature $kT_c \approx 150$ MeV (Fukugita 1988, Quashnock *et al.* 1989). During this period, hadronic bubbles are formed out of the quark-gluon plasma on scales $\mathcal{O}\,10^{1\pm1}$ cm, and the nucleation sites are separated by *ca.*10× that scale. The characteristic temperature (T_c) permits a postulated mechanism for supersonic shock heating from the newly formed hadronic bubbles. The bubbles could form a

Cosmological time table

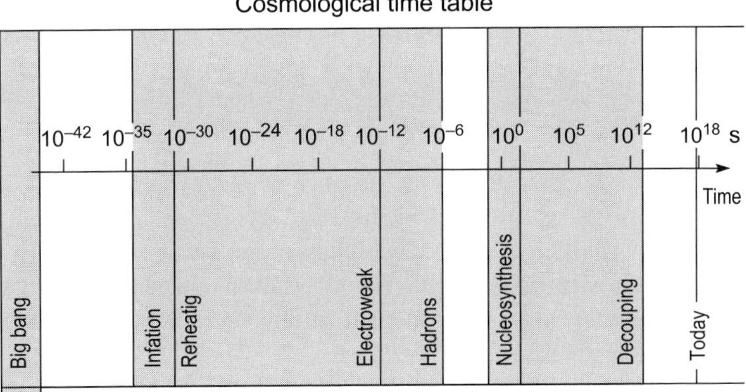

Figure 13.1. Periods of cosmic time (**T**) from the assumed beginning, at $\approx 10^{-44}$ s, to the present time. Fiducial transition phases are shown according to the Standard Model. (Adapted from Widrow *et al.* (2012).)

deflagration front, which releases enormous latent heat into the quark-gluon plasma. This heat compensates for cooling due to expansion during the few-microseconds of the QCD phase transition. At that point, around a Hubble time of ~10 μs, the quarks and gluons have been transformed into mesons and baryons (Kajantie & Kurki-Suonio 1986). Currents could have been set up, due to a slight charge difference between positively charged quarks and coexisting negatively charged leptons, and these are governed by different equations of state Quashnock *et al.* 1989, see also Vachaspati (2003).

The result is an electric field that is associated with subsonically moving hadronic shock fronts. It has been argued that, in this model, the collision of these shock fronts and the consequent vorticity will, via a Biermann battery-like mechanism, produce $\mathcal{O}$ 5 G magnetic fields on a scale of the separation between the bubbles, which is $\approx 10^{2\pm1}$ cm. By invoking some further assumptions, including the local scale-related field diffusion time, a field strength of ~2 $\times$ 10^{-17} G on a scale of ~5 $\times$ 10^{10} cm would have appeared at the Recombination epoch. However, this is only of order 1 AU at the present epoch, very small compared to galaxy scales, unless they are stretched in one dimension by many orders of magnitude. In the absence of such "bubble stretching" the question remains open on whether turbulent fields on the correspondingly high k-scales could effectively serve as seed fields in protogalaxy systems. The very small bubble dimensions require that some mechanism stretch the bubbles by $\approx 10^{10}\times$ in order to make them detectable at the last scattering surface. Without such stretching, they would need an unrealistic number of e-folding times of dynamo amplification to reach observed field values in the IGM, galaxy clusters and galaxies. On the other hand, the higher end of the estimated range above (ca. 10^{-10} G), may be consistent with current observational limits. Some newer observations suggest that some of the first stellar and galaxy systems were magnetised at $\approx\mu$G level as early as at **T** ~ 10^6 year.

13.5.2 Coupling of an inflation scalar potential to a Maxwellian gauge field

The generation of more detectable inflation-era seed magnetic fields has been discussed by Ratra (1992a,b), based on an exponential potential model introduced by Ratra & Peebles (1988). This model included a sequence of scenarios beginning at the inflation and radiation eras, and extended the modified hot big bang model. The important modification adds a coupling during inflation between a scalar field (Φ) chosen as an exponential to define inflation, and an Abelian (Maxwellian) gauge field A_μ, where

$$A_\mu \propto e^{\alpha\Phi} F_{\mu\nu} F^{\mu\nu}. \tag{13.3}$$

$F_{\mu\nu} F^{\mu\nu}$ represents the Lagrangian density of an Abelian vector field. α is a (significant) scaling number, and $F_{\mu\nu}$ is the electromagnetic field strength tensor of A_μ ($=\partial_\mu A_\nu - \partial_\nu A_\mu$). Given the less than fully clarified physics of the inflation process, it will be understood that this formulation of the exponential scaling should not be taken as an explicit derivation from basic principles of quantum physics and general relativity. But it represents what has been thought to be a reasonable approximation to Nature.

Ratra's model covers the epoch of scalar field dominance, the radiation dominant era, and the baryon era (Ratra 1992a). Allowing for various uncertainties in the physics, especially at transition points, the model arrived at a range of present-epoch fields ranging from 10^{-65} to $\sim 10^{-10}$ G on a scale of a few Mpc. The higher fields, associated with relatively large α, arise from models close to the de Sitter inflation model, and these might be strong enough to provide seed fields for subsequent B regeneration during star and galaxy formation, e.g. through infall, or some subsequent dynamo amplification due to outflow, rotation, etc., as discussed in earlier chapters.

13.5.3 *Primeval turbulence, bubbles, and forcing*

The postulated rapid rise of turbulence during Inflation, and its slow decay, are believed to be associated with the inflation-generated magnetic fields (Widrow *et al.* 2012, Kahniashvili *et al.* 2012). It is proposed that these fields can emerge at the electroweak phase transition (EWPT), or the QCD phase transition (QCTPT) – e.g. Kahniashvili *et al.* (2013). In both cases the presence of significant helicity (as distinct from no helicity) favours stronger fields, and the ranges of resulting field strengths *and* their scales are thought relatively higher for the QCDPT-produced fields by factors of ≈ 10–100 (Kahniashvili *et al.* 2013). The range of strongest present-epoch PT-generated fields covers 10^{-12} to 10^{-9} G, on scales ranging from ~ 1 to ~ 100 kpc at $z = 0$.

The very strong turbulence of the primordial fluid during phase transitions, associated with an enormous release of energy, is also a likely source of *gravitational* waves. (For additional reading on these points the interested reader could consult e.g. Ichiki *et al.* 2006, Caprini & Durrer 2001, Wang 2010, Kahniashvili *et al.* 2011, Widrow *et al.* 2012, and Binétruy *et al.* 2012). A gravitational wave signal, possibly related to an anisotropic magnetic stress during inflation, may, at this time of writing have been detected in CMB B-mode signals. Whether the B-mode signals contain a magnetic field component and how strong it is relative to the gravitational wave component of foregrounds and detected B-modes remain as currently open questions.

During the inflation phase the conductivity is low, but at the end of this phase (Fig. 13.2) the newly emerged plasma has a very high conductivity. This means that the EMF term $V(\Phi)$ is greatly reduced after the inflation phase, nearly shorting out the electric field. However, the currents, and their coupled magnetic fields remain. At this point the radiation energy density, ρ_{rad}, might be comparable to the thermal energy density of the plasma ρ_{th}. This

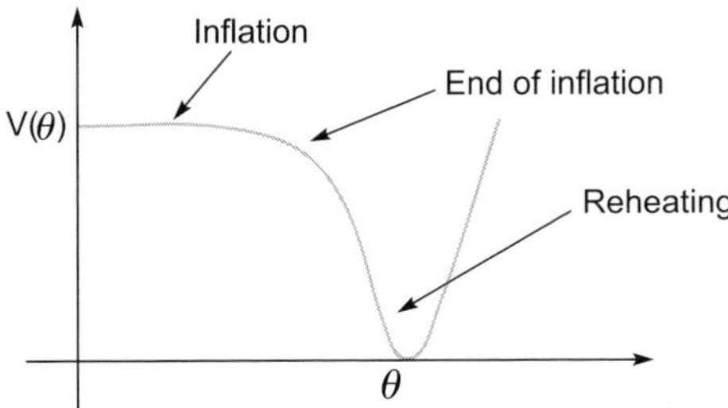

Figure 13.2. A simple model of the modification and evolution of the inflation scalar field, showing the transition of the Φ – dependent EMF, $V(\Phi)$, as Φ proceeds from the end of inflation to reheating and beyond. This assumes a spatially homogeneous background $\Phi(t)$. (Courtesy of Bharat Ratra, 2014.)

implies that a characteristic Alfvén velocity can be defined in terms of the radiation energy density ρ_{rad}, rather than the usual gas density:

$$v_A = \frac{B}{\sqrt{(16\pi\rho_{\mathrm{rad}})/3}}. \tag{13.4}$$

Here, the cosmological parameters become Ω_{rad} (density parameter) and h_{100} (Hubble constant normalised to $H_0 = 100$ kms^{-1} Mpc^{-1}). This gives $(\rho_{\mathrm{rad}}\, h_{100}^2) = 2.56 \times 10^{-5}$, conforming to a CMB temperature of 2.74 K (Kahniashvili *et al.* 2012). From the point at which reheating occurs (Fig. 13.2), the hot plasma remains ionised until Recombination at $\mathbf{T} \sim 4 \times 10^5$ year.

An important element of this theory is that the phase transition bubble size is thought to impose a forcing scale which sets the kinetic energy spectrum. This "forcing" in a phase transition bubble can be either irrotational or vortical, that is, described either by a scalar potential forcing of the type $f(x,t) = \nabla\phi$ where $\phi(x,t)$ is a random scalar function, or, by generating a random vector potential $\boldsymbol{\Psi}$, such that $f = \nabla \times \boldsymbol{\Psi}$.

Different configurations of both forcing types can be considered; These are, plane wave irrotational forcing on a smooth bubble, and vortical forcing (with differing vorticity). These have been explored in some detail by Kahniashvili *et al.* (2012), who conclude that the forcing is the source of MHD turbulence around the phase transitions. The bubbles rise to large scales during the forcing after only a few turnover times. This is followed by a period of much slower decay – of both the magnetic and fluid energies. Exactly which of the forcing scenarios operates is not greatly affected by the helicity, but in this paradigm the decay stage *is* strongly affected by magnetic helicity. As above, the scale of the forcing during the inflation phase is generally that of the bubble size. In all of the above studies, the inflation-generated magnetic field during this highly energetic phase (Fig. 13.2) is tightly coupled to the fluid. If this is true, the original big question in the first sentence of Section 13.1 would seem to be answered; that is, magnetic fields seeded at inflation, and possibly at the electroweak phase transition, will likely leave a detectable post-Recombination imprint.

13.6 Summary and conclusion

The preceding few sections have led to territory unfamiliar to many, when the Universe had an early form – before the appearance of commonly familiar sub-atomic particles, atoms, stars and galaxies, and with an unimaginably large infusion of energy. The physics of the primordial Universe appears to involve gravity, probably quantised. An important outcome of this is that gravity waves may constitute the most revealing probe of the primordial fluid, or fireball, since they can penetrate its electromagnetic opaqueness.

Magnetism appears closely connected to the origin of the energy of the Universe, and by virtue of that, of course, to our very existence. The discussion in this final chapter points to the likelihood that magnetic energy is one of the fundamental early components of the "proto-Universe" that long preceded the Universe that we know – and that is the connecting point to the theme of this book.

References

Biermann, L. 1950, Über den Ursprung der Magnetfelder auf Sternen und im interstellaren Raum (mit einem Anhang von A. Schlüter), *Zeitschrift für Naturforschung*, 5a, 65

Binétruy, P., Bohe, A., Caprini, C., & Dufaux, J.-F. 2012, Cosmological Backgrounds of Gravitational Waves and eLISA/NGO: Phase Transitions, Cosmic Strings and Other Sources, *J. Cosmol. Astropart. Phys.*, 06, 027

Bond, J. R. & Efstathiou, G. 1984, Cosmic Background Radiation Anisotropies in Universes Dominated by Nonbaryonic Dark Matter, *Astrophys. J.*, 285, L45

Dyer, C. C. & Shaver, E. G. 1992, On the Rotation of Polarization by a Gravitational Lens, *Astrophys. J.*, 390, L5

Hanson, D. *et al.* 2013, Detection of B-Mode Polarization in the Cosmic Microwave Background with Data from the South Pole Telescope, *Phys. Rev. Lett.*, 111, 141301

Hu, W., Scott, D., Sugiyama, N., & White, M. 1995, Effect of Physical Assumptions on the Calculation of Microwave Background Anisotropies, *Phys. Rev. D.*, 52, 5498

Hu, W. & Okamoto, T. 2002, Mass Reconstruction with Cosmic Microwave Background Polarization, *Astrophys. J.*, 574, 566

Hu, W. & White, M. 1997, A CMB Polarization Primer, *New Astron.*, 2, 323

Ichiki, K., Takahashi, K., Ohno, H., Hanayama, H., & Sugiyama, N. 2006, Cosmological Magnetic Field: A Fossil of Density Perturbations in the Early Universe, *Science*, 311, 827

Kahniashvili, T., Brandenburg, A., Campanelli, A., Ratra, B., & Tevadze, A. G. 2012, Evolution of Inflation-Generated Magnetic Field through Phase Transitions, *Phys. Rev. D.*, 86, 103005

Kahniashvili, T., Tevadze, A. G., Brandenburg, A., & Neronov, A. 2013, Evolution of Primordial Magnetic Fields from Phase Transitions, *Phys. Rev. D.*, 87, 083007

Kamionkowski, M., Kosowsky, A., & Stebbins, A. 1997, Statistics of Cosmic Microwave Background Polarization, *Phys. Rev. D.*, 55, 7368

Knox, L. & Song, Y. 2002, Limit on the Detectability of the Energy Scale of Inflation, *Phys. Rev. Lett.*, 89, 011303

Kosowsky, A. & Loeb, A. 1996, Faraday Rotation of Microwave Background Polarization by a Primordial Magnetic Field, *Astrophys. J.*, 469, 1

Kovac, J. M., Leitch, E. M., Pryke, C., Carlstrom, J. E., Halverson, N. W., & Holzapfel, W. L. 2002, Detection of Polarization in the Cosmic Microwave Background Using DASI, *Nature*, 420, 772

Mason, B., Pearson, T. J., Readhead, A. C. S., Shepherd, M. C., Sievers, J., Udomprasert, P. S., Cartwright, J. K., Farmer, A. J., Padin, S., Myers, S. T., *et al.* 2003, The Anisotropy of the Microwave Background to l = 3500: Deep Field Observations with the Cosmic Background Imager, *Astrophys. J.*, 591, 540

Polnarev, A. G. 1985, Polarization and Anisotropy Induced in the Microwave Background by Cosmological Gravitational Waves, *Sov. Astron.*, 29, 607

Ratra, B. 1992a, Cosmological "Seed" Magnetic Field from Inflation, *Astrophys. J. Lett.*, 391, L1

Ratra, B. 1992b, Inflation in an Exponential-Potential Scalar Field Model, *Phys. Rev. D.*, 45, 1913

Ratra, B. & Peebles, P. J. E. 1988, Cosmological Consequences of a Rolling Homogeneous Scalar Field, *Phys. Rev. D.*, 37, 3406

Rees, M. J. 1968, Polarization and Spectrum of the Primeval Radiation in an Anisotropic Universe, *Astrophys. J.*, 153, L1

Seljak, U. & Zaldarriaga, M. 1997, Signature of Gravity Waves in the Polarization of the Microwave Background, *Phys. Rev. Lett.*, 78, 2054

Turner, M. S. & Widrow, L. M. 1988, Inflation-Produced, Large-Scale Magnetic Fields, *Phys. Rev. D.*, 37, 2743

Vachaspati, T. 1991, Magnetic Fields From Cosmological Phase Transitions, *Phys. Lett.*, 265, 258

Vachaspati, T. 2003, The Bubbling Universe, *Int. J. Mod. Phys. D.*, 12, 1783

Wang, S. 2010, New Primordial Magnetic Field Limit from the Latest LIGO S5 data, *Phys. Rev. D.*, 81, 023002

Widrow, L. M., Ryu, D., Schleicher, D. R. G., Subrahmanian, K., Tsagas, C. G., & Treumann, R. A. 2012, The First Magnetic Fields, *Space Sci. Rev.*, 166, 37

Abbreviations

AGN, active galactic nucleus/nuclei, 29
BAO, baryonic acoustic oscillations, 260
CNM, cold neutral medium, 11
DM, dispersion measure, 15
EM, electromagnetic, 9
EMF, electromotive force, 36
ICM, intracluster medium, 29
LOFAR, Low Frequency Array, 21
RM, rotation measure, 9
SNR, supernova remnant, 21
UHECR, ultra-high energy cosmic ray, 30
VLBI, Very Long Baseline Interferometry, 11

Symbols

S, modified entropy, 122

T, temperature, 183

t, time, 15, 34, 63, 91, 109, 120–1, 253

t_D, diffusion time, 108

T_{rec}, cosmic temperature at Recombination, 263

t_{sp}, differential stopping time, 108

u, a bulk streaming velocity, 258

U, linearly polarised component (Stokes parameter), 10

V_F, volume filled, 110

V, circularly polarised component (Stokes parameter), 10

W, line equivalent widths, 236

W_{mag}, magnetic energy released in one turn at the inner accretion disc radius, 135

x, dimensionless frequency, 185

x, magnetic energy, 183

x_{loss}, loss length, 214

y, dimensionless Compton y- parameter, 183

z, redshift, 253

Z_0, impedance of free space, 140

Z_i, ion charge, 49

α, alpha effect in a dynamo, 91

$\gamma(\beta)$, parameter defined by Felten (1966), 172

η, energy conversion efficiency factor, 152

θ_j, jet opening angle, 200

π^0, neutral pion, 166

ρ_{BH}, black hole density, 156

τ, time of arrival delay, 212

v_T, turbulent velocity, 91

χ, orientation of the polarisation plane, 25

ω_{LHW}, LHW frequency, 67

τ, magnetic field decay time, 39

$\Sigma_a (z)$, intervenor cross-section for spectral absorption, 237

ε, internal energy, 122

ε_B, magnetic energy density, 171

ε_{cr}, energy densities of interstellar synchrotron-emitting cosmic ray gas, 89

ε_{th}, thermal energy density, 171

ε^{tot}_{er}, energy densities of relativistic electrons, 148

$\varepsilon(v)$, emissivity, 1

κ, elasticity, 214

μ_i, ion magnetic moment, 49

v, frequency or neutrino, 10, 214

v_C, critical frequency, 28

ρ, density, 120, 128, 253

ρ, mass density, 34

ρ^c, co-moving galaxy density, 111

$\tau(E_{\gamma 0})$, energy dependence of optical depth for primary photons, 199

Object Index

Subject Index